*Springer Monographs in Mathematics*

R. Stekolshchik

# Notes on Coxeter Transformations and the McKay Correspondence

Springer

Rafael Stekolshchik
Str. Kehilat Klivlend 7
Tel-Aviv
Israel
rs2@biu.013.net.il

ISBN 978-3-540-77398-6          e-ISBN 978-3-540-77399-3

DOI 10.1007/978-3-540-77399-3

Springer Monographs in Mathematics ISSN 1439-7382

Library of Congress Control Number: 2007941499

Mathematics Subject Classification (2000): 20F55, 15A18, 17B20, 16G20

*Cover design:* WMXDesign GmbH, Heidelberg

Printed on acid-free paper

9 8 7 6 5 4 3 2 1

springer.com

# Summary

We consider the Coxeter transformation in the context of the McKay correspondence, representations of quivers, and Poincaré series.

We study in detail the Jordan forms of the Coxeter transformations and prove splitting formulas due to Subbotin and Sumin for the characteristic polynomials of the Coxeter transformations. Using splitting formulas we calculate characteristic polynomials of the Coxeter transformation for the diagrams $T_{2,3,r}, T_{3,3,r}, T_{2,4,r}$, prove J. S. Frame's formulas, and generalize R. Steinberg's theorem on the spectrum of the affine Coxeter transformation for the multiply-laced diagrams. This theorem is the key statement in R. Steinberg's proof of the McKay correspondence. For every extended Dynkin diagram, the spectrum of the Coxeter transformation is easily obtained from R. Steinberg's theorem.

In the study of representations $\pi_n$ of $SU(2)$, we extend B. Kostant's construction of a vector-valued generating function $P_G(t)$. B. Kostant's construction appears in the context of the McKay correspondence and gives a way to obtain multiplicities of irreducible representations $\rho_i$ of the binary polyhedral group $G$ in the decomposition of $\pi_n|G$. In the case of multiply-laced graphs, instead of irreducible representations $\rho_i$ we use restricted representations and induced representations of $G$ introduced by P. Slodowy. Using B. Kostant's construction we generalize to the case of multiply-laced graphs W. Ebeling's theorem which connects the Poincaré series $[P_G(t)]_0$ and the Coxeter transformations. According to W. Ebeling's theorem

$$[P_G(t)]_0 = \frac{\mathcal{X}(t^2)}{\tilde{\mathcal{X}}(t^2)},$$

where $\mathcal{X}$ is the characteristic polynomial of the Coxeter transformation and $\tilde{\mathcal{X}}$ is the characteristic polynomial of the corresponding affine Coxeter transformation.

Using the Jordan form of the Coxeter transformation we prove a criterion of V. Dlab and C. M. Ringel of the regularity of quiver representations, consider necessary and sufficient conditions of this criterion for extended Dynkin diagrams and for diagrams with indefinite Tits form.

We prove one more observation of McKay concerning the Kostant generating functions $[P_G(t)]_i$:

$$(t + t^{-1})[P_G(t)]_i = \sum_{j \leftarrow i} [P_G(t)]_j,$$

where $j$ runs over all successor vertices to $i$.

A connection between fixed and anti-fixed points of the powers of the Coxeter transformations and Chebyshev polynomials of the first and second kind is established.

In memory of V. F. Subbotin

# Contents

# List of Figures

# List of Tables

# List of Notions

# 1

## Introduction

...A second empirical procedure for finding the exponents was discovered by H. S. M. Coxeter. He recognized that the exponents can be obtained from a particular transformation $\gamma$ in the Weyl group, which he had been studying, and which we take the liberty of calling a Coxeter-Killing transformation...

B. Kostant, [Kos59, p.974], 1959.

### 1.1 The three historical aspects of the Coxeter transformation

The three areas, where the Coxeter transformation plays a dramatic role, are:
- the theory of Lie algebras of the compact simple Lie groups;
- the representation theory of algebras and quivers;
- the McKay correspondence.

A *Coxeter transformation* or a *Coxeter element* is defined as the product of all the reflections in the simple roots. Neither the choice of simple roots nor the ordering of reflections in the product affects its conjugacy class, see [Bo, Ch.5, §6], see also Remark B.6. H. S. M. Coxeter studied these elements and their eigenvalues in [Cox51].

Let $h$ be the order of the Coxeter transformation (called the *Coxeter number*), $|\Delta|$ the number of roots in the corresponding root system $\Delta$, and $l$ the number of eigenvalues of the Coxeter transformation, i.e., the rank of the Cartan subalgebra. Then

$$hl = |\Delta|. \tag{1.1}$$

This fact was observed by H. S. M. Coxeter in [Cox51] and proved by B. Kostant in [Kos59, p.1021].

H. S. M. Coxeter also observed that the order of the Weyl group is equal to

$$(m_1 + 1)(m_2 + 1)...(m_l + 1),$$

where the $m_i$ are the exponents of the eigenvalues of the Coxeter transformation, and the factors $m_i + 1$ are the degrees of $l$ basic polynomial invariants of the Weyl group. Proofs of these facts were obtained by C. Chevalley [Ch55] and other authors; for historical notes, see [Bo], [Kos59]; for details, see §2.3.2.

Let $\Delta_+ \subset \Delta$ be the subset of simple positive roots, let

$$\beta = n_1\alpha_1 + \cdots + n_l\alpha_l,$$

where $\alpha_i \in \Delta_+$, be the highest root in the root system $\Delta$. The Coxeter number $h$ from (1.1) and coordinates $n_i$ of $\beta$ are related as follows:

$$h = n_1 + n_2 + \cdots + n_l + 1. \tag{1.2}$$

Observation (1.2) is due to H. S. M. Coxeter [Cox49, p.234], see also [Stb59, Th.1.4.], [Bo, Ch.6, 1, §11,Prop.31].

The Coxeter transformation is important in the study of representations of algebras, quivers, partially ordered sets (posets) and lattices. The distinguished role of the Coxeter transformation in these areas is related to the construction of the Coxeter functors given by I. N. Bernstein, I. M. Gelfand, V. A. Ponomarev in [BGP73]. Further revelation of the role of Coxeter functors for representations of algebras is due to V. Dlab and C. M. Ringel [DR76]; for a construction of the functor $D\mathrm{Tr}$ — an analog of the Coxeter functor for hereditary Artin algebras, see M. Auslander, M. I. Platzeck and I. Reiten [AuPR79], [AuRS95, Ch.8,§2]. For an application of the Coxeter functors in the representations of posets, see [Drz74]; for their applications in the representations of the modular lattices, see [GP74], [GP76], [St07].

Another area where the affine Coxeter transformations appeared is the McKay correspondence — a one-to-one correspondence between finite subgroups of $SU(2)$ and simply-laced extended Dynkin diagrams. Affine Coxeter transformations play the principal role in R. Steinberg's work [Stb85] on the proof of the McKay correspondence, see §6.2. B. Kostant ([Kos84]) obtains multiplicities of the representations related to the concrete nodes of the extended Dynkin diagram from the orbit structure of the affine Coxeter transformation, see §A.4, §5.5.

We only consider two areas of application of the Coxeter transformation: representations of quivers and the McKay correspondence. We do not consider other areas where the Coxeter transformation plays an important role, such as the theory of singularities of differentiable maps (the monodromy operator coincides with a Coxeter transformation, see, e.g., [A'C75], [Gu76], [ArGV86], [Il87], [Il95], [EbGu99]), Alexander polynomials, pretzel knots, Lehmer's problem, growth series of Coxeter groups (see, e.g., [Lev66], [Hir02], [GH01], [McM02]).

## 1.2 A brief review of this work

In Ch.2, we recall some common definitions and notions.

In Ch.3, we establish general results about the Jordan form and the spectrum of the Coxeter transformation.

In Ch.4, we give the eigenvalues of the affine Coxeter transformation. After that we prove some splitting formulas concerning the characteristic polynomials of the Coxeter transformation. The main splitting formula is the Subbotin-Sumin *splitting along the edge* formula which is extended in this chapter to the multiply-laced case. One of applications of splitting formulas is a construction of characteristic polynomials of the Coxeter transformation for the hyperbolic Dynkin diagrams $T_{2,3,r}$, $T_{3,3,r}$ and $T_{2,4,r}$. Two *Frame formulas* from [Fr51] (see Remark 4.14 and Proposition 5.2) are easily obtained from the splitting formulas.

In Ch.5, we generalize a number of results appearing in a context of the McKay correspondence to multiply-laced diagrams. First, we consider R. Steinberg's theorem playing the key role in his proof of the McKay correspondence. Essentially, R. Steinberg observed that the orders of eigenvalues of the affine Coxeter transformation corresponding to the *extended* Dynkin diagram $\tilde{\Gamma}$ coincide with the lengths of branches of the corresponding Dynkin diagram $\Gamma$, [Stb85, p.591,(*)].

Further, in Ch.5, we move on to B. Kostant's construction of a vector-valued *generating function* $P_G(t)$ [Kos84]. Let $G$ be a binary polyhedral group, let $\rho_i$, where $i = 0, \ldots, r$, be irreducible representations of $G$ corresponding by the McKay correspondence to simple roots $\alpha_i$ of the extended Dynkin diagram, let $\pi_n$, where $n = 0, 1, \ldots$, be irreducible representations of $SU(2)$ in the symmetric algebra $\mathrm{Sym}^n(\mathbb{C}^2)$, where $\mathrm{Sym}^n(V)$ is the $n$th symmetric power of $V$. Representations $\{\pi_n \mid n = 0, 1, \ldots \}$ constitute the set of all irreducible representations of $SU(2)$, [Sp77, Ch. 3.2]. Let $m_i(n)$ be *multiplicities* in the decomposition

$$\pi_n | G = \sum_{i=0}^{r} m_i(n)\rho_i;$$

set

$$v_n = \sum_{i=0}^{r} m_i(n)\alpha_i.$$

Then the Kostant generating function is defined as the following vector-valued function:

$$P_G(t) = \begin{pmatrix} [P_G(t)]_0 \\ [P_G(t)]_1 \\ \ldots \\ [P_G(t)]_r \end{pmatrix} := \sum_{n=0}^{\infty} v_n t^n. \tag{1.3}$$

In particular, $[P_G(t)]_0$ is the Poincaré series of the algebra of invariants $\mathrm{Sym}(\mathbb{C}^2)^G$, i.e.,

$$[P_G(t)]_0 = P(\mathrm{Sym}(\mathbb{C}^2)^G, t). \qquad (1.4)$$

B. Kostant obtained an explicit expression for the series $[P_G(t)]_i$, and the multiplicities $m_i(n)$, see [Kos84, Ch. 6.1.]. In Ch.5, we extend B. Kostant's construction to the case of multiply-laced graphs. For this purpose, we use P. Slodowy's generalization [Sl80, App.III] of the McKay correspondence to the multiply-laced case. The main idea of P. Slodowy is to consider the pair of binary polyhedral groups $H \lhd G$ and their *restricted representations* $\rho \downarrow_H^G$ and *induced representations* $\tau \uparrow_H^G$ instead of the representations $\rho_i$.

In Appendix A, we study in detail P. Slodowy's generalization for the case of the binary octahedral group $\mathcal{O}$ and the binary tetrahedral group $\mathcal{T}$, where $\mathcal{T} \lhd \mathcal{O}$. The generalization of the McKay correspondence to the multiply-laced case is said to be the *Slodowy correspondence*.

Generally, one can speak about the *McKay-Slodowy correspondence*.

In Ch.5 we generalize W. Ebeling's theorem [Ebl02] to the multiply-laced case: it relates the Poincaré series $[P_G(t)]_0$ and the Coxeter transformations. According to W. Ebeling's theorem,

$$[P_G(t)]_0 = \frac{\mathcal{X}(t^2)}{\tilde{\mathcal{X}}(t^2)},$$

where $\mathcal{X}$ is the characteristic polynomial of the Coxeter transformation $\mathbf{C}$ and $\tilde{\mathcal{X}}$ is the characteristic polynomial of the corresponding affine Coxeter transformation $\mathbf{C}_a$, see Theorem 5.12.

In §5.5 we prove one more observation of McKay concerning the Kostant generating functions $[P_G(t)]_i$, see (1.3):

$$(t + t^{-1})[P_G(t)]_i = \sum_{j \leftarrow i} [P_G(t)]_j,$$

where $j$ runs over all successor[1] vertices to $i$, and $[P_G(t)]_0$ related to the affine vertex $\alpha_0$ occurs in the right side only: $i = 1, 2, \ldots, r$.

The results of Ch.4 and Ch.5 are published for the first time.

In Ch.6, we study the affine Coxeter transformation $\mathbf{C}_a$. We discuss details of the original proof of R. Steinberg's theorem and its generalization. In §6.3 we consider the important linear form concerning the affine Coxeter transformation: the *defect* $\delta(z)$. It was introduced by Dlab and Ringel in [DR76] for the classification of tame type quivers in the representation theory of quivers. The following remarkable formula is due to V. Dlab and C. M. Ringel, see [DR76]:

$$\mathbf{C}_a^h z = z + h\delta(z)z^1,$$

---

[1] For a definition of the successor vertex, see Remark 5.16.

where $h$ is the Coxeter number, $z^1$ is the eigenvector corresponding to eigenvalue 1 of the affine Coxeter transformation $\mathbf{C}_a$. This formula is proved in §6.3.

In Ch.B, using results on the Jordan form of the Coxeter transformation we prove a criterion of V. Dlab and C.M. Ringel of the regularity of quiver representations, we consider necessary and sufficient conditions of this criterion for extended Dynkin diagrams and for diagrams with indefinite Tits form.

In §C.7 we establish a connection between fixed and anti-fixed points of the powers of the Coxeter transformations and Chebyshev polynomials of the first and second kind.

**Acknowledgements.** I am most grateful to John McKay for helpful comments.

I am extremely thankful to Dimitry Leites for careful editing, encouragement and advice that helped to improve the text, and MPIMiS, Leipzig, for the most creative environment which enabled him to do this job.

Many thanks to Curt McMullen, Chris Smyth, Vlad Kolmykov, Alexei Lebedev for notes they kindly sent to me during preparation of this text.

**About V. F. Subbotin.**

V. F. Subbotin was the head of an algebraic seminar of the department of Algebra and Topological Methods of Analysis of Voronezh University. At that time, the chairman of department was Professor Y. G. Borisovich, who, together with Subbotin, was my scientific advisor and to whom I am also grateful.

In 1973, Subbotin brought to me the then just published and now well-known paper *Coxeter functors and Gabriel's theorem* [BGP73]. It was the remarkable concurrence from all points of view and it was an illustration of his pedagogical talent. My research and our joint work with Subbotin began with it.

Subbotin tragically died in May 1998.

**On bibliography.** The results published in difficult to access VINITI depositions (useless to the Western reader) are listed, separately, to give due credit to the priority of the results.

The results of Ch.3 about the Jordan normal form of the Coxeter transformation are obtained by V. F. Subbotin and the author, [SuSt75], [SuSt78], [St82a], [St85].

The results of Ch.4 on the "splitting along the edge" obtained by V. F. Subbotin and M. V. Sumin, [SuSum82]. The formula of splitting along the weighted edge is obtained by the author in [St05]. An explicit calculation of eigenvalues for extended Dynkin diagrams is carried out by V. F. Subbotin and the author [SuSt75], [St82a], [St81]. Formulas of the characteristic polynomials for some diagrams $T_{p,q,r}$ are obtained by the author [St05]. R. Steinberg's theorem [Stb85] and W. Ebeling's theorem [Ebl02] from Ch.5

were generalized by the author for the multiply-laced case in [St05]. The proof of the observation of McKay [McK99] was obtained by the author [St06].

Necessary regularity conditions for the extended Dynkin diagrams from Ch.6 were obtained by V. Dlab and C. M. Ringel, [DR76]. Necessary and sufficient regularity conditions for the extended Dynkin diagrams for an arbitrary orientation were obtained by V. F. Subbotin and the author [SuSt79], [St82]. These conditions are obtained by means of a careful study of conjugations in the Weyl group which connect Coxeter transformations related to different orientations [St82], [St85].

## 1.3 The spectrum and the Jordan form

### 1.3.1 The Jordan form and reduction to the golden pair of matrices

In this work we review the research started more than 30 years ago in the teamwork with V.F. Subbotin and discuss the Jordan form and the spectrum of the Coxeter transformation. We show that the study of eigenvalues of the Coxeter transformation reduces to the study of the *golden pair* of matrices satisfying conditions of the Perron-Frobenius theorem [MM64], [Ga90] and having other nice properties.

In §3.1.3 we recall (see (3.2), (3.4), (3.5)) that any Cartan matrix $K$ associated with a tree graph can be represented in the form

$$K = \begin{cases} 2\mathbf{B} & \text{for } K \text{ symmetric, see (3.2)} \\ U\mathbf{B} & \text{for } K \text{ symmetrizable, see (3.4)}, \end{cases}$$

where $U$ is a diagonal matrix, $\mathbf{B}$ is a symmetric matrix. For simply-laced (resp. multiply-laced) diagrams[1], the *golden pair* of matrices is

$$DD^t \text{ and } D^t D \text{ (resp. } DF \text{ and } FD),$$

where the matrices $D$ and $D^t$ are found from the expressions

$$\mathbf{B} = \begin{pmatrix} I_m & D \\ D^t & I_k \end{pmatrix} \tag{1.5}$$

in the simply-laced case, and $D$ and $F$ are found from the expressions

$$K = \begin{pmatrix} 2I_m & 2D \\ 2F & 2I_k \end{pmatrix} \tag{1.6}$$

for the multiply-laced case; for details, see §2.1.1 and §3.1.3.

The above reduction method works only for trees (they have symmetrizable Cartan matrices) and for graphs with even cycles. In §4.2 we give some bibliographical remarks regarding generalized Cartan matrices.

---

[1] For a definition of simply-laced and multiply-laced diagrams, see §2.1.1.

*Remark 1.1.* We associate the Cartan matrix $K$ with every tree graph $\Gamma$, §2.1.1. Given a Cartan matrix $K$, we construct the Weyl group $W$, §2.2.1. From every orientation $\Omega$ of the graph $\Gamma$ we construct the Coxeter transformation $\mathbf{C}_\Omega$ in the Weyl group $W$, §2.2.6.

Let the $\varphi_i$ be the eigenvalues of $DD^t$ and $D^tD$. The corresponding eigenvalues $\lambda_{1,2}^{\varphi_i}$ of the Coxeter transformations are (according to Proposition 3.4, relation (3.13))

$$\lambda_{1,2}^{\varphi_i} = 2\varphi_i - 1 \pm 2\sqrt{\varphi_i(\varphi_i - 1)}.$$

Let $\mathcal{B}$ be the quadratic form corresponding to the matrix $\mathbf{B}$. This quadratic form is called the *Tits quadratic form*. For details, see §2.1.1.

One of the central results of this work is Theorem 3.15 (§3, Fig. 3.3) on Jordan form [SuSt75, SuSt78, St85]:

1) *The Jordan form of the Coxeter transformation is diagonal if and only if the Tits form is non-degenerate.*

2) *If $\mathcal{B}$ is non-negative definite (i.e., $\Gamma$ is an extended Dynkin diagram), then the Jordan form of the Coxeter transformation contains one $2 \times 2$ Jordan block. All other Jordan blocks are of size $1 \times 1$. All eigenvalues $\lambda_i$ lie on the unit circle centered at the origin.*

3) *If $\mathcal{B}$ is indefinite and degenerate, then the number of $2 \times 2$ Jordan blocks coincides with* $\dim \ker \mathbf{B}$. *All other Jordan blocks are of size $1 \times 1$. The maximal $\lambda_1^{\varphi_1}$ and the minimal $\lambda_2^{\varphi_1}$ eigenvalues such that*

$$\lambda_1^{\varphi_1} > 1, \quad \lambda_2^{\varphi_1} < 1$$

are simple.

C. M. Ringel [Rin94] generalized the result of Theorem 3.15 for non-symmetrizable Cartan matrices, see §4.2.

### 1.3.2 An explicit construction of eigenvectors. The eigenvalues are roots of unity

An important point of this work is an explicit construction of eigenvectors and adjoint vectors of the Coxeter transformation — the vectors that form a Jordan basis, see Proposition 3.10, [SuSt75, SuSt78]. This construction is used to obtain the necessary and sufficient regularity conditions of representations [SuSt75, SuSt78]. This condition was also found by V. Dlab and C.M. Ringel, [DR76], see §6.3.1.

The eigenvalues for all cases of extended Dynkin diagrams are easily calculated using a generalized R. Steinberg theorem (Theorem 5.5) and Table 1.2.

Theorem 4.1 and Table 4.1 summarize this calculation as follows ([SuSt79, St81, St82a, St85]):

*the eigenvalues of the Coxeter transformation for any extended Dynkin diagram are* **roots of unity.**

The case of $\widetilde{A}_n$ is considered[1] in the §4.2. According to eq. (4.3), the characteristic polynomial of the Coxeter transformation for $\widetilde{A}_n$ is as follows:

$$(\lambda^{n-k+1} - 1)(\lambda^k - 1),$$

where $k$ is the number characterizing the conjugacy class of the Coxeter transformation, see [MeSu82], [Men85], [Col89], [Shi00], [BT97]. For $\widetilde{A}_n$, there are $[n/2]$ characteristic polynomials [Col89], (see §4.2), but, in all these cases, the eigenvalues of the Coxeter transformation are roots of unity.

### 1.3.3 Study of the Coxeter transformation and the Cartan matrix

To study of the Coxeter transformation is almost the same as to study of the Cartan matrix. The Cartan matrix and the matrix of the Coxeter transformation (more precisely, the bicolored representative[2] of the conjugacy class of the Coxeter transformation, §3.1.3) are constructed from the same blocks, see relations (3.2), (3.4).

By Proposition 3.2, the eigenvalues $\lambda$ of the Coxeter transformation and the eigenvalues $\gamma$ of the matrix **B** of the quadratic Tits form[3] are related as follows:

$$\frac{(\lambda + 1)^2}{4\lambda} = (\gamma - 1)^2.$$

By Corollary 3.11, the Jordan form of the Coxeter transformation is diagonal if and only if the Tits form is nondegenerate.

The Coxeter transformation contains more information than the Cartan matrix, namely, the Coxeter transformation contains additional information about orientation, see §2.2.6 and considerations on the graphs containing cycles in §4.2.

The Coxeter transformation and the Cartan matrix are also related to the *fixed points* and *anti-fixed points* of the powers of the Coxeter transformation. This connection is given by means of the Chebyshev polynomials in Theorem C.19 ([SuSt82]).

### 1.3.4 Monotonicity of the dominant eigenvalue of the golden pair

According to §2.1.1 we associate the Cartan matrix $K$ to every tree graph $\Gamma$, and by (1.6) we can associate the matrices $DD^t$ and $D^tD$ to the graph $\Gamma$. According to Corollary 3.8, the matrices $DD^t$ and $D^tD$ have a common simple

---

[1] For notation of the extended Dynkin diagrams, see §2.1.6.

[2] For a definition of the bicolored representative of the conjugacy class of the Coxeter transformation, i.e., the bicolored Coxeter transformation, see §3.1.2.

[3] For a definition of the Tits form, see §1.3.1 or §2.1.1.

positive eigenvalue $\varphi_1$, the maximal eigenvalue. This eigenvalue is said to be the *dominant* eigenvalue. Thus, the dominant eigenvalue $\varphi_1$ of the matrices $DD^t$ and $D^tD$ is a certain characteristic of the graph $\Gamma$. In Proposition 3.12, we show that if any edge is added to $\Gamma$, then this characteristic only grows. The same is true for the maximal eigenvalue $\lambda_1^{\varphi_1}$, see Proposition 3.4, relation (3.13).

**Problem.** Is the dominant eigenvalue $\varphi_1$ an invariant of the graph $\Gamma$ in the class of trees, $\mathbb{T}$? I.e., is there a one-to-one correspondence between the dominant eigenvalue and the graph provided the Tits form of the graph is indefinite?

If there exist two graphs $\Gamma_1$ and $\Gamma_2$ with the same dominant eigenvalue $\varphi_1$, and $\theta$ is the assignment of the dominant eigenvalue to a graph, then what class of graphs $\mathbb{T}/\theta$ do we obtain modulo the relation given by $\theta$?

This problem is solved in this work for the diagrams $T_{2,3,r}$, $T_{3,3,r}$, $T_{2,4,r}$, see Propositions 4.16, 4.17, 4.19 and Tables 4.4, 4.5, 4.6 in §4.4.

For a definition of the diagrams $T_{p,q,r}$, see §2.1.8 and Fig. 2.1. Some of the Dynkin diagrams (as well as the extended Dynkin diagrams[1], §2.1.6, and the hyperbolic Dynkin diagrams, §2.1.8) are subclasses of the diagrams of the form $T_{p,q,r}$.

## 1.4 Splitting formulas and the diagrams $T_{p,q,r}$

### 1.4.1 Splitting formulas for the characteristic polynomial

There is a number of recurrence formulas used to calculate the characteristic polynomial of the Coxeter transformation of a given graph in terms of characteristic polynomials of the Coxeter transformation of components of the graph. Subbotin and Sumin proved the formula of *splitting along the edge* [SuSum82]:

$$\mathcal{X}(\Gamma, \lambda) = \mathcal{X}(\Gamma_1, \lambda)\mathcal{X}(\Gamma_2, \lambda) - \lambda\mathcal{X}(\Gamma_1\backslash\alpha, \lambda)\mathcal{X}(\Gamma_2\backslash\beta, \lambda). \qquad (1.7)$$

The proof of the Subbotin-Sumin formula is given in Proposition 4.8.

Another formula ([KMSS83]) is given in Proposition 4.11.

V. Kolmykov kindly informed me that the following statement holds:

---

[1] For notation of extended Dynkin diagrams, see §2.1.6. There we give two different notation: one used in the context of representations of quivers, and another one used in the context of affine Lie algebras, see Table 2.1, Table 2.2. For example, we have:

$$\widetilde{A}_n \text{ vs. } A_n^{(1)}.$$

**Proposition 1.2.** *If $\lambda$ is the eigenvalue of Coxeter transformations for graphs $\Gamma_1$ and $\Gamma_2$, then $\lambda$ is also the eigenvalue of the Coxeter transformation for the graph $\Gamma$ obtained by gluing $\Gamma_1$ and $\Gamma_2$.*

For details, see Proposition 4.12.

In Proposition 4.9 we generalize eq. (1.7) to the multiply-laced case. The formula of *splitting along the weighted edge*[1] holds:

$$\mathcal{X}(\Gamma, \lambda) = \mathcal{X}(\Gamma_1, \lambda)\mathcal{X}(\Gamma_2, \lambda) - \rho\lambda\mathcal{X}(\Gamma_1 \backslash \alpha, \lambda)\mathcal{X}(\Gamma_2 \backslash \beta, \lambda), \qquad (1.8)$$

where the factor $\rho$ is as follows:

$$\rho = k_{\alpha\beta}k_{\beta\alpha},$$

and $k_{ij}$ are elements of the Cartan matrix. Corollary 4.10 deals with the case where $\Gamma_2$ contains a single point. In this case, we have

$$\mathcal{X}(\Gamma, \lambda) = -(\lambda + 1)\mathcal{X}(\Gamma_1, \lambda) - \rho\lambda\mathcal{X}(\Gamma_1 \backslash \alpha, \lambda). \qquad (1.9)$$

A formula similar to (1.8) can be proved not only for the Coxeter transformation, but, for example, for the Cartan matrix, see È. B. Vinberg's paper [Vin85, Lemma 5.1]. See also Remark 4.14 concerning the works of J. S. Frame and S. M. Gussein-Zade.

### 1.4.2 An explicit calculation of characteristic polynomials

We use recurrence formulas (1.7), (1.8) and (1.9) to calculate the characteristic polynomials of the Coxeter transformation for the Dynkin diagrams and extended Dynkin diagrams.

The characteristic polynomials of the Coxeter transformations for the Dynkin diagrams are presented in Table 1.1. For calculations, see §5.2.

The characteristic polynomials of the Coxeter transformations for the extended Dynkin diagrams are presented in Table 1.2. The polynomials $\mathcal{X}_n$ from Table 1.2 are, up to a sign, equal to the characteristic polynomials of the Coxeter transformation for the Dynkin diagram $A_n$:

$$\mathcal{X}_n = (-1)^n \mathcal{X}(A_n),$$

defined (see §4.3 and Remark 4.13) to be

$$\mathcal{X}_n = \frac{\lambda^{n+1} - 1}{\lambda - 1} = \lambda^n + \lambda^{n-1} + \ldots + \lambda^2 + \lambda + 1.$$

For calculations of characteristic polynomials for the extended Dynkin diagrams, see §5.1 and §5.3.2.

---

[1] For a definition of the weighted edge, simply-laced and multiply-laced diagrams, see §2.1.1.

**Table 1.1.**    The characteristic polynomials, the Dynkin diagrams

| Dynkin diagram | Characteristic polynomial | Form with denominator |
|---|---|---|
| $A_n$ | $\lambda^n + \lambda^{n-1} + \cdots + \lambda + 1$ | $\dfrac{\lambda^{n+1} - 1}{\lambda - 1}$ |
| $B_n, C_n$ | $\lambda^n + 1$ | |
| $D_n$ | $\lambda^n + \lambda^{n-1} + \lambda + 1$ | $(\lambda + 1)(\lambda^{n-1} + 1)$ |
| $E_6$ | $\lambda^6 + \lambda^5 - \lambda^3 + \lambda + 1$ | $\dfrac{(\lambda^6 + 1)}{(\lambda^2 + 1)} \dfrac{(\lambda^3 - 1)}{(\lambda - 1)}$ |
| $E_7$ | $\lambda^7 + \lambda^6 - \lambda^4 - \lambda^3 + \lambda + 1$ | $\dfrac{(\lambda + 1)(\lambda^9 + 1)}{(\lambda^3 + 1)}$ |
| $E_8$ | $\lambda^8 + \lambda^7 - \lambda^5 - \lambda^4 - \lambda^3 + \lambda + 1$ | $\dfrac{(\lambda^{15} + 1)(\lambda + 1)}{(\lambda^5 + 1)(\lambda^3 + 1)}$ |
| $F_4$ | $\lambda^4 - \lambda^2 + 1$ | $\dfrac{\lambda^6 + 1}{\lambda^2 + 1}$ |
| $G_2$ | $\lambda^2 - \lambda + 1$ | $\dfrac{\lambda^3 + 1}{\lambda + 1}$ |

**Table 1.2.**    The characteristic polynomials, the extended Dynkin diagrams

| Extended Dynkin diagram | Characteristic polynomial | Form with $\mathcal{X}_i$ | Class $g$ |
|---|---|---|---|
| $\widetilde{D}_4$ | $(\lambda - 1)^2(\lambda + 1)^3$ | $(\lambda - 1)^2\mathcal{X}_1^3$ | 0 |
| $\widetilde{D}_n$ | $(\lambda^{n-2} - 1)(\lambda - 1)(\lambda + 1)^2$ | $(\lambda - 1)^2\mathcal{X}_{n-3}\mathcal{X}_1^2$ | 0 |
| $\widetilde{E}_6$ | $(\lambda^3 - 1)^2(\lambda + 1)$ | $(\lambda - 1)^2\mathcal{X}_2^2\mathcal{X}_1$ | 0 |
| $\widetilde{E}_7$ | $(\lambda^4 - 1)(\lambda^3 - 1)(\lambda + 1)$ | $(\lambda - 1)^2\mathcal{X}_3\mathcal{X}_2\mathcal{X}_1$ | 0 |
| $\widetilde{E}_8$ | $(\lambda^5 - 1)(\lambda^3 - 1)(\lambda + 1)$ | $(\lambda - 1)^2\mathcal{X}_4\mathcal{X}_2\mathcal{X}_1$ | 0 |
| $\widetilde{CD}_n, \widetilde{DD}_n$ | $(\lambda^{n-1} - 1)(\lambda^2 - 1)$ | $(\lambda - 1)^2\mathcal{X}_{n-2}\mathcal{X}_1$ | 1 |
| $\widetilde{F}_{41}, \tilde{F}_{42}$ | $(\lambda^2 - 1)(\lambda^3 - 1)$ | $(\lambda - 1)^2\mathcal{X}_2\mathcal{X}_1$ | 1 |
| $\widetilde{B}_n, \widetilde{C}_n, \widetilde{BC}_n$ | $(\lambda^n - 1)(\lambda - 1)$ | $(\lambda - 1)^2\mathcal{X}_{n-1}$ | 2 |
| $\widetilde{G}_{21}, \widetilde{G}_{22}$ | $(\lambda - 1)^2(\lambda + 1)$ | $(\lambda - 1)^2\mathcal{X}_1$ | 2 |
| $A_{11}, A_{12}$ | $(\lambda - 1)^2$ | $(\lambda - 1)^2$ | 3 |

### 1.4.3 Formulas for the diagrams $T_{2,3,r}, T_{3,3,r}, T_{2,4,r}$

For the three classes of diagrams — $T_{2,3,r}, T_{3,3,r}, T_{2,4,r}$ — explicit formulas of characteristic polynomials of the Coxeter transformations are obtained, see §2.1.8, Fig. 2.1.

The case of $T_{2,3,r}$, where $r \geq 2$, contains diagrams $D_5, E_6, E_7, E_8, \widetilde{E}_8, E_{10}$, and so we call these diagrams the $E_n$-series, where $n = r + 3$. The diagram $T_{2,3,7}$ is *hyperbolic*, (see §2.1.8) and, for all $r \geq 3$, we have

$$\mathcal{X}(T_{2,3,r}) = \lambda^{r+3} + \lambda^{r+2} - \sum_{i=3}^{r} \lambda^i + \lambda + 1,$$

see (4.15) and Table 4.4.

The spectral radius of $\mathcal{X}(T_{2,3,r})$ converges to the *smallest Pisot number*[1].

$$\sqrt[3]{\frac{1}{2} + \sqrt{\frac{23}{108}}} + \sqrt[3]{\frac{1}{2} - \sqrt{\frac{23}{108}}} \approx 1.324717...,$$

as $r \to \infty$, see Proposition 4.16.

The case of $T_{3,3,r}$, where $r \geq 2$, contains diagrams $E_6, \widetilde{E}_6$, and so we call these diagrams the $E_{6,n}$-*series*, where $n = r - 2$. The diagram $T_{3,3,4}$ is *hyperbolic*, (see §2.1.8) and, for all $r \geq 3$, we have

$$\mathcal{X}(T_{3,3,r}) = \lambda^{r+4} + \lambda^{r+3} - 2\lambda^{r+1} - 3\sum_{i=4}^{r} \lambda^i - 2\lambda^3 + \lambda + 1,$$

see (4.19) and Table 4.5. The spectral radius of $\mathcal{X}(T_{3,3,r})$ converges to the *golden ratio*[2]

$$\frac{\sqrt{5}+1}{2} \approx 1.618034...,$$

as $r \to \infty$, see Proposition 4.17.

The case of $T_{2,4,r}$, where $r \geq 2$, contains diagrams $D_6, E_7, \widetilde{E}_7$, and so we call these diagrams the $E_{7,n}$-*series*, where $n = r - 3$. The diagram $T_{2,4,5}$ is *hyperbolic*, (see §2.1.8) and, for all $r \geq 4$, we have

$$\mathcal{X}(T_{2,4,r}) = \lambda^{r+4} + \lambda^{r+3} - \lambda^{r+1} - 2\sum_{i=4}^{r} \lambda^i - \lambda^3 + \lambda + 1,$$

see (4.22) and Table 4.6. The spectral radius of $\mathcal{X}(T_{2,4,r})$ converges to

$$\frac{1}{3} + \sqrt[3]{\frac{58}{108} + \sqrt{\frac{31}{108}}} + \sqrt[3]{\frac{58}{108} - \sqrt{\frac{31}{108}}} \approx 1.465571...$$

as $r \to \infty$, see Proposition 4.19.


# 1.5 Coxeter transformations and the McKay correspondence

## 1.5.1 The generalized R. Steinberg theorem

Here we generalize R. Steinberg's theorem concerning the mysterious connection between lengths of branches of any Dynkin diagram and orders of

---

[1] Hereafter we give all such numbers with six decimal points. About Pisot numbers see §C.2, about the smallest Pisot number 1.324717... see Remark C.4.

[2] The *golden ratio* is a well-known mathematical constant $\varphi$, expressed as follows:

$$\varphi = \frac{a+b}{a} = \frac{a}{b}.$$

eigenvalues of the affine Coxeter transformation. R. Steinberg proved this theorem for the simply-laced case in [Stb85, p.591,(∗)]; it was a key statement in his explanation of the phenomena of the McKay correspondence. We prove the R. Steinberg theorem for the simply-laced case in §5.1, Theorem 5.1. The multiply-laced case (generalized R. Steinberg's theorem) is proved in §5.3.2, Theorem 5.5. Essentially, the generalized R. Steinberg theorem immediately follows from Table 1.2.

We introduce the *class number g* for the extended Dynkin diagram $\tilde{\Gamma}$, see §5.3. The class number $g$ is defined by the number of weighted edges of the diagram, see (5.14), and $g$ may take values $0, 1, 2, 3$. Let $p, q, r$ be the lengths of branches of the diagram.

**Theorem 1.3 (The generalized R. Steinberg theorem).** *The affine Coxeter transformation with the extended Dynkin diagram $\tilde{\Gamma}$ has the same eigenvalues as the product of $3 - g$ factors, each of which is the Coxeter transformation of type $A_i$, where $i \in \{p - 1, q - 1, r - 1\}$. In other words,*

$$\text{For } g = 0, \text{ the product } \mathcal{X}_{p-1}\mathcal{X}_{q-1}\mathcal{X}_{r-1} \text{ is taken.}$$
$$\text{For } g = 1, \text{ the product } \mathcal{X}_{p-1}\mathcal{X}_{q-1} \text{ is taken.}$$
$$\text{For } g = 2, \text{ the product consists only of } \mathcal{X}_{p-1}.$$
$$\text{For } g = 3, \text{ the product is trivial } (= 1).$$

For details, see Theorem 5.5.

### 1.5.2 The Kostant generating functions and W. Ebeling's theorem

Now we consider B. Kostant's construction of the vector-valued *generating function* $P_G(t)$ [Kos84]. Let $G$ be a binary polyhedral group, and $\rho_i$, where $i = 0, \ldots, r$, irreducible representations of $G$ corresponding (by the McKay correspondence) to simple roots $\alpha_i$ of the extended Dynkin diagram; let $\pi_n$, where $n = 0, 1, \ldots$, be irreducible representations[1] of $SU(2)$ in $\mathrm{Sym}^n(\mathbb{C}^2)$. Let $m_i(n)$ be *multiplicities* in the decomposition

$$\pi_n | G = \sum_{i=0}^{r} m_i(n)\rho_i,$$

and so $m_i(n) = \langle \pi_n | G, \rho_i \rangle$, where $\langle \cdot, \cdot \rangle$ is the inner product of the characters corresponding to the representations, see §5.4.2; set

$$v_n = \sum_{i=0}^{r} m_i(n)\alpha_i = \begin{pmatrix} m_0(n) \\ \ldots \\ m_r(n). \end{pmatrix}$$

---

[1] Representations $\{\pi_n \mid n = 0, 1, \ldots\}$ constitute the set of all irreducible representations of $SU(2)$, [Sp77, Ch. 3.2].

Then

$$P_G(t) = \begin{pmatrix} [P_G(t)]_0 \\ \cdots \\ [P_G(t)]_r \end{pmatrix} := \sum_{n=0}^{\infty} v_n t^n = \begin{pmatrix} \sum_{n=0}^{\infty} m_0(n)t^n \\ \cdots \\ \sum_{n=0}^{\infty} m_r(n)t^n \end{pmatrix} \qquad (1.10)$$

is a vector-valued series. In particular, $[P_G(t)]_0$ is the Poincaré series of the algebra of invariants $\mathrm{Sym}(\mathbb{C}^2)^G$, i.e.,

$$[P_G(t)]_0 = P(\mathrm{Sym}(\mathbb{C}^2)^G, t).$$

B. Kostant obtained an explicit formulas for the series $[P_G(t)]_i$, $i = 0, 1, \ldots, r$, the multiplicities $m_i(n)$, see [Kos84]. B. Kostant's construction is generalized in Ch. 5 to the multiply-laced case. For this purpose, we use the P. Slodowy generalization [Sl80, App.III] of the McKay correspondence to the multiply-laced case.

The main idea of P. Slodowy is to consider a pair of binary polyhedral groups $H \triangleleft G$ and their *restricted representations* $\rho \downarrow_H^G$ and *induced representations* $\tau \uparrow_H^G$ instead of representations $\rho_i$. In Appendix A, we study in detail P. Slodowy's generalization for the pair $\mathcal{T} \triangleleft \mathcal{O}$, where $\mathcal{O}$ is the binary octahedral group and $\mathcal{T}$ is the binary tetrahedral group. We call the generalization of the McKay correspondence to the multiply-laced case the *Slodowy correspondence*. Finally, in Ch. 5, we generalize W. Ebeling's theorem [Ebl02] which relates the Poincaré series $[P_G(t)]_0$ and the Coxeter transformations to the multiply-laced case.

First, we prove the following proposition due to B. Kostant [Kos84]. It holds for the McKay operator and also for the Slodowy operator.

**Proposition 1.4 (B. Kostant [Kos84]).** *If $B$ is either the McKay operator $A$ or the Slodowy operator $\widetilde{A}$ or $\widetilde{A}^{\vee}$, then*

$$Bv_n = v_{n-1} + v_{n+1}.$$

For details, see Proposition 5.10 in §5.4.2.

**Theorem 1.5 (The generalized W. Ebeling theorem, [Ebl02]).** *Let $G$ be a binary polyhedral group and $[P_G(t)]_0$ the Poincaré series (5.29) of the algebra of invariants $\mathrm{Sym}(\mathbb{C}^2)^G$. Then*

$$[P_G(t)]_0 = \frac{\det M_0(t)}{\det M(t)},$$

*where*

$$\det M(t) = \det |t^2 I - \mathbf{C}_a|, \qquad \det M_0(t) = \det |t^2 I - \mathbf{C}|,$$

*and where $\mathbf{C}$ is the Coxeter transformation and $\mathbf{C}_a$ is the corresponding affine Coxeter transformation.*

In Theorem 1.5 the Coxeter transformation $\mathbf{C}$ and the affine Coxeter transformation $\mathbf{C}_a$ are related to the binary polyhedral group $G$. In the multiply-laced case, we consider a pair of binary polyhedral groups $H \triangleleft G$ and again the operators $\mathbf{C}$ and $\mathbf{C}_a$ are related to the group $G$. We generalize W. Ebeling's theorem for the multiply-laced case, see Theorem 5.12. For a definition of the Poincaré series for the multiply-laced case, see (5.30) from §5.4.1.

## 1.6 The affine Coxeter transformation

### 1.6.1 The R. Steinberg trick

We consider the original proof of R. Steinberg's theorem based on the careful study of the affine Coxeter transformation[1]. Let $\mathbf{C}$ be a Coxeter transformation represented in the form

$$\mathbf{C} = w_2 w_1,$$

corresponding to a bicolored partition[2] of a given Dynkin diagram, see §3.1.2. Let $\alpha_0$ be the additional ("affine") vertex, the one that extends the Dynkin diagram to the extended Dynkin diagram, see §6.1.3. Then

$$\mathbf{C}_a = s_{\alpha_0} w_2 w_1$$

is the affine Coxeter transformation. Let $\beta$ be a root such that

$$\beta = \omega - \alpha_0,$$

where $\omega$ is the nil-root[3], i.e., the vector from the one-dimensional kernel of the Tits form $\mathcal{B}$, see (2.26). The root $\beta$ is the highest root in the root system $\Delta$ corresponding to the given Dynkin diagram. For any vector $z \in V$, let $t_\lambda$ be the translation connected to $\lambda \in V$:

$$t_\lambda(z) = z - 2\frac{(\lambda, z)}{(\lambda, \lambda)}\omega,$$

see (6.7). We have

$$t_\beta = s_{\alpha_0} s_\beta, \quad s_{\alpha_0} = t_\beta s_\beta, \quad t_{-\beta} = t_{\alpha_0}$$

where $s_{\alpha_0}$ and $s_\beta$ are reflections corresponding to the roots $\alpha_0$ and $\beta$, see Proposition 6.8. The R. Steinberg trick is to take, instead of the affine Coxeter transformation, the so-called *linear part* of the affine Coxeter transformation

$$\mathbf{C}' = s_\beta w_2 w_1$$

---

[1] For a definition of the affine Coxeter transformations, see §6.1.4.
[2] For a definition of the bicolored partitions, see §3.1.1.
[3] The nil-root $\omega$ coincides (up to a factor) with fixed point $z^1$ of the Coxeter transformation, see §2.2.1.

having the same spectrum:

**Proposition (see Proposition 6.11)** *1) The affine Coxeter transformation* $\mathbf{C}_a$ *and the linear part of the affine Coxeter transformation* $\mathbf{C}'$ *are related by a translation* $t_{\alpha_0}$ *as follows:*

$$\mathbf{C}' = t_{\alpha_0} \mathbf{C}_a.$$

*2) Let* $W_a$ *be the affine Weyl group that acts on the linear space* $V$, *and let* $V' \subset V$ *be the hyperplane of vectors orthogonal to* $\omega$. *The spectrum of* $\mathbf{C}_a$ *with deleted eigenvalue* 1 *coincides with the spectrum of the operator* $\mathbf{C}'$ *restricted onto* $V'$.

Let $\tau^{(n)}$ be alternating products given as follows:

$$\tau^{(1)} = w_1,$$
$$\tau^{(2)} = w_2 w_1,$$
$$\tau^{(3)} = w_1 w_2 w_1,$$
$$\dots$$
$$\tau^{(2n)} = (w_2 w_1)^n,$$
$$\tau^{(2n+1)} = w_1 (w_2 w_1)^n,$$
$$\dots$$

see (5.60). The highest root $\beta$ and the branch root $b$ (corresponding to the branch point) are conjugate by means of the alternating product $\tau^{(g-1)}$:

$$b = \tau^{(g-1)} \beta,$$

where $h = 2g$ is the Coxeter number, see Remark 6.13 and Proposition 6.21. Then corresponding reflections $s_\beta$ and $s_b$ are conjugate as follows:

$$s_b = w s_\beta w^{-1}, \text{ where } w = \tau^{(g-1)},$$

see Corollary 6.26.

**Proposition (see Proposition 6.27)** *The linear part* $\mathbf{C}'$ *is conjugate to* $w_2 w_1$ *(and also* $w_1 w_2$*) with canceled reflection* $s_b$ *corresponding to the branch point* $b$.

From this proposition R. Steinberg's theorem (Theorem 5.1) immediately follows:

*The affine Coxeter transformation for the extended Dynkin diagram* $\tilde{\Gamma}$ *has the same eigenvalues as the product of three Coxeter transformations of types* $A_n$, *where* $n = p - 1$, $q - 1$, *and* $r - 1$, *corresponding to the branches of the Dynkin diagram* $\Gamma$.

The generalized R. Steinberg theorem (Theorem 1.3) is also proved in this way, see §6.2.5.

### 1.6.2 The defect and the Dlab-Ringel formula

There is the important characteristic connected with the affine Coxeter transformation. It is a linear form called *defect*. For the extended Dynkin diagram, it defines the hyperplane of regular representations.

In [DR76], V. Dlab and C. M. Ringel introduced the *defect* $\delta_\Omega$ obtained as the solution to the following equation, see (6.36):

$$\mathbf{C}_\Omega^* \delta_\Omega = \delta_\Omega,$$

where $\mathbf{C}_\Omega^*$ means the dual operator to the Coxeter transformation $\mathbf{C}_\Omega$, see (6.37). Dlab and Ringel in [DR76] used the defect for the classification of *tame type quivers* in the representation theory of quivers. For the case of the extended Dynkin diagram $\widetilde{D}^4$, the defect $\delta$ was applied by Gelfand and Ponomarev in [GP72] in the study of quadruples of subspaces.

In [St75], [SuSt75], [SuSt78], the linear form $\rho_\Omega(z)$ was considered, see Definition 6.29. For the simply-laced case, it is defined as

$$\rho_\Omega(z) = \langle Tz, \tilde{z}^1 \rangle,$$

and, for the multiply-laced case, as

$$\rho_\Omega(z) = \langle Tz, \tilde{z}^{1\vee} \rangle.$$

Here $z^1$ is the eigenvector of the Coxeter transformation corresponding to eigenvalue 1, and $z^{1\vee}$ is the eigenvector corresponding to eigenvalue 1 of the Coxeter transformation for the dual diagram $\Gamma^\vee$;

$\tilde{v}$ denotes the *dual* vector to $v$ obtained from $v$ by changing the sign of the $\mathbb{Y}$-component (for a definition of $\mathbb{X}$- and $\mathbb{Y}$-components, see Remark 3.1, see also Definition 6.31);

$T$ is one of the elements interrelating Coxeter transformations for different orientations:

$$\mathbf{C}_\Omega = T^{-1}\mathbf{C}_\Lambda T,$$

where $\Lambda$ is the bicolored orientation, see §3.1.1 and (6.38).

The linear form $\rho_\Omega(z)$ is said to be the $\Omega$-*defect* of the vector $z$, or the *defect of the vector $z$ in the orientation* $\Omega$.

In Proposition 6.35, [St85], we show that the Dlab-Ringel defect $\delta_\Omega$ coincides with the $\Omega$-defect $\rho_\Omega$.

The following formula is due to V. Dlab and C. M. Ringel, see [DR76]:

$$\mathbf{C}_\Omega^h z = z + h\delta_\Omega(z)z^1,$$

where $h$ is the Coxeter number, see (6.40). This formula is proved in Proposition 6.34.

For some applications of the defect not directly connected to representation theory of quivers, see [Rin80], [St84], [SW00].

## 1.7 The regular representations of quivers

In the category of all representation of a given quiver, the regular representations are the most complicated ones, see §6.3.2. Essentially, the regularity of quiver representations is defined by means of the Coxeter transformation and depends on the orientation of the quiver, see §6.3.2, Definition 6.30. For every Dynkin diagram, there are only finite number representations, and they all are non-regular (P. Gabriel's theorem, Th. 2.14). The regular representations are completely described only for the extended Dynkin diagrams, ([Naz73], [DR76]), see §2.2.3.

### 1.7.1 The regular and non-regular representations of quivers

A given representation $V$ is said to be *regular* if

$$(\Phi^+)^k V \neq 0 \text{ and } (\Phi^-)^k V \neq 0 \text{ for every } k \in \mathbb{Z},$$

where $\Phi^+$, $\Phi^-$ are Coxeter functors[1].

Thanks to Lemma B.2, we can use another regularity condition. Namely, the representation $V$ is *regular* if and only if

$$\mathbf{C}^k(\dim V) > 0 \text{ for all } k \in \mathbb{Z}, \tag{1.11}$$

where $\mathbf{C}$ is the Coxeter transformation (2.27) associated with a given orientation, see [DR76], [St75], [SuSt75], [SuSt78]. The representations, which does not satisfy (1.11), are *non-regular* representation, i.e.,

$$\mathbf{C}^k(\dim V) \not> 0 \text{ for some } k \in \mathbb{Z},$$

see §6.3.1. Of course, the definition of regular and non-regular representation must include dependence on the orientation, see (6.41), (6.42).

Essentially, the non-regular representations constitute the simple part of representations of quivers of any *representation type* – finite-type, tame or wild. The non-regular representations can be constructed by means of repeated applications of the Coxeter functor to simple representations corresponding to simple roots in the root system.

Though, according to Kac's theorem, §2.2.4, the set of dimensions of indecomposable representations is independent of the orientation and coincides with the set of positive roots of the corresponding quiver, the partition "regular – non-regular" depends on the orientation. The vector-dimension of a regular representation in one orientation can be the dimension of a non-regular representation in another orientation and vice versa.

Consider, for example, regular representations of $\widetilde{D}_4$ in the bicolored orientation[2] $\Lambda$ and in the orientation $\Lambda''$ depicted in Fig. B.1. The regular representations in these orientations satisfy the following relations:

---

[1] For a definition of Coxeter functors, see §B.1.

[2] For a definition of the bicolored partition and the bicolored orientation, see §3.1.1.

$$y_1 + y_2 + y_3 + y_4 - 2x_0 \qquad \text{for the orientation } \Lambda,$$

$$y_1 + y_2 = y_3 + y_4 \qquad \text{for the orientation } \Lambda'',$$

see (B.45) and (B.46).

The vector-dimension $v_1$ (resp. $v_2$) from (1.12) is regular in $\Lambda$ (resp. in $\Lambda''$) and is non-regular in $\Lambda''$ (resp. in $\Lambda$).

$$v_1 = \begin{pmatrix} 2n+3 \\ n+2 \\ n+2 \\ n+1 \\ n+1 \end{pmatrix} \begin{matrix} x_0 \\ y_1 \\ y_2 \\ y_3 \\ y_4 \end{matrix}, \qquad v_2 = \begin{pmatrix} 2n+3 \\ n+1 \\ n+1 \\ n+1 \\ n+1 \end{pmatrix} \begin{matrix} x_0 \\ y_1 \\ y_2 \\ y_3 \\ y_4 \end{matrix}. \qquad (1.12)$$

### 1.7.2 The necessary and sufficient regularity conditions

In Theorem 6.33 proved by Dlab-Ringel [DR76] and by Subbotin-Stekolshchik [SuSt75], [SuSt78], we show the necessary condition of the regularity of the representation $V$:

*If* $\dim V$ *is a regular vector for an extended Dynkin diagram* $\Gamma$
*in an orientation* $\Omega$*, then*

$$\rho_\Omega(\dim V) = 0. \qquad (1.13)$$

In Proposition B.4 (for the bicolored orientation) and in Proposition B.9 (for an arbitrary orientation $\Omega$) we show that (see [St82])

*the condition (1.13) is also sufficient if* $\dim V$ *is a positive root in the root system associated with the extended Dynkin diagram* $\Gamma$*.*

To prove the sufficient regularity condition, we study the transforming elements interrelating Coxeter transformations for different orientations. Here, Theorem B.7, [St82], plays a key role. Proposition B.5 and Theorem B.7 yield the following:

1) Let $\Omega', \Omega''$ be two arbitrary orientations of the graph $\Gamma$ that differ by the direction of $k$ edges. Consider a chain of orientations, in which every two adjacent orientations differ by the direction of one edge:

$$\Omega' = \Lambda_0, \Lambda_1, \Lambda_2, \dots, \Lambda_{k-1}, \Lambda_k = \Omega''.$$

Then, in the Weyl group, there exist elements $P_i$ and $S_i$, where $i = 1, 2, ..., k$, such that

$$\mathbf{C}_{\Lambda_0} = P_1 S_1,$$
$$\mathbf{C}_{\Lambda_1} = S_1 P_1 = P_2 S_2,$$
$$\dots$$
$$\mathbf{C}_{\Lambda_{k-1}} = S_{k-1} P_{k-1} = P_k S_k,$$
$$\mathbf{C}_{\Lambda_k} = S_k P_k.$$

2) $T^{-1}\mathbf{C}_{\Omega'}T = \mathbf{C}_{\Omega''}$ for the following $k+1$ transforming elements $T := T_i$:

$$T_1 = P_1 P_2 P_3 ... P_{k-2} P_{k-1} P_k,$$

$$T_2 = P_1 P_2 P_3 ... P_{k-2} P_{k-1} S_k^{-1},$$

$$T_3 = P_1 P_2 P_3 ... P_{k-2} S_{k-1}^{-1} S_k^{-1},$$

$$...$$

$$T_{k-1} = P_1 P_2 S_3^{-1} ... S_{k-2}^{-1} S_{k-1}^{-1} S_k^{-1},$$

$$T_k = P_1 S_2^{-1} S_3^{-1} ... S_{k-2}^{-1} S_{k-1}^{-1} S_k^{-1},$$

$$T_{k+1} = S_1^{-1} S_2^{-1} S_3^{-1} ... S_{k-2}^{-1} S_{k-1}^{-1} S_k^{-1}.$$

In addition, for each reflection $\sigma_\alpha$, there exists a $T_i$ whose decomposition does not contain this reflection.

3) The following relation holds:

$$T_p T_q^{-1} = \mathbf{C}_{\Omega'}^{q-p}.$$

For each graph $\Gamma$ with indefinite Tits form $\mathcal{B}$, we prove the following *necessary regularity condition*, see Theorem B.3 [SuSt75], [SuSt78]:

If $z$ is the regular vector in the orientation $\Omega$, then

$$\langle Tz, \tilde{z}_1^m \rangle \quad \leq \quad 0, \qquad \langle Tz, \tilde{z}_2^m \rangle \quad \geq \quad 0,$$

where $\tilde{z}_1^m$ and $\tilde{z}_2^m$ are the *dual vectors* (Definition 6.31) to the vectors $z_1^m$ and $z_2^m$ correspond to the maximal eigenvalue $\varphi^m = \varphi^{max}$ of $DD^t$ and $D^t D$, respectively, see §B.2.

Similar results were obtained by Y. Zhang in [Zh89, Prop.1.5], and by J.A. de la Peña and M. Takane in [PT90, Th.2.3].

For an application of this necessary condition to the star graph, see §B.4.4.

# 2

# Preliminaries

...Having computed the $m$'s several years earlier ..., I recognized them[1] in the Poincaré polynomials while listening to Chevalley's address at the International Congress in 1950. I am grateful to A. J. Coleman for drawing my attention to the relevant work of Racah, which helps to explain the "coincidence"; also, to J. S. Frame for many helpful suggestions...

H. S. M. Coxeter, [Cox51, p.765], 1951

## 2.1 The Cartan matrix and the Tits form

### 2.1.1 The generalized and symmetrizable Cartan matrix

Let $K$ be an $n \times n$ matrix with the following properties: [Mo68], [Kac80]

$$(C1) \quad k_{ii} = 2 \text{ for } i = 1, .., n,$$
$$(C2) \quad -k_{ij} \in \mathbb{Z}_+ = \{0, 1, 2, ...\} \text{ for } i \neq j, \qquad (2.1)$$
$$(C3) \quad k_{ij} = 0 \text{ implies } k_{ji} = 0 \text{ for } i, j = 1, ..., n.$$

Such a matrix is called a *generalized Cartan matrix*. A generalized Cartan matrix $M$ is said to be *symmetrizable* if there exists an invertible diagonal matrix $U$ with positive integer coefficients and a symmetric matrix $\mathbf{B}$ such that $M = U\mathbf{B}$.

---

[1] The numbers $m_i$ mentioned by H. S. M. Coxeter in the epigraph are the exponents of the eigenvalues of the Coxeter transformation, they are called the *exponents* of the Weyl group, see §2.3.2.

*Remark 2.1.* According to the classical Bourbaki definition a *Cartan matrix* is the matrix satisfying the condition (2.1), such that the non-diagonal entries constitute a certain subset of the set $\{-1, -2, -3, -4\}$, see ([Bo, Ch.6.1.5, Def. 3]). The following inclusions hold between three classes of Cartan matrices:

$$\{M \mid M \text{ is a Cartan matrix by Bourbaki}\} \subset$$
$$\{M \mid M \text{ is a symmetrizable matrix satisfying } (2.1)\} \subset \qquad (2.2)$$
$$\{M \mid M \text{ is a generalized Cartan matrix}\}.$$

According to V. Kac [Kac80], [Kac82], a matrix (2.1) is referred to a *Cartan matrix*, but in the first edition of Kac's book [Kac93] the matrix satisfying (2.1) is already called a *generalized Cartan matrix*. We are interested in the case of symmetrizable generalized Cartan matrices, so when no confusion is possible, "the Cartan matrix" means "the symmetrizable generalized Cartan matrix".  □

Let $\Gamma_0$ (resp. $\Gamma_1$) be the set of vertices (resp. edges) of a given graph $\Gamma$. A *valued graph* or *diagram* $(\Gamma, d)$ ([DR76], [AuRS95, p. 241]) is a finite set $\Gamma_1$ (of edges) rigged with numbers $d_{ij}$ for all pairs $i, j \in \partial \Gamma_1 \subset \Gamma_0$ of the endpoints of the edges in such a way that

$(D1)$   $d_{ii} = 2$ for $i = 1, .., n$,

$(D2)$   $d_{ij} \in \mathbb{Z}_+ = \{0, 1, 2, ...\}$ for $i \neq j$,

$(D3)$   $d_{ij} = 0$ implies $d_{ji} = 0$ for $i, j = 1, ..., n$.

The rigging of the edges of $\Gamma_1$ is depicted by symbols

$$i \xrightarrow{(d_{ij}, d_{ji})} j$$

If $d_{ij} = d_{ji} = 1$, we simply write

$$i \underline{\hspace{3cm}} j$$

There is, clearly, a one-to-one correspondence between valued graphs and generalized Cartan matrices, see [AuRS95]. The following relation holds:

$$d_{ij} = |k_{ij}| \text{ for } i \neq j,$$

where $k_{ij}$ are elements of the Cartan matrix (2.1), see [AuRS95, p. 241]. The integers $d_{ij}$ of the valued graph are called *weights*, and the corresponding edges are called *weighted edges*. If

$$d_{ij} = d_{ji} = 1, \qquad (2.3)$$

we say that the corresponding edge is not weighted. A diagram is called *simply-laced* (resp. *multiply-laced*) if it does not contain (resp. contains) weighted edges.

*Remark 2.2.* Any generalized Cartan matrix $K$ whose diagram contains no cycles is symmetrizable ([Mo68, §3]). Any generalized Cartan matrix whose diagram is a simply-laced diagram (even with cycles) is symmetrizable because it is symmetric. In particular, $\widetilde{A}_n$ has a symmetrizable Cartan matrix. In this work we consider only diagrams without cycles, so the diagrams we consider correspond to symmetrizable Cartan matrices.

For notation of diagrams with the non-negative Tits form, i.e., extended Dynkin diagrams, see §2.1.6. The notation of $\widetilde{A}_n$ in the context of affine Lie algebras is $A_n^{(1)}$, see Table 2.1, Table 2.2.  □

Let $\mathcal{B}$ be the quadratic form associated with matrix $\mathbf{B}$, and let $(\cdot, \cdot)$ be the corresponding symmetric bilinear form. The quadratic form $\mathcal{B}$ is called the *quadratic Tits form*. We have

$$\mathcal{B}(\alpha) = (\alpha, \alpha).$$

In the simply-laced case,

$$K = 2\mathbf{B}$$

with the symmetric Cartan matrix $K$. So, in the simply-laced case, the Tits form is the Cartan-Tits form. In the multiply-laced case, the symmetrizable matrix $K$ factorizes

$$K = U\mathbf{B}, \tag{2.4}$$

where $U$ is a diagonal matrix with positive integers on the diagonal, and $\mathbf{B}$ is a symmetric matrix. It is easy to see that the matrix $U$ is unique up to a factor.

## 2.1.2 The Tits form and diagrams $T_{p,q,r}$

Consider the diagram $T_{p,q,r}$ depicted in Fig. 2.1. The *diagram* $T_{p,q,r}$ is defined as the tree graph with three arms of lengths $p, q, r$ having one common vertex. On Fig. 2.1 this common vertex is

$$x_p = y_q = z_r.$$

**Proposition 2.3.** *Let $\mathcal{B}$ be the quadratic Tits form connected to the diagram $T_{p,q,r}$. Then*

$$2\mathcal{B} = U + (\mu - 1)u_0^2, \tag{2.5}$$

*where $U$ is a non-negative quadratic form, and*

$$\mu = \frac{1}{p} + \frac{1}{q} + \frac{1}{r}. \tag{2.6}$$

*Proof.* By eq.(3.3) the quadratic Tits form $\mathcal{B}$ can be expressed in the following form

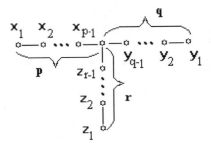

**Fig. 2.1.**    The diagram $T_{p,q,r}$

$$
\begin{aligned}
\mathcal{B}(z) =\, & x_1^2 + x_2^2 + \cdots + x_{p-1}^2 - x_1 x_2 - \cdots - x_{p-2} x_{p-1} - x_{p-1} u_0 + \\
& y_1^2 + y_2^2 + \cdots + y_{q-1}^2 - y_1 y_2 - \cdots - y_{q-2} y_{q-1} - y_{q-1} u_0 + \\
& z_1^2 + z_2^2 + \cdots + z_{r-1}^2 - z_1 z_2 - \cdots - z_{r-2} z_{r-1} - z_{r-1} u_0 + u_0^2,
\end{aligned}
\tag{2.7}
$$

where $u_0$ is the coordinate of the vector $z$ corresponding to the branch point of the diagram. Let

$$
\begin{aligned}
V(x_1, ..., & x_{p-1}, u_0) = \\
& x_1^2 + x_2^2 + \cdots + x_{p-1}^2 - x_1 x_2 - \cdots - x_{p-2} x_{p-1} - x_{p-1} u_0, \\
V(y_1, ..., & y_{q-1}, u_0) = \\
& y_1^2 + y_2^2 + \cdots + y_{q-1}^2 - y_1 y_2 - \cdots - y_{q-2} y_{q-1} - y_{q-1} u_0, \\
V(z_1, ..., & z_{r-1}, u_0) = \\
& z_1^2 + z_2^2 + \cdots + z_{r-1}^2 - z_1 z_2 - \cdots - z_{r-2} z_{r-1} - z_{r-1} u_0.
\end{aligned}
$$

Then

$$
\mathcal{B}(z) = V(x_1, ..., x_{p-1}) + V(y_1, ....y_{q-1}) + V(z_1, ....z_{r-1}) + u_0^2.
$$

It is easy to check that

$$
\begin{aligned}
2V(x_1, ..., x_{p-1}, u_0) &= \sum_{i=1}^{p-1} \frac{i+1}{i} \left( x_i - \frac{i}{i+1} x_{i+1} \right)^2 - \frac{p-1}{p} u_0^2, \\
2V(y_1, ..., y_{q-1}, u_0) &= \sum_{i=1}^{q-1} \frac{i+1}{i} \left( y_i - \frac{i}{i+1} y_{i+1} \right)^2 - \frac{q-1}{q} u_0^2, \\
2V(z_1, ..., z_{r-1}, u_0) &= \sum_{i=1}^{r-1} \frac{i+1}{i} \left( z_i - \frac{i}{i+1} z_{i+1} \right)^2 - \frac{r-1}{r} u_0^2.
\end{aligned}
\tag{2.8}
$$

Denote by $U(x)$, $U(y)$, $U(z)$ the corresponding sums in (2.8):

$$2V(x_1, ..., x_{p-1}, u_0) = U(x) - \frac{p-1}{p}u_0^2,$$

$$2V(y_1, ..., y_{q-1}, u_0) = U(y) - \frac{q-1}{q}u_0^2,$$

$$2V(z_1, ..., z_{r-1}, u_0) = U(z) - \frac{r-1}{r}u_0^2.$$

Thus,

$$2\mathcal{B}(z) = U(x) + U(y) + U(z) - \frac{p-1}{p}u_0^2 - \frac{q-1}{q}u_0^2 - \frac{r-1}{r}u_0^2 + 2u_0^2,$$

or

$$2\mathcal{B}(z) = U(x) + U(y) + U(z) + \left(\frac{1}{p} + \frac{1}{q} + \frac{1}{r} - 1\right)u_0^2.$$

Set

$$\mu = \frac{1}{p} + \frac{1}{q} + \frac{1}{r};$$

then we have

$$2\mathcal{B}(z) = U(x) + U(y) + U(z) + (\mu - 1)u_0^2. \qquad \square \qquad (2.9)$$

### 2.1.3 The simply-laced Dynkin diagrams

The diagram with the positive definite Tits form is said to be the *Dynkin diagram*. In §2.1.3, §2.1.4, §2.1.5 we find full list of Dynkin diagrams. In the other words, we prove the following theorem:

**Theorem 2.4.** *The Tits form* $\mathcal{B}$ *of the diagram* $\Gamma$ *is positive definite if and only if* $\Gamma$ *is one of diagrams in Fig. 2.3.*

*Remark 2.5.* 1) Let $K$ be the Cartan matrix associated with the tree $\Gamma$. If $d_{ij}$ and $d_{ji}$ are distinct non-zero elements of $K$, let us multiply the $i$th row by $\frac{d_{ji}}{d_{ij}}$, so the new element $d'_{ij}$ is equal to $d_{ji}$.

Let $k$ be a certain index, $k \neq i, j$. If $d_{ik} \neq 0$, i.e., there exists an edge $\{i, k\}$, then $d_{jk} = 0$, otherwise $\{i, k, j\}$ is a loop. In the next step, if $d_{ik}$ and $d_{ki}$ are distinct, then we multiply the $k$th row by $\frac{d_{ik}}{d_{ki}}$. Then $d'_{ki}$ is equal to $d_{ik}$, whereas the $i$th and $j$th rows did not change. We continue the process in this way until the Cartan matrix $K$ becomes symmetric. Since $\Gamma$ is a tree, this process terminates.

2) Let the Tits form be positive definite. Since this property is true for all values of vectors $z \in \mathcal{E}_\Gamma$, we can select some coordinates of $z$ to be zero. In this case, it suffices to consider only the submatrix of the Cartan matrix corresponding to non-zero coordinates.

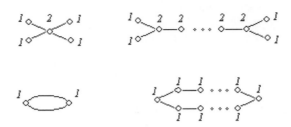

**Fig. 2.2.**    The Tits form is not positive definite. The simply-laced case

**Proposition 2.6.** *If diagram* $\Gamma$ *contains one of subdiagrams of Fig. 2.2 then the corresponding Tits form is not positive definite.*

*Proof.* Let $\Gamma_S$ be one of subdiagrams of Fig. 2.2, and let $\Gamma\backslash\Gamma_S$ be the part complementary to $\Gamma_S$. The numbers depicted in Fig. 2.2 are the coordinates of the vector corresponding to $\Gamma_S$. Complete the coordinates of $z \in \mathcal{E}_\Gamma$ corresponding to the vertices of $\Gamma\backslash\Gamma_S$ by zeros. Then $\mathcal{B}(z) \leq 0$.  □

**Proposition 2.7.** *The simply-laced Dynkin diagrams are diagrams* $A_n$, $D_n$, *and* $E_n$, *where* $n = 6, 7, 8$, *see Fig. 2.3.*

*Proof.* From Proposition 2.6 we see that the simply-laced Dynkin diagrams are only the diagrams $T_{p,q,r}$. The chains constitute a particular case of $T_{p,q,r}$ with $p = 1$.

From Proposition 2.3, (2.6), we see that $\mathcal{B}$ is positive definite for

$$\mu = \frac{1}{p} + \frac{1}{q} + \frac{1}{r} > 1,$$

i.e., the triples $p, q, r$ can only be as follows:

1) $(1, q, r)$ for any $q, r \in \mathbb{N}$, i.e., the diagrams $A_n$,
2) $(2, 2, r)$, i.e., the diagrams $D_n$,
3) $(2, 3, r)$, where $r = 3, 4, 5$, i.e., the diagrams $E_6, E_7, E_8$.

In all cases $p + q + r = n + 2$, where $n$ is the number or vertices of the diagram.  □

### 2.1.4 The multiply-laced Dynkin diagrams. Possible weighted edges

In order to prove Proposition 2.8, Proposition 2.10 and Proposition 2.11 we use the common arguments of Remark 2.5.

**Proposition 2.8.** *Let $\Gamma$ be a tree.*

*1) Consider a weighted edge $(d_{ij}, d_{ji})$, i.e., one of $d_{ij}, d_{ji}$ differs from 1. If $\mathcal{B}$ is positive definite, then*

$$d_{ij}d_{ji} < 4, \tag{2.10}$$

*i.e., only the following pairs $(d_{ij}, d_{ji})$ are possible:*

$$(1,2), (2,1), (1,3), (3,1). \tag{2.11}$$

*2) If the weighted edge is $(2,2)$, then the form $\mathcal{B}$ is not positive definite.*

*3) If the weighted edge is $(1,3)$, then the diagram consists of only one edge, and we have the diagram $G_2$, see Fig. 2.3.*

*Proof.* 1) According to Remark 2.5 factorize the Cartan matrix as follows:

$$K = \begin{pmatrix} 2 & -d_{ij} \\ -d_{ji} & 2 \end{pmatrix} = \begin{pmatrix} d_{ij} & 0 \\ 0 & d_{ji} \end{pmatrix} \begin{pmatrix} \dfrac{2}{d_{ij}} & -1 \\ -1 & \dfrac{2}{d_{ji}} \end{pmatrix} \tag{2.12}$$

and, for all $k \neq i, j$, set $x_k = 0$. Then

$$\mathcal{B}(z) = \frac{2}{d_{ij}}x_i^2 - 2x_ix_j + \frac{2}{d_{ji}}x_j^2,$$

or, up to a non-zero factor

$$\mathcal{B}(z) = d_{ji}x_i^2 - d_{ji}d_{ij}x_ix_j + d_{ij}x_j^2, \tag{2.13}$$

The discriminant $(d_{ji}d_{ij})^2 - 4d_{ji}d_{ij}$ should be negative, i.e., $d_{ij}d_{ji} < 4$.

2) Let $d_{ij} = d_{ji} = 2$. For all $k \neq i, j$, in (2.13) set $x_k = 0$. Then

$$\mathcal{B}(z) = 2x_i^2 + 2x_j^2 - 4x_ix_j = (x_i - x_j)^2 \tag{2.14}$$

For $x_i = x_j$, we have $\mathcal{B}(z) = 0$.

3) Let the diagram $\Gamma$ have two edges $\{l, k\}$ and $\{k, j\}$, and

$$d_{lk}d_{kl} = 3. \tag{2.15}$$

a) Let $\{l, k\}$ be the weighted edge $(3, 1)$ and $\{k, j\}$ be the weight edge $(d_{kj}, d_{jk})$. Let us factorize the component of the Cartan matrix corresponding to the two edges $\{l, k\}$ and $\{k, j\}$ as follows:

$$K = \begin{pmatrix} 2 & -d_{kj} & -3 \\ -d_{jk} & 2 & 0 \\ -1 & 0 & 2 \end{pmatrix} = \begin{pmatrix} 3 & 0 & 0 \\ 0 & \dfrac{3d_{jk}}{d_{kj}} & 0 \\ 0 & 0 & 1 \end{pmatrix} \begin{pmatrix} \dfrac{2}{3} & -\dfrac{d_{kj}}{3} & -1 \\ -\dfrac{d_{kj}}{3} & \dfrac{2d_{kj}}{3d_{jk}} & 0 \\ -1 & 0 & 2 \end{pmatrix} \begin{matrix} k \\ j \\ l \end{matrix}$$

and, for all $i \neq l, k, j$, set $x_i = 0$. Then

$$\mathcal{B}(z) = d_{jk}x_k^2 + d_{kj}x_j^2 + 3d_{jk}x_l^2 - d_{kj}d_{jk}x_jx_k - 3d_{jk}x_kx_l.$$

Set $x_l = 1$, $x_k = 2$, $x_j = 1$. Then

$$\mathcal{B}(z) = 4d_{jk} + d_{kj} + 3d_{jk} - 2d_{kj}d_{jk} - 6d_{jk} =$$
$$d_{kj} + d_{jk} - 2d_{kj}d_{jk} < 0.$$

b) Let $\{l, k\}$ be the weighted edge $(1, 3)$ and let $\{k, j\}$ be the weighted edge $(d_{kj}, d_{jk})$. As above, we factorize the component of the Cartan matrix corresponding to the two edges $\{l, k\}$ and $\{k, j\}$ as follows:

$$K = \begin{pmatrix} 2 & -d_{kj} & -1 \\ -d_{jk} & 2 & 0 \\ -3 & 0 & 2 \end{pmatrix} = \begin{pmatrix} 1 & 0 & 0 \\ 0 & \dfrac{d_{jk}}{d_{kj}} & 0 \\ 0 & 0 & 3 \end{pmatrix} \begin{pmatrix} 2 & -d_{kj} & -1 \\ -d_{kj} & \dfrac{2d_{kj}}{d_{jk}} & 0 \\ -1 & 0 & \dfrac{2}{3} \end{pmatrix} \begin{matrix} k \\ j \\ l \end{matrix}$$

and, for all $i \neq l, k, j$, set $x_i = 0$. Then

$$\mathcal{B}(z) = 3d_{jk}x_k^2 + 3d_{kj}x_j^2 + d_{jk}x_l^2 - 3d_{kj}d_{jk}x_jx_k - 3d_{jk}x_kx_l.$$

Set $x_l = 3$, $x_k = 2$, $x_j = 1$. Then

$$\mathcal{B}(z) = 12d_{jk} + 3d_{kj} + 9d_{jk} - 6d_{kj}d_{jk} - 18d_{jk} =$$
$$3d_{kj} + 3d_{jk} - 6d_{kj}d_{jk} < 0.$$

c) The diagram consisting of only on weighted edge $(1, 3)$ has the positive definite Tits form

$$\mathcal{B}(z) = x_l^2 - 3x_lx_k + 3x_k^2.$$

This is the diagram $G_2$.  □

**Corollary 2.9.** *If $\mathcal{B}$ is the positive definite Tits form associated with the diagram $\Gamma$ having more than one edge, then the only possible weighted edges are $(1, 2)$ and $(2, 1)$.*

□

In what follows, the only weighted edges we consider are $(1, 2)$ and $(2, 1)$.

## 2.1.5 The multiply-laced Dynkin diagrams. A branch point

**Proposition 2.10.** *If the diagram $\Gamma$ has two adjacent weighted edges, then the corresponding Tits form is not positive definite.*

*Proof.* Let $\{l, k\}$ be the weighted edge $(1, 2)$ and $\{k, j\}$ be the weighted edge $(d_{jk}, d_{kj})$. As in Proposition 2.8, b) above, we have

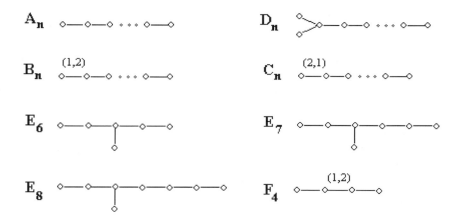

**Fig. 2.3.**    The Dynkin diagrams

$$\mathcal{B}(z) = d_{jk}x_k^2 + d_{kj}x_j^2 + 2d_{jk}x_l^2 - d_{kj}d_{jk}x_jx_k - 2d_{jk}x_kx_l.$$

For $(d_{jk}, d_{kj}) = (2,1)$, we have

$$\mathcal{B}(z) = 2x_k^2 + x_j^2 + 4x_l^2 - 2x_jx_k - 4x_kx_l = (x_k - x_j)^2 + (x_k - 2x_l)^2,$$

and $\mathcal{B}(z)$ is not positive definite. For $(d_{jk}, d_{kj}) = (1, 2)$, we have

$$\mathcal{B}(z) = x_k^2 + 2x_j^2 + 2x_l^2 - 2x_jx_k - 2x_kx_l = \frac{1}{2}(x_k - 2x_j)^2 + \frac{1}{2}(x_k - 2x_l)^2,$$

and $\mathcal{B}(z)$ is also not positive definite.

Now, let $\{l, k\}$ be the weighted edge $(2, 1)$ and $\{k, j\}$ be the weighted edge $(d_{jk}, d_{kj})$. Here,

$$\mathcal{B}(z) = 2d_{jk}x_k^2 + 2d_{kj}x_j^2 + d_{jk}x_l^2 - 2d_{kj}d_{jk}x_jx_k - 2d_{jk}x_kx_l.$$

For $(d_{jk}, d_{kj}) = (2, 1)$, we have

$$\mathcal{B}(z) = 4x_k^2 + 2x_j^2 + 2x_l^2 - 4x_jx_k - 4x_kx_l = 2(x_k - x_j)^2 + 2(x_k - 2x_l)^2,$$

and $\mathcal{B}(z)$ is not positive definite. For $(d_{jk}, d_{kj}) = (1, 2)$, we have

$$\mathcal{B}(z) = 2x_k^2 + 4x_j^2 + x_l^2 - 4x_jx_k - 2x_kx_l = (x_k - 2x_j)^2 + (x_k - x_l)^2,$$

and $\mathcal{B}(z)$ is also not positive definite.    □

**Proposition 2.11.** *Let the diagram* $\Gamma$ *have only one branch point.*

*1) If one of edges ending in the branch point is a weighted edge, then the corresponding Tits form is not positive definite, see Fig. 2.4, a), b).*

*2) If the diagram* $\Gamma$ *has a weighted edge, then the corresponding Tits form is not positive definite, see Fig. 2.4, c), d).*

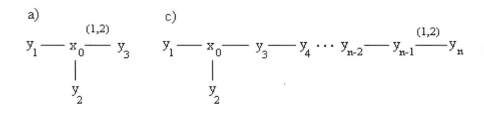

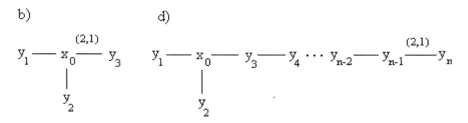

**Fig. 2.4.**     The Tits form is not positive definite. Cases with a branch point

*Proof.* 1) Case a) in Fig. 2.4. The Tits form is

$$B(z) = x_0^2 + y_1^2 + y_2^2 + 2y_3^2 - x_0 y_1 - x_0 y_2 - 2x_0 y_3.$$

Set $x_0 = 2, y_1 = 1, y_2 = 1, y_3 = 1$. We have

$$B(z) = 4 + 1 + 1 + 2 - 2 - 2 - 4 = 0.$$

Case b) in Fig. 2.4. The Tits form is

$$B(z) = 2x_0^2 + 2y_1^2 + 2y_2^2 + y_3^2 - 2x_0 y_1 - 2x_0 y_2 - 2x_0 y_3.$$

Set $x_0 = 2, y_1 = 1, y_2 = 1, y_3 = 2$. We have

$$B(z) = 8 + 2 + 2 + 4 - 4 - 4 - 8 = 0.$$

*Remark 2.12.* If the weighted edge $(1,2)$ (resp. $(2,1)$) in cases a) c), (resp. b), d)) is not a terminal edge from the right side, see Fig. 2.4, complete all remaining coordinates corresponding to vertices till the terminal edge by zeros, see Remark 2.5, 2).

2) Case c) in Fig. 2.4. The Cartan matrix is

$$
K = \begin{pmatrix}
2 & -1 & -1 & -1 & & & & & & \\
-1 & 2 & & & & & & & & \\
-1 & & 2 & & & & & & & \\
-1 & & & 2 & -1 & & & & & \\
& & & -1 & 2 & -1 & & & & \\
& & & & -1 & 2 & & & & \\
& & \cdots & & & & \cdots & & & \\
& & & & & & & 2 & -1 & \\
& & & & & & & -1 & 2 & -2 \\
& & & & & & & & -1 & 2
\end{pmatrix} , \tag{2.16}
$$

the matrix $U$ from the factorization (2.4) and the matrix of the Tits form are as follows:

$$
U = \mathrm{diag} \begin{pmatrix} 1 \\ 1 \\ 1 \\ 1 \\ 1 \\ 1 \\ \cdots \\ 1 \\ 1 \\ \frac{1}{2} \end{pmatrix} , \qquad
\mathbf{B} = \begin{pmatrix}
2 & -1 & -1 & -1 & & & & & \\
-1 & 2 & & & & & & & \\
-1 & & 2 & & & & & & \\
-1 & & & 2 & -1 & & & & \\
& & & -1 & 2 & -1 & & & \\
& & & & -1 & 2 & & & \\
& & \cdots & & & & \cdots & & \\
& & & & & & 2 & -1 & \\
& & & & & & -1 & 2 & -2 \\
& & & & & & & -2 & 4
\end{pmatrix} . \tag{2.17}
$$

Up to a factor 2, the Tits form is

$$
\mathcal{B}(z) = x_0^2 + \sum_{i=1}^{n-1} y_i^2 + 2y_n^2 - x_0(y_1 + y_2 + y_3) - \sum_{i=3}^{n-2} y_i y_{i+1} - 2 y_{n-1} y_n.
$$

Set $x_0 = 2, y_1 = 1, y_2 = 1, y_3 = \cdots = y_{n-1} = 2, y_n = 1$. Then

$$
\mathcal{B}(z) = 4 + 1 + 1 + 4(n-3) + 2 - 2(1+1+2) - 4(n-4) - 4 = 0.
$$

Case d) in Fig. 2.4. The Cartan matrix is

$$
K = \begin{pmatrix}
2 & -1 & -1 & -1 & & & & & & \\
-1 & 2 & & & & & & & & \\
-1 & & 2 & & & & & & & \\
-1 & & & 2 & -1 & & & & & \\
& & & -1 & 2 & -1 & & & & \\
& & & & -1 & 2 & & & & \\
& & \cdots & & & & \cdots & & & \\
& & & & & & & 2 & -1 & \\
& & & & & & & -1 & 2 & -1 \\
& & & & & & & & -2 & 2
\end{pmatrix} , \tag{2.18}
$$

the matrix $U$ from the factorization (2.4) and the matrix of the Tits form are as follows:

$$
U = \operatorname{diag}
\begin{pmatrix}
1 \\ 1 \\ 1 \\ 1 \\ 1 \\ 1 \\ \cdots \\ 1 \\ 1 \\ 2
\end{pmatrix}, \quad
B =
\begin{pmatrix}
2 & -1 & -1 & -1 & & & & & \\
-1 & 2 & & & & & & & \\
-1 & & 2 & & & & & & \\
-1 & & & 2 & -1 & & & & \\
& & & -1 & 2 & -1 & & & \\
& & & & -1 & 2 & & & \\
& & & \cdots & & & \cdots & & \\
& & & & & & 2 & -1 & \\
& & & & & & -1 & 2 & -1 \\
& & & & & & & -1 & 1
\end{pmatrix}. \quad (2.19)
$$

The Tits form is

$$
\mathcal{B}(z) = 2x_0^2 + 2 \sum_{i=1}^{n-1} y_i^2 + y_n^2 - 2x_0(y_1 + y_2 + y_3) - 2 \sum_{i=3}^{n-1} y_i y_{i+1}.
$$

Set $x_0 = 2, y_1 = 1, y_2 = 1, y_3 = \cdots = y_{n-1} = y_n = 2$. Then

$$
\mathcal{B}(z) = 8 + 2 + 2 + 8(n-3) + 4 - 4(1+1+2) - 8(n-3) = 0. \quad \square
$$

**Corollary 2.13.** *If the diagram $\Gamma$ has a weighted edge, and the corresponding Tits form is positive definite, then the diagram $\Gamma$ is a chain.*

*To finish the proof of Theorem 2.4* it remains to show that the diagram $\Gamma$ is a chain with a weighted edge only if $\Gamma$ is $B_n$ or $C_n$ or $F_4$, see Fig. 2.3. In order to prove this fact, it suffices to prove, that diagrams $\widetilde{B}_n$, $\widetilde{C}_n$, $\widetilde{BC}_n$, $\widetilde{F}_{41}$, $\widetilde{F}_{42}$ have the non-negative Tits form[1].

As above in Proposition 2.11, 2), we write down the corresponding Tits form $\mathcal{B}$. For every mentioned diagram from Fig. 2.6, we insert the vector $z$ with coordinates depicted in the figure, into the form $\mathcal{B}$. In all these cases, we have $\mathcal{B}(z) = 0$. Consider, for example, the diagrams $\widetilde{F}_{41}$ and $\widetilde{F}_{42}$, see Fig. 2.5.

a) Diagram $\widetilde{F}_{41}$. Here, the Cartan matrix is

$$
K =
\begin{pmatrix}
2 & -1 & -2 & & \\
-1 & 2 & & -1 & \\
-1 & & 2 & & -1 \\
& -1 & & 2 & \\
& & -1 & & 2
\end{pmatrix}
\begin{matrix}
x_0 \\ y_1 \\ y_2 \\ y_3 \\ y_4
\end{matrix} \quad (2.20)
$$

---

[1] For notation of diagrams with the non-negative Tits form, i.e., extended Dynkin diagrams, see §2.1.6. There we give two different notation: one used in the context of representations of quivers, and another one used in the context of affine Lie algebras, see Table 2.1, Table 2.2. For example, two notations:

$$
\widetilde{F}_{41} \text{ vs. } F_4^{(1)}, \quad \widetilde{F}_{42} \text{ vs. } E_6^{(2)}, \quad \widetilde{B}_{n+1} \text{ vs. } D_{n+1}^{(2)}.
$$

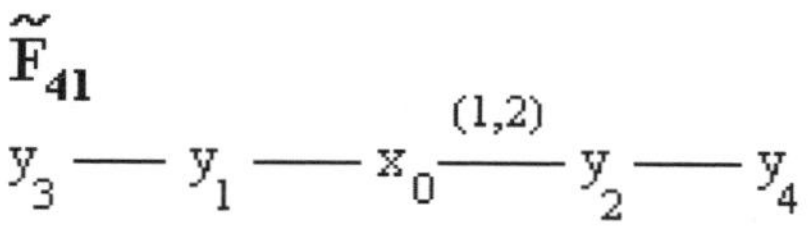

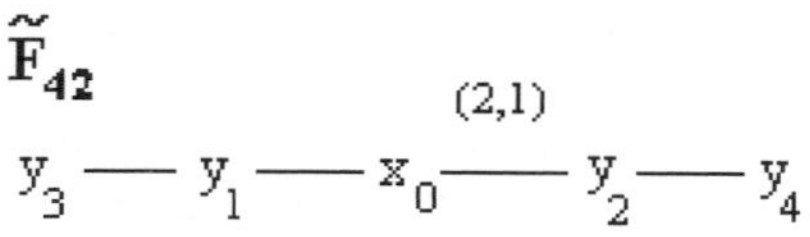

$$\widetilde{F}_{41}$$

$$\underset{3}{y} \; \text{---} \; \underset{1}{y} \; \text{---} \; \underset{0}{x} \overset{(1,2)}{\text{---}} \; \underset{2}{y} \; \text{---} \; \underset{4}{y}$$

$$\widetilde{F}_{42}$$

$$\underset{3}{y} \; \text{---} \; \underset{1}{y} \; \text{---} \; \underset{0}{x} \overset{(2,1)}{\text{---}} \; \underset{2}{y} \; \text{---} \; \underset{4}{y}$$

**Fig. 2.5.**    The diagrams $\widetilde{F}_{41}$ and $\widetilde{F}_{42}$

the matrix $U$ from (2.4) and the matrix of the Tits form are as follows:

$$U = \operatorname{diag} \begin{pmatrix} 1 \\ 1 \\ 1/2 \\ 1 \\ 1/2 \end{pmatrix}, \quad B = \begin{pmatrix} 2 & -1 & -2 & & \\ -1 & 2 & & -1 & \\ -2 & & 4 & & -2 \\ & -1 & & 2 & \\ & & -2 & & 4 \end{pmatrix}. \quad (2.21)$$

Up to a factor 2, the Tits form is

$$\mathcal{B}(z) = x_0^2 + y_1^2 + 2y_2^2 + y_3^2 + 2y_2^4 - y_1 x_0 - 2y_2 x_0 - 2y_2 y_4 - y_3 y_1.$$

Set $y_3 = 1, y_1 = 2, x_0 = 3, y_2 = 2, y_4 = 1$. Then

$$\mathcal{B}(z) = 9 + 4 + 8 + 1 + 2 - 6 - 12 - 4 - 2 = 0.$$

a) Diagram $\widetilde{F}_{42}$. The Cartan matrix is

$$K = \begin{pmatrix} 2 & -1 & -1 & & \\ -1 & 2 & & -1 & \\ -2 & & 2 & & -1 \\ & -1 & & 2 & \\ & & -1 & & 2 \end{pmatrix} \begin{matrix} x_0 \\ y_1 \\ y_2 \\ y_3 \\ y_4 \end{matrix}, \quad (2.22)$$

the matrix $U$ from (2.4) and the matrix of the Tits form are as follows:

$$U = \operatorname{diag} \begin{pmatrix} 1 \\ 1 \\ 2 \\ 1 \\ 2 \end{pmatrix}, \quad B = \begin{pmatrix} 2 & -1 & -1 & & \\ -1 & 2 & & -1 & \\ -1 & & 1 & & -\dfrac{1}{2} \\ & -1 & & 2 & \\ & & -\dfrac{1}{2} & & 1 \end{pmatrix}. \quad (2.23)$$

The Tits form is

$$\mathcal{B}(z) = 2x_0^2 + 2y_1^2 + y_2^2 + 2y_3^2 + y_2^4 - 2y_1x_0 - 2y_2x_0 - y_2y_4 - 2y_3y_1.$$

Set $y_3 = 1, y_1 = 2, x_0 = 3, y_2 = 4, y_4 = 2$. Then

$$\mathcal{B}(z) = 18 + 8 + 16 + 2 + 4 - 12 - 24 - 8 - 4 = 0. \quad \square$$

## 2.1.6 The extended Dynkin diagrams. Two different notation

Any diagram $\Gamma$ with a non-negative definite Tits form $\mathcal{B}$ is said to be an *extended Dynkin diagram*. All extended Dynkin diagrams are listed in Fig. 2.6, see [Bo], [DR76], [Kac80], [Kac83].

There are two different notation systems of the extended Dynkin diagrams: one used in the context of representations of quivers, and another one used in the context of twisted affine Lie algebras, see Table 2.1, Table 2.2, Remark 4.4, Remark 6.5 and Table 4.2.

**Table 2.1.**  Notation of extended Dyknin diagrams.

| In the context of representations of quivers | In the context of twisted affine Lie algebras | In the context of representations of quivers | In the context of twisted affine Lie algebras |
|---|---|---|---|
| $\widetilde{E}_6$ | $E_6^{(1)}$ | $\widetilde{G}_{22}$ | $G_2^{(1)}$ |
| $\widetilde{E}_7$ | $E_7^{(1)}$ | $\widetilde{G}_{21}$ | $D_4^{(3)}$ |
| $\widetilde{E}_8$ | $E_8^{(1)}$ | $\widetilde{F}_{42}$ | $F_4^{(1)}$ |
| $\widetilde{D}_n$ | $D_n^{(1)}$ $(n \geq 4)$ | $\widetilde{F}_{41}$ | $E_6^{(2)}$ |
| $\widetilde{A}_{11}$ | $A_2^{(2)}$ | $\widetilde{C}_n$ | $C_n^{(1)}$ $(n \geq 2)$ |
| $\widetilde{A}_{12}$ | $A_1^{(1)}$ | $\widetilde{B}_n$ | $D_{n+1}^{(2)}$ $(n \geq 2)$ |
| $\widetilde{BC}_n$ | $A_{2n}^{(2)}$ $(n \geq 2)$ | $\widetilde{A}_n$ | $A_n^{(1)}$ $(n \geq 2)$ |
| $\widetilde{CD}_n$ | $B_n^{(1)}$ $(n \geq 3)$ | $\widetilde{DD}_n$ | $A_{2n-1}^{(2)}$ $(n \geq 3)$ |

According to Proposition 2.3, we see that the simply-laced diagrams with only one branch point and with the non-negative definite Tits form are characterized as follows:

$$\mu = \frac{1}{p} + \frac{1}{q} + \frac{1}{r} = 1,$$

i.e., the following triples $p, q, r$ are possible:

**Table 2.2.**    Notation of affine Lie algebras. The upper index $r$ in the notation of twisted affine Lie algebras has an invariant sense: it is the order of the diagram automorphism $\mu$ of $\mathfrak{g}$. In this table, if $\mathfrak{g}$ is a complex simple finite dimensional Lie algebra of type $X_n$, then the corresponding affine Lie algebra is of type $X_n^{(r)}$, see Remark 4.4.

| Rang | Affine Lie algebra of type $X_n^{(r)}$ | | | | | Note |
|------|------|------|------|------|------|------|
| Aff1 | $A_1^{(1)}$, | $A_n^{(1)}$, | $B_n^{(1)}$, | $C_n^{(1)}$, | $D_n^{(1)}$, | Non-twisted, |
|      | $G_2^{(1)}$, | $F_4^{(1)}$, | $E_6^{(1)}$, | $E_7^{(1)}$ | $E_8^{(1)}$ | $r = 1$ |
| Aff2 | $A_2^{(2)}$ | $A_{2n}^{(2)}$ | $A_{2n-1}^{(2)}$ | $D_{n+1}^{(2)}$ | $E_6^{(2)}$ | Twisted, |
|      | | | | | | $r = 2$ |
| Aff3 | | | $D_4^{(3)}$ | | | Twisted, |
|      | | | | | | $r = 3$ |

1) $(3, 3, 3)$, i.e., the diagram $\widetilde{E}_6$,
2) $(2, 4, 4)$, i.e., the diagram $\widetilde{E}_7$,
3) $(2, 3, 6)$, i.e., the diagram $\widetilde{E}_8$,
see Fig. 2.6.

The multiply-laced extended Dynkin diagrams are: $\widetilde{A}_{12}$, $\widetilde{A}_{11}$, $\widetilde{BC}_n$, $\widetilde{B}_n$, $\widetilde{C}_n$, $\widetilde{CD}_n$, $\widetilde{DD}_n$, $\widetilde{F}_{41}$, $\widetilde{F}_{42}$, $\widetilde{G}_{21}$, $\widetilde{G}_{22}$, see Fig. 2.6.

### 2.1.7 Three sets of Tits forms

1) All Tits forms $\mathcal{B}$ fall into 3 non-intersecting sets:
   a) $\{\mathcal{B} \mid \mathcal{B}$ is positive definite, i.e., $\Gamma$ is the Dynkin diagram$\}$,
   b) $\{\mathcal{B} \mid \mathcal{B}$ is non-negative definite, i.e., $\Gamma$ is the extended Dynkin diagram$\}$,
   c) $\{\mathcal{B} \mid \mathcal{B}$ is indefinite$\}$.

Consider two operations:

$\wedge$ : *Add* a vertex and connect it with $\Gamma$ by only one edge. We denote the new graph $\overset{\wedge}{\Gamma}$.

$\vee$ : *Remove* a vertex and all incident edges (the new graph may contain more than one component). We denote the new graph $\overset{\vee}{\Gamma}$.

2) It is easy to see that
   a) The set $\{\mathcal{B} \mid \mathcal{B}$ is positive definite$\}$ is stable under $\vee$ (*Remove*) and is not stable under $\wedge$ (*Add*) ,

$\wedge$ : $\{\mathcal{B} \mid \mathcal{B}$ is positive definite$\}$ $\Longrightarrow$ $\begin{cases} \{\mathcal{B} \mid \mathcal{B}$ is non-negative definite$\}$ $\coprod \\ \{\mathcal{B} \mid \mathcal{B}$ is indefinite$\}, \end{cases}$

$\vee$ : $\{\mathcal{B} \mid \mathcal{B}$ is positive definite$\}$ $\Longrightarrow \{\mathcal{B} \mid \mathcal{B}$ is positive definite$\}$.

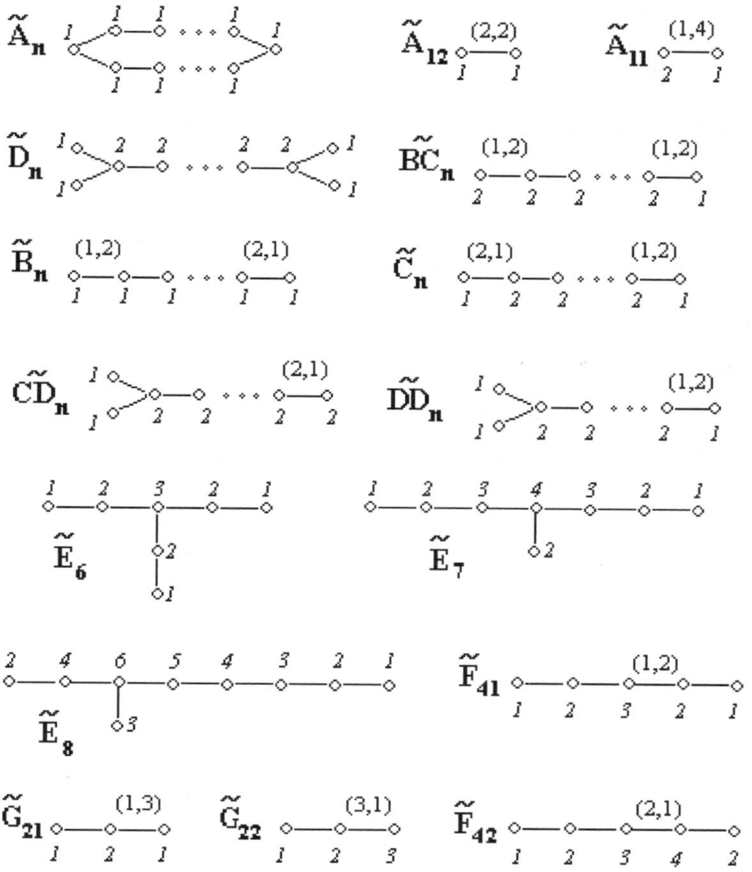

$\tilde{A}_n$

$\tilde{A}_{12}$  (2,2)

$\tilde{A}_{11}$  (1,4)

$\tilde{D}_n$

$\tilde{BC}_n$  (1,2)  (1,2)

$\tilde{B}_n$  (1,2)  (2,1)

$\tilde{C}_n$  (2,1)  (1,2)

$\tilde{CD}_n$  (2,1)

$\tilde{DD}_n$  (1,2)

$\tilde{E}_6$

$\tilde{E}_7$

$\tilde{E}_8$

$\tilde{F}_{41}$  (1,2)

$\tilde{G}_{21}$  (1,3)

$\tilde{G}_{22}$  (3,1)

$\tilde{F}_{42}$  (2,1)

**Fig. 2.6.** The extended Dynkin diagrams. The numerical labels at the vertices are the coefficients of the imaginary root which coincides with the fixed point $z^1$ of the Coxeter transformation, see §3.3.1, (3.23).

We consider the vector space $\mathcal{E}_\Gamma$ over $\mathbb{Q}$; set

$$\dim \mathcal{E}_\Gamma = |\Gamma_0|.$$

Let

$$\alpha_i = \{0, 0, ...0, \overset{i}{1}, 0, ...0, 0\} \in \mathcal{E}_\Gamma$$

be the basis vector corresponding to the vertex $v_i \in \Gamma_0$. The space $\mathcal{E}_\Gamma$ is spanned by the vectors $\{\alpha_i \mid i \in \Gamma_0\}$; the vectors $\alpha_1, ..., \alpha_n$ form a basis in $\mathcal{E}_\Gamma$. Let

$$\mathcal{E}_+ = \{\alpha = \sum k_i \alpha_i \in \mathcal{E}_\Gamma \mid k_i \in \mathbb{Z}_+, \ \sum k_i > 0\}$$

be the set of all non-zero elements in $\mathcal{E}_\Gamma$ with non-negative integer coordinates in the basis $\{\alpha_1, ..., \alpha_n\}$.

Define the linear functions $\phi_1, \ ..., \ \phi_n$ on $\mathcal{E}_\Gamma$ by means of the elements of the Cartan matrix (2.4):

$$\phi_i(\alpha_j) = k_{ij} \ .$$

The *positive root system* $\Delta_+$ associated to the Cartan matrix $K$ is a subset in $\mathcal{E}_+$ defined by the following properties (R1)–(R3):

(R1) $\alpha_i \in \Delta_+$ and $2\alpha_i \notin \Delta_+$ for $i = 1, ..., n$.

(R2) If $\alpha \in \Delta_+$ and $\alpha \neq \alpha_i$, then $\alpha + k\alpha_i \in \Delta_+$ for $k \in \mathbb{Z}$ if and only if $-p \leq k \leq q$ for some non-negative integers $p$ and $q$ such that $p - q = \phi_i(\alpha)$.

(R3) Any $\alpha \in \Delta_+$ has a connected support.

We define endomorphisms $\sigma_1, \ ..., \ \sigma_n$ of $\mathcal{E}_\Gamma$ by the formula

$$\sigma_i(x) = x - \phi_i(x)\alpha_i. \tag{2.24}$$

Each endomorphism $\sigma_i$ is the reflection in the hyperplane $\phi_i = 0$ such that $\sigma_i(\alpha_i) = -\alpha_i$. These reflections satisfy the following relations[1]:

$$\sigma_i^2 = 1, \quad (\sigma_i\sigma_j)^{n_{ij}} = 1,$$

where the exponents $n_{ij}$, corresponding to $k_{ij}k_{ji}$, are given in the Table 2.3, taken from [Kac80, p.63].

**Table 2.3.**    The exponents $n_{ij}$

| $k_{ij}k_{ji}$ | 0 | 1 | 2 | 3 | $\geq 4$ |
|---|---|---|---|---|---|
| $n_{ij}$ | 2 | 3 | 4 | 6 | $\infty$ |

The group $W$ generated by the reflections $\sigma_1, \ ..., \ \sigma_n$ is called the *Weyl group*. The vectors $\alpha_1, \ ..., \ \alpha_n$ are called *simple roots*; we denote by $\Pi$ the set of all simple roots. Let $W(\Pi)$ be the orbit of $\Pi$ under the $W$-action.

Following V. Kac [Kac80, p.64], set

$$M = \{\alpha \in \mathcal{E}_+ \mid \phi_i(\alpha) \leq 0 \text{ for } i = 1, ..., n, \text{ and } \alpha \text{ has a connected support }\}.$$

The set $M$ is called the *fundamental set*. Let $W(M)$ be the orbit of $M$ under the $W$-action. We set

---

[1] $\sigma^\infty = 1$ means that no power of $\sigma$ is equal to 1, i.e., $\sigma$ is free.

$$\Delta_+^{re} = \bigcup_{w \in W} (w(\Pi) \cap \mathcal{E}_+), \qquad \Delta_+^{im} = \bigcup_{w \in W} (w(M)). \qquad (2.25)$$

The elements of the set $\Delta_+^{re}$ are called *real roots* and the elements of the set $\Delta_+^{im}$ are called *imaginary roots*. By [Kac80], the *system of positive roots* $\Delta_+$ is the disjoint union of the sets $\Delta_+^{re}$ and $\Delta_+^{im}$:

$$\Delta_+ = \Delta_+^{re} \coprod \Delta_+^{im}.$$

We denote by $\Delta$ the set of all roots. It consists of the set of positive roots $\Delta_+$ and the set of negative root $\Delta_-$ obtained from $\Delta_+$ by multiplying by $-1$:

$$\Delta = \Delta_+ \coprod \Delta_-.$$

If the Tits form $\mathcal{B}$ (associated with the Cartan matrix $K$) is positive definite, then the root system is finite, it corresponds to a simple finite dimensional Lie algebra. In this case, the root system consists of real roots.

If the Tits form $\mathcal{B}$ is non-negative definite, we have an infinite root system whose imaginary root system is one-dimensional:

$$\Delta_+^{im} = \{\omega, 2\omega, 3\omega, ...\}, \qquad \text{where} \quad \omega = \sum_i k_i \alpha_i, \qquad (2.26)$$

the coefficients $k_i$ being the labels of the vertices from Fig. 2.6.

The elements $k\omega$ ($k \in \mathbb{N}$) are called *nil-roots*. Every nil-root is a fixed point for the Weyl group.

## 2.2.2 A category of representations of quivers and the P. Gabriel theorem

A *quiver* $(\Gamma, \Omega)$ is a connected graph $\Gamma$ with an orientation $\Omega$. Let $\Gamma_0$ be the set of all vertices of $\Gamma$. Any orientation $\Omega$ is given by a set of arrows $\Gamma_1$ that constitute $\Gamma_1$. Every arrow $\alpha \in \Gamma_1$ is given by its source point $s(\alpha) \in \Gamma_0$ and its target point $t(\alpha) \in \Gamma_0$. In [Gab72], P. Gabriel introduced the notion of representations of quivers in order to formulate a number of problems of linear algebra in a general way. P. Gabriel consider graphs without weighted edges.

Let $k$ be a field. A *representation* of the quiver $(\Gamma, \Omega)$ over $k$ is a set of spaces and linear maps between them $(V_i, \phi_\alpha)$, where to any vertex $i \in \Gamma_0$ a finite-dimensional space $V_i$ over $k$ is assigned, and to any arrow

$$i \xrightarrow{\alpha} j$$

a linear operator $\phi_\alpha : V_i \to V_j$ corresponds. All representations $(V_i, \phi_\alpha)$ of the quiver $(\Gamma, \Omega)$ constitute the category $\mathfrak{L}(\Gamma, \Omega)$. In this category, a *morphism*

$$\eta : (V_i, \phi_\alpha) \to (V_i', \phi_\alpha')$$

is a collection of linear maps $\eta_i : V_i \to V_i'$ for $i \in \Gamma_0$ such that for each arrow $\alpha : s(\alpha) \to t(\alpha)$, we have

$$\eta_{t(\alpha)}\phi_\alpha = \phi_\alpha'\eta_{s(\alpha)},$$

or, equivalently, the following square is commutative

$$
\begin{array}{ccc}
V_{s(\alpha)} & \xrightarrow{\ \phi_\alpha\ } & V_{t(\alpha)} \\
\downarrow{\scriptstyle \eta_{s(\alpha)}} & & \downarrow{\scriptstyle \eta_{t(\alpha)}} \\
V_{s(\alpha)}' & \xrightarrow{\ \phi_\alpha'\ } & V_{t(\alpha)}'
\end{array}
$$

A morphism $\eta : (V_i, \phi_\alpha) \to (V_i', \phi_\alpha')$ is an *isomorphism* in the category $\mathfrak{L}(\Gamma\Omega)$ if there exists a morphism $\eta' : (V_i', \phi_\alpha') \to (V_i, \phi_\alpha)$ such that $\eta\eta' = \mathrm{Id}_{(V_i', \phi_\alpha')}$ and $\eta'\eta = \mathrm{Id}_{(V_i, \phi_\alpha)}$.

Objects of the category $\mathfrak{L}(\Gamma, \Omega)$ are said to be *representations of the quiver* $(\Gamma, \Omega)$ considered up to an isomorphism. The *dimension* $\dim V$ of the representation $(V_i, \phi_\alpha)$ (or, which is the same, the *vector-dimension*) is an element of $\mathcal{E}_\Gamma$, with $(\dim V)_i = \dim V_i$ for all $i \in \Gamma_0$.

The *direct sum* of objects $(V, f)$ and $(U, g)$ in the category $\mathfrak{L}(\Gamma, \Omega)$ is the object

$$(W, h) = (V, f) \oplus (U, g),$$

where $W_i = V_i \oplus U_i$ and $h_\alpha = f_\alpha \oplus g_\alpha$ for all $i \in \Gamma_0$ and $\alpha \in \Gamma_1$. The nonzero object $(V, f)$ is said to be *indecomposable*, if it can not be represented as a direct sum of two nonzero objects, see [Gab72], [BGP73]. As usual in the representation theory, the main question is the description of all indecomposable representations in the category $\mathfrak{L}(\Gamma, \Omega)$.

**Theorem 2.14.** (P. Gabriel, [Gab72]) *1) A given quiver has only a finite number of indecomposable representations if and only if it is a simply-laced Dynkin diagram with arbitrary orientations of the edges.*

*2) Vector-dimensions of indecomposable representations coincide with positive roots in the root system of the corresponding Dynkin diagram.*

V. Dlab and C. M. Ringel [DR76] and L. A. Nazarova, S. A. Ovsienko and A. V. Roiter [NaOR77], [NaOR78] extended the P. Gabriel theorem to the multiply-laced case.

### 2.2.3 Finite-type, tame and wild quivers

In [BGP73], Bernstein, Gelfand and Ponomarev introduced the *Coxeter functors* $\Phi^+$, $\Phi^-$ leading to a new proof and new understanding of the P. Gabriel theorem. They also introduced *regular representations* of quivers, i.e., representations that never vanish under the Coxeter functors.

The Dynkin diagrams do not have regular representations; in the category of quiver representations[1], only a finite set of indecomposable representations is associated to any Dynkin diagram. Quivers with such property are called *finite-type quivers*. According to P. Gabriel's theorem [Gab72], *a quiver is of finite-type if and only if it is a (simply-laced) Dynkin diagram.*

In the category of all representations of a given quiver, the regular representations are the most complicated ones; they have been completely described only for the extended Dynkin diagrams which for this reason, in the representation theory of quivers, were called *tame quivers*. The following theorem is an extension of the P. Gabriel theorem to extended Dynkin diagrams and is due to L. A. Nazarova [Naz73], P. Donovan, and M. R. Freislich [DF73], V. Dlab and C. M. Ringel [DR74], [DR74a], [DR76].

**Theorem 2.15.** (Nazarova, Donovan-Freislich, Dlab-Ringel) *1) A quiver $\Gamma$ is of tame type if and only if it is a simply-laced extended Dynkin diagram with arbitrary orientations of the edges.*

*2) Dimensions of indecomposable representations coincide with positive roots in the root system of the corresponding diagram. For each positive real root $\alpha^{re}$ of $\Gamma$, there exists a unique indecomposable $V$ (up to isomorphism) with $\dim V = \alpha^{re}$. For each positive imaginary root $\alpha^{im}$, the isomorphism classes of indecomposable representations $V$ with $\dim V = \alpha^{im}$ are parameterized by an infinite subset of $k\mathbb{P}^1$.*

Every quiver with indefinite Tits form $\mathcal{B}$ is *wild*, i.e., the description of its representations contains the problem of classifying pairs of matrices up to simultaneous similarity; this classification is hopeless in a certain sense, see [GP69], [Naz73], [Drz80], [Kac83].

### 2.2.4 The V. Kac theorem on the possible dimension vectors

According to the P. Gabriel theorem (Th.2.14) and the Nazarova, Donovan-Freislich, Dlab-Ringel theorem (Th.2.15) all quivers, which are neither Dynkin diagram nor extended Dynkin diagram are wild. H. Kraft and Ch. Riedtmann write in [KR86, p.109] : *"Since all remaining quivers are wild, there was little hope to get any further, except maybe in some special case. Therefore Kac's spectacular paper [Kac80], where he describes the dimension types of all indecomposables of arbitrary quivers, came as a big surprise. In [Kac82] and [Kac83] Kac improved and completed his first results"*. Note, that the work [KR86] of H. Kraft and Ch. Riedtmann is a nice guide to the Kac theorem.

Following V. Kac [Kac82, p.146] (see also [Kac83, p.84], [KR86, p.125]) the integers $\mu_\alpha$ an $r_\alpha$ are introduced as follows:

$\mu_\alpha$ *is the maximal dimension of an irreducible component in the set of isomorphism classes of indecomposable representations of dimension $\alpha$ and $r_\alpha$ is the number of such components.*

The following theorem was first proved in [Kac80]. It is given here in a more consolidated version [Kac83, §1.10].

---

[1] Sometimes, one says about *graph representations* meaning *quiver representations*.

**Theorem 2.16 (V. Kac).** *Let the base field $k$ be algebraically closed. Let $(\Gamma, \Omega)$ be a quiver. Then*

*a) There exists an indecomposable representation of dimension $\alpha \in \mathcal{E}_\Gamma \backslash 0$ if and only if $\alpha \in \Delta_+(\Gamma)$.*

*b) There exists a unique indecomposable representation of dimension $\alpha$ if and only if $\alpha \in \Delta_+^{re}(\Gamma)$.*

*c) If $\alpha \in \Delta_+^{im}(\Gamma)$, then*

$$\mu_\alpha(\Gamma, \Omega) = 1 - (\alpha, \alpha) > 0, \qquad r_a = 1,$$

*where $(\cdot, \cdot)$ is the quadratic Tits form, see §2.1.1.*

The integer $\mu_\alpha$ is called the *number of parameters*, [KR86]. For example, let $V$ be an indecomposable representation of a tame quiver and $\dim V$ an imaginary root, i.e., $(\dim V, \dim V) = 0$. Then $\mu_\alpha = 1$ and the irreducible component of indecomposable representations is one-parametric as in Theorem 2.15. Let $V$ be an indecomposable representation of a wild quiver and $(\dim V, \dim V) < 0$. Then $\mu_\alpha > 1$, and the irreducible component containing $V$ is at least two-parametric.

For a wild quiver with the indefinite Tits form, we have

The following lemma [Kac83, §1.10] is one of the crucial moments in the proof of Kac's theorem.

**Lemma 2.17 (V. Kac).** *The number of indecomposable representations of dimension $\alpha$ (if it is finite) and $\mu_\alpha(\Gamma, \Omega)$ are independent of the orientation $\Omega$.*

The proof of this lemma and further details of Kac's theorem can be also found in the work [KR86, §5] of H. Kraft and Ch. Riedtmann and in works [CrW93, p.32], [CrW99], [CrW01, p.40] of W. Crawley-Boevey.

The main part in the proof of this lemma is counting points of certain varieties for the case of algebraic closure $\overline{\mathbb{F}}_p$ of the finite field $\mathbb{F}_p$, for any prime $p$, see [KR86, §5.6], with subsequent transfer of the result to fields of characteristic zero [KR86, §5.6]. *"So this is one of the examples where the only proof known for a result about fields of characteristic zero passes via fields of positive characteristic."* [KR86, §5.1]

### 2.2.5 The quadratic Tits form and vector-dimensions of representations

In the simply-laced case, by (1.5), (3.3), the quadratic Tits form on the space of vectors $\{x_v\}_{v \in \Gamma_0}$ associated with quiver $(\Gamma, \Omega)$ can be expressed as follows:

$$\mathcal{B}(x) = \sum_{v \in \Gamma_0} x_v^2 - \sum_{l \in \Gamma_1} x_{v(l)} x_{u(l)},$$

where $v(l)$ and $u(l)$ are the endpoints of the edge $l \in \Gamma_1$, see, e.g., (2.7). By Kac's theorem (Theorem 2.16, b)), the representation $V$ of $(\Gamma, \Omega)$ is a unique indecomposable representation of the given dimension if and only if

$$\mathcal{B}(\dim V) = 1.$$

It is so because such dimensions are real vectors lying by eq. (2.25) on the orbit of simple positive roots under the action of the Weyl group, which preserves the quadratic form $\mathcal{B}$, see, e.g., [BGP73].

An analogous proposition takes place for representations of posets. Let $(\mathfrak{D}, \geq)$ be a finite partially ordered set (*poset*). Let $\mathfrak{D} = \{1, 2, \ldots, n\}$ (not necessary with the natural order) and denote:

$$\widehat{\mathfrak{D}} = \mathfrak{D} \coprod \{0\}.$$

A poset is called *primitive* if it is a disjoint unit of several ordered chains such that the elements of different chains are non-comparable. We denote such a poset by $(n_1, n_2, \ldots, n_s)$, where $n_i$ are the lengths of the chains. A *representation* $V$ of $\mathfrak{D}$ over a field $k$ is an order-preserving map $\mathfrak{D}$ into the set of subspaces of a finite dimensional vector space $V(0)$ over $k$, see [DrK04]. The category of representations of $\mathfrak{D}$ and indecomposable representations are defined in the natural way [NR72]. The *dimension* $\dim(u)$ of an element $u \in \mathfrak{D}$ is the vector whose components are

$$\dim(0) = \dim V(0), \quad \dim(a) = \dim(V(a)/\sum_{b \leq a} V(b)), \quad a \in \mathfrak{D}.$$

The quadratic form $\mathcal{B}_{\mathfrak{D}}$ associated to a poset $\mathfrak{D}$ is the quadratic form

$$\mathcal{B}_{\mathfrak{D}}(x_0, x_1, \ldots, x_n) = \sum_{a \in \widehat{\mathfrak{D}}} x_a^2 + \sum_{\substack{a,b \in \mathfrak{D} \\ a \leq b}} x_a x_b - \sum_{a \in \mathfrak{D}} x_0 x_a.$$

In [DrK04], Yu. A. Drozd and E. Kubichka consider the *dimensions of finite type* of representations of a partially ordered set, i.e., such that there are only finitely many isomorphism classes of representations of this dimension. In particular, they show that an element $u \in \mathfrak{D}$ is indecomposable if and only if $\mathcal{B}_{\mathfrak{D}}(\dim(u)) = 1$. For primitive posets, this theorem was deduced by P. Magyar, J. Weyman, A. Zelevinsky in [MWZ99] from the results of Kac [Kac80]. This approach cannot be applied in general case, see [DrK04, p.5], so Yu. A. Drozd and E. Kubichka used the original technique of *derivations* (or *differentiations*), which is due to L. A. Nazarova and A. V. Roiter, [NR72].

### 2.2.6 Orientations and the associated Coxeter transformations

Here, we follow the definitions of Bernstein-Gelfand-Ponomarev [BGP73]. Let us consider the graph $\Gamma$ endowed with an orientation $\Omega$.

The vertex $v_i$ is called *sink-admissible* (resp. *source-admissible*) in the orientation $\Omega$ if every arrow containing $v_i$ ends in (resp. starts from) this vertex. The reflection $\sigma_i$ is applied to the vector-dimensions by (2.24) and can be applied only to vertex which is either sink-admissible or source-admissible. The reflection $\sigma_i$ acts on the orientation $\Omega$ by reversing all arrows containing the vertex $v_i$.

Consider now a sequence of vertices and the corresponding reflections. A sequence of vertices

$$\{v_{i_n}, v_{i_{n-1}}, \; ..., \; v_{i_3}, v_{i_2}, v_{i_1}\}$$

is called *sink-admissible*, if the vertex $v_{i_1}$ is sink-admissible in the orientation $\Omega$, the vertex $v_{i_2}$ is sink-admissible in the orientation $\sigma_{i_1}(\Omega)$, the vertex $v_{i_3}$ is sink-admissible in the orientation $\sigma_{i_2}\sigma_{i_1}(\Omega)$, and so on. *Source-admissible sequences* are similarly defined.

A sink-admissible (resp. source-admissible) sequence

$$\mathcal{S} = \{v_{i_n}, v_{i_{n-1}}, \; ..., \; v_{i_3}, v_{i_2}, v_{i_1}\}$$

is called *fully sink-admissible* (resp. *fully source-admissible*) if $\mathcal{S}$ contains every vertex $v \in \Gamma_0$ exactly once. Evidently, the inverse sequence $\mathcal{S}^{-1}$ of a fully sink-admissible sequence $\mathcal{S}$ is fully source-admissible and vice versa.

Every tree has a fully sink-admissible sequence $\mathcal{S}$. To every sink-admissible sequence $\mathcal{S}$, we assign the Coxeter transformation depending on the order of vertices in $\mathcal{S}$:

$$\mathbf{C} = \sigma_{i_n}\sigma_{i_{n-1}}...\sigma_{i_2}\sigma_{i_1}. \tag{2.27}$$

For every orientation $\Omega$ of the tree, every fully sink-admissible sequence gives rise to the same Coxeter transformation $\mathbf{C}_\Omega$, and every fully source-admissible sequence gives rise to $\mathbf{C}_\Omega^{-1}$. Thus, *to every orientation $\Omega$ of the tree, we assign two Coxeter transformations:* $\mathbf{C}_\Omega$ *and* $\mathbf{C}_\Omega^{-1}$.

Obviously, $\mathbf{C}_\Omega$ acts trivially on the orientation $\Omega$ because every edge of the tree is twice reversed. However, the Coxeter transformation does not act trivially on the space of vector-dimensions, see (2.24).

## 2.3 The Poincaré series

### 2.3.1 The graded algebras, symmetric algebras, algebras of invariants

Let $k$ be a field. We define a *graded $k$-algebra* to be a finitely generated $k$-algebra $A$ (associative, commutative, and with unit), together with a direct sum decomposition (as vector space)

$$A = A_0 \oplus A_1 \oplus A_2 \oplus \dots,$$

such that $A_0 = k$ and $A_i A_j \subset A_{i+j}$. The component $A_n$ is called the $n$th *homogeneous part of $A$* and any element $x \in A_n$ is said to be *homogeneous of degree $n$*, notation: $\deg x = n$.

We define a *graded $A$-module* to be a finitely generated $A$-module, together with a direct sum decomposition

$$M = M_0 \oplus M_1 \oplus M_2 \oplus \dots,$$

such that $A_i M_j \subset M_{i+j}$.

The *Poincaré series* of a graded algebra $A = \overset{\infty}{\underset{n=0}{\oplus}} A_i$ is the formal series

$$P(A, t) = \sum_{n=0}^{\infty} (\dim A_n) t^n.$$

The *Poincaré series* of a graded $A$-module $M = \overset{\infty}{\underset{n=0}{\oplus}} M_i$ is the formal series

$$P(M, t) = \sum_{n=0}^{\infty} (\dim M_n) t^n,$$

see [Sp77], [PV94].

**Theorem 2.18.** (Hilbert, Serre, see [AtMa69, p.117, Th.11.1]) *The Poincaré series $P(M, t)$ of a finitely generated graded $A$-module is a rational function in $t$ of the form*

$$P(M, t) = \frac{f(t)}{\prod\limits_{i=1}^{s} (1 - t^{k_i})}, \quad \text{where } f(t) \in \mathbb{Z}[t].$$

In what follows in this section, any algebraically closed field $k$ can be considered instead of $\mathbb{C}$. Set

$$R = \mathbb{C}[x_1, \dots, x_n]. \tag{2.28}$$

The set $R_d$ of homogeneous polynomials of degree $d$ is a finite dimensional subspace of $R$, and $R_0 = \mathbb{C}$. Moreover, $R_d R_e \subset R_{d+e}$, and $R$ is a graded $\mathbb{C}$-algebra with a direct sum decomposition (as a vector space)

$$R = R_0 \oplus R_1 \oplus R_2 \oplus \dots \tag{2.29}$$

Let $V = \mathrm{Span}(x_1, \dots, x_n)$. Let $f_1, \dots, f_n \in V^* = \mathrm{Hom}_{\mathbb{C}}(V, \mathbb{C})$ be the linear forms defined by $f_i(x_j) = \delta_{ij}$, i.e.,

$$f_i(\lambda_1 x_1 + \dots \lambda_n x_n) = \lambda_i, \quad \text{where } \lambda_i \in \mathbb{C}.$$

b) The set $\{\mathcal{B} \mid \mathcal{B}$ is non-negative definite$\}$ is not stable under $\vee$ (*Remove*) and $\wedge$ (*Add*),

$\wedge$ : $\{\mathcal{B} \mid \mathcal{B}$ is non-negative definite$\}$ $\implies$ $\{\mathcal{B} \mid \mathcal{B}$ is indefinite$\}$,

$\vee$ : $\{\mathcal{B} \mid \mathcal{B}$ is non-negative definite$\}$ $\implies$ $\{\mathcal{B} \mid \mathcal{B}$ is positive definite$\}$.

c) The set $\{\mathcal{B} \mid \mathcal{B}$ is indefinite$\}$ is stable under $\wedge$ (*Add*) but not stable under $\vee$ (*Remove*):

$\wedge$ : $\{\mathcal{B} \mid \mathcal{B}$ is indefinite$\}$ $\implies$ $\{\mathcal{B} \mid \mathcal{B}$ is indefinite$\}$,

$$\vee : \{\mathcal{B} \mid \mathcal{B} \text{ is indefinite}\} \implies \left\{ \begin{array}{l} \{\mathcal{B} \mid \mathcal{B} \text{ is positive definite}\} \coprod \\ \{\mathcal{B} \mid \mathcal{B} \text{ is non-negative definite}\} \coprod \\ \{\mathcal{B} \mid \mathcal{B} \text{ is indefinite}\}. \end{array} \right.$$

3) If the graph $\Gamma$ with indefinite form $\mathcal{B}$ is obtained from any Dynkin diagram by adding an edge, then the same graph $\Gamma$ can be obtained by adding an edge (or maybe several edges) to an extended Dynkin diagram.

### 2.1.8 The hyperbolic Dynkin diagrams and hyperbolic Cartan matrices

A connected graph $\Gamma$ with indefinite Tits form is said to be a *hyperbolic Dynkin diagram* (resp. *strictly hyperbolic Dynkin diagram*) if every subgraph $\Gamma' \subset \Gamma$ is a Dynkin diagram or an extended Dynkin diagram (resp. Dynkin diagram). The corresponding Cartan matrix $K$ is said to be *hyperbolic* (resp. *strictly hyperbolic*), see [Kac93, exs. of §4.10]. The corresponding Weyl group is said to be a *hyperbolic Weyl group* (resp. *compact hyperbolic Weyl group*), see [Bo, Ch.5, exs. of §4].

The Tits form $\mathcal{B}$ is indefinite for $\mu < 1$. Since $U(x)$, $U(y)$, $U(z)$ in (2.9) are non-negative definite, we see that, in this case, the signature of $\mathcal{B}$ is equal to $(n-1,1)$, see [Kac93, exs.4.2]. It is easy to check that the triples $p, q, r$ corresponding to hyperbolic graphs $T_{p,q,r}$ are only the following ones:

1) $(2, 3, 7)$, i.e., the diagram $E_{10}$, and $1 - \mu = \dfrac{1}{42}$,

2) $(2, 4, 5)$, and $1 - \mu = \dfrac{1}{20}$,

3) $(3, 3, 4)$, and $1 - \mu = \dfrac{1}{12}$.

## 2.2 Representations of quivers

### 2.2.1 The real and imaginary roots

Recall now the definitions of *imaginary* and *real* roots in the infinite root system associated with infinite dimensional Kac-Moody Lie algebras. We mostly follow V. Kac's definitions [Kac80], [Kac82], [Kac93].

Then the $f_i$, where $i = 1, \ldots, n$, generate a *symmetric algebra*[1] $\mathrm{Sym}(V^*)$ isomorphic to $R$.

Define an isomorphism $\Phi : \mathrm{Sym}(V^*) \to R$ by setting

$$\Phi f_i = x_i \text{ for all } i.$$

Let $\mathrm{Sym}^m(V^*)$ denote the *mth symmetric power* of $V^*$, which consists of the homogeneous polynomials of degree $m$ in $x_1, \ldots, x_n$. Thus,

$$\mathrm{Sym}^0(V^*) = \mathbb{C},$$
$$\mathrm{Sym}^1(V^*) = V^*,$$
$$\mathrm{Sym}^2(V^*) = \{x_i x_j \mid 1 \le i \le j \le n\}, \quad \dim \mathrm{Sym}^2(V^*) = \binom{n+1}{2},$$
$$\mathrm{Sym}^3(V^*) = \{x_i x_j x_k \mid 1 \le i \le j \le k \le n\}, \quad \dim \mathrm{Sym}^3(V^*) = \binom{n+2}{3},$$

$$\cdots$$

$$\mathrm{Sym}^m(V^*) = \{x_{i_1} \ldots x_{i_m} \mid 1 \le i_1 \le \cdots \le i_m \le n\},$$
$$\dim \mathrm{Sym}^m(V^*) = \binom{n+m-1}{m},$$

see [Sp77], [Ben93].

For $n = 2$, $m \ge 1$, we have

$$\mathrm{Sym}^m(V^*) = \{x^m, x^{m-1}y, \ldots, xy^{m-1}, y^m\}$$

is the space of homogeneous polynomials of degree $m$ in two variables $x$, $y$, and

$$\dim \mathrm{Sym}^m(V^*) = \binom{m+1}{m} = m.$$

For any $a \in GL(V)$ and $f \in R$, define $af \in R$ by the rule

$$(af)(v) = f(a^{-1}v) \text{ for any } v \in V.$$

Then, for any $a, b \in G$ and $f \in R$, we have

$$a(bf) = bf(a^{-1}v) = f(b^{-1}a^{-1}v) = f((ab)^{-1}v) = ((ab)f)(v),$$

and $aR_d = R_d$.

Let $G$ be a subgroup in $GL(V)$. We say that $f \in R$ is *G-invariant* if $af = f$ for all $a \in G$. The G-invariant polynomial functions form a subalgebra $R^G$ of $R$, which is a graded, or, better say, homogeneous, subalgebra, i.e.,

$$R^G = \bigoplus R^G \cap R_i.$$

---

[1] The symmetric algebra $\mathrm{Sym}(V)$ is the quotient ring of the tensor algebra $T(V)$ by the ideal generated by elements $vw - wv$ for $v$ and $w$ in $V$, where $vw := v \otimes w$.

The algebra $R^G$ is said to be the *algebra of invariants* of the group $G$. If $G$ is finite, then

$$P(R^G, t) = \frac{1}{|G|} \frac{1}{\sum\limits_{g \in G} \det(1 - tg)} . \tag{2.30}$$

Eq. (2.30) is a classical theorem of Molien (1897), see [PV94, §3.11], or [Bo, Ch.5, §5.3].

### 2.3.2 The invariants of finite groups generated by reflections

Let $G$ be a finite subgroup of $GL(V)$ and $g \in G$. Then $g$ is called a *pseudo-reflection* if precisely one eigenvalue of $g$ is not equal to 1. Any pseudo-reflection with determinant $-1$ is called a *reflection*. For example, the element

$$g = \begin{pmatrix} -1 & 0 & 0 \\ 0 & 1 & 1 \\ 0 & 0 & 1 \end{pmatrix}$$

has infinite order and does not belong to any finite group $G$, i.e., $g$ is not reflection.

**Theorem 2.19 (Shephard-Todd, Chevalley, Serre).** *Let $G$ be a finite subgroup of $GL(V)$. There exist $n = \dim V$ algebraically independent homogeneous invariants $\theta_1, \ldots, \theta_n$ such that*

$$R^G = \mathbb{C}[\theta_1, \ldots, \theta_n]$$

*if and only if $G$ is generated by pseudo-reflections.*

For references, see [Stn79, Th.4.1], [Ch55, Th.A, p.778].

Shephard and Todd [ShT54] explicitly determined all finite subgroups of $GL(V)$ generated by pseudo-reflections and verified the sufficient condition of Theorem 2.19. Chevalley [Ch55] found the classification-free proof of this sufficient condition for a particular case where $G$ is generated by reflections. Serre observed that Chevalley's proof is also valid for groups generated by pseudo-reflections. Shephard and Todd ([ShT54]) proved the necessary condition of the theorem by a strong combinatorial method; see also Stanley [Stn79, p.487].

The coefficients of the Poincaré series are called the *Betti numbers*, see [Ch50], [Col58]. The Poincaré polynomial of the algebra $R^G$ is

$$(1 + t^{2p_1+1})(1 + t^{2p_2+1}) \ldots (1 + t^{2p_n+1}), \tag{2.31}$$

where $p_i + 1$ are the degrees of homogeneous basis elements of $R^G$, see [Cox51], and [Ch50]. See Table 2.4 taken from [Cox51, p.781,Tab.4].

Let $\lambda_1, \ldots, \lambda_n$ be the eigenvalues of a Coxeter transformation in a finite Weyl group. These eigenvalues can be given in the form

**Table 2.4.**    The Poincaré polynomials for the simple compact Lie groups

| Dynkin diagram | Poincaré polynomial |
|---|---|
| $A_n$ | $(1+t^3)(1+t^5)\ldots(1+t^{2n+1})$ |
| $B_n$ or $C_n$ | $(1+t^3)(1+t^7)\ldots(1+t^{4n-1})$ |
| $D_n$ | $(1+t^3)(1+t^7)\ldots(1+t^{4n-5})(1+t^{2n-1})$ |
| $E_6$ | $(1+t^3)(1+t^9)(1+t^{11})(1+t^{15})(1+t^{17})(1+t^{23})$ |
| $E_7$ | $(1+t^3)(1+t^{11})(1+t^{15})(1+t^{19})(1+t^{23})(1+t^{27})(1+t^{35})$ |
| $E_8$ | $(1+t^3)(1+t^{15})(1+t^{23})(1+t^{27})(1+t^{35})(1+t^{39})(1+t^{47})(1+t^{59})$ |
| $F_4$ | $(1+t^3)(1+t^{11})(1+t^{15})(1+t^{23})$ |
| $G_2$ | $(1+t^3)(1+t^{11})$ |

$$\omega^{m_1}, \ldots, \omega^{m_n},$$

where $\omega = \exp^{2\pi i/h}$ is a primitive root of unity. The numbers $m_1, ..., m_n$ are called the *exponents* of the Weyl group. H. S. M. Coxeter observed that the *exponents* $m_i$ and numbers $p_i$ in (2.31) coincide (see the epigraph to this chapter).

**Theorem 2.20 (Coxeter, Chevalley, Coleman, Steinberg).** *Let*

$$u_1, \ldots, u_n$$

*be homogeneous elements generating the algebra of invariants* $R^G$, *where* $G$ *is the Weyl group corresponding to a simple compact Lie group. Let* $m_i + 1 = \deg u_i$, *where* $i = 1, 2, ...n$. *Then the exponents of the group* $G$ *are*

$$m_1, \ldots, m_n.$$

For more details, see [Bo, Ch.5, §6.2, Prop.3] and historical notes in [Bo], and [Ch50], [Cox51], [Col58], [Stb85].

# The Jordan normal form of the Coxeter transformation

> It turned out that most of the classical concepts of the Killing-Cartan-Weyl theory can be carried over to the entire class of Kac-Moody algebras, such as the Cartan subalgebra, the root system, the Weyl group, etc. ... I shall only point out that $\mathfrak{g}'(K)$ [1] does not always possess a nonzero invariant bilinear form. This is the case if and only if the matrix $K$ is *symmetrizable* ...

V. Kac, [Kac93, p.XI], 1993

## 3.1 The Cartan matrix and the Coxeter transformation

In this subsection a graph $\Gamma$ and a partition $S = S_1 \coprod S_2$ of its vertices are fixed.

### 3.1.1 A bicolored partition and a bipartite graph

A partition $S = S_1 \coprod S_2$ of the vertices of the graph $\Gamma$ is said to be *bicolored* if all edges of $\Gamma$ lead from $S_1$ to $S_2$. A *bicolored partition* exists if and only if all cycles in $\Gamma$ are of even length. The graph $\Gamma$ admitting a bicolored partition is said to be *bipartite* [McM02]. An orientation $\Lambda$ is said to be *bicolored*, if there is the corresponding sink-admissible sequence

---

[1] Here, $\mathfrak{g}'(K)$ is the subalgebra $[\mathfrak{g}(K), \mathfrak{g}(K)]$ of the Kac-Moody algebra $\mathfrak{g}(K)$ associated with the generalized Cartan matrix $K$. One has

$$\mathfrak{g}(K) = \mathfrak{g}'(K) + \mathfrak{h},$$

where $\mathfrak{h}$ is the Cartan subalgebra, $\mathfrak{g}(K) = \mathfrak{g}'(K)$ if and only if $\det K \neq 0$, [Kac93, §1.3 and Th.2.2].

$$\{v_1, v_2, \ldots, v_m, v_{m+1}, v_{m+2}, \ldots v_{m+k}\}$$

of vertices in this orientation $\Lambda$, such that subsequences

$$S_1 = \{v_1, v_2, \ldots, v_m\} \text{ and } S_2 = \{v_{m+1}, v_{m+2}, \ldots, v_{m+k}\}$$

form a bicolored partition, the set $S_1$ (resp. $S_2$) contains all sources (resp. sinks) of $S$. In other words, all arrows go from $S_1$ to $S_2$.

Let $W(S)$ (resp. $W(S_1)$, resp. $W(S_2)$) be the Coxeter group generated by all reflections corresponding to vertices of $S$ (resp. $S_1$, resp. $S_2$). Any two generators $g'$ and $g''$ of the Coxeter group $W(S)$ (contained in the same subpartition) commute and the subgroups $W(S_1)$ and $W(S_2)$ are abelian. So, the products $w_i \in W(S_i)$ for $i = 1, 2$ of generators of $W(S_i)$ are involutions, i.e.,

$$w_1^2 = 1, \quad w_2^2 = 1 .$$

For the first time (as far as I know), the technique of bipartite graphs was used by R. Steinberg in [Stb59], where he gave classification-free proofs for some results of H. S. M. Coxeter [Cox34] concerning properties of the order of the Coxeter transformation; see also R. Carter's paper [Car70].

### 3.1.2 Conjugacy of Coxeter transformations

All Coxeter transformations are conjugate for any tree or forest $\Gamma$ [Bo, Ch.5, §6]; see also Proposition B.5 and Remark B.6. The Coxeter transformations for the graphs with cycles are studied in [Col89], [Rin94], [Shi00], [BT97], and in the works by Menshikh and Subbotin of 1982–1985, see §4.2. Here, we consider only trees.

As it is mentioned in §2.2.6, there are two Coxeter transformations corresponding to every orientation of the tree. Two Coxeter transformations corresponding to the bicolored orientation are called *bicolored Coxeter transformations*. We choose one of two bicolored Coxeter transformations as very simple to study. Here,

$$\mathbf{C} = w_1 w_2 \quad \text{or} \quad \mathbf{C}^{-1} = w_2 w_1. \tag{3.1}$$

From now on we assume that $S_1$ contains $m$ elements and $S_2$ contains $k$ elements, we denote by $a_1, \ldots, a_m$ (resp. $b_1, \ldots, b_k$) basis vectors corresponding to vertices $v_1, \ldots, v_m$ of $S_1$ (resp. vertices $v_{m+1} \ldots, v_{m+k}$ of $S_2$). We denote by $\mathcal{E}_{\Gamma_a}$ (resp. $\mathcal{E}_{\Gamma_b}$) the vector space generated by the $a_i$, where $i = 1, \ldots, m$ (resp. by the $b_i$, where $i = 1, \ldots, k$). So,

$$\dim \mathcal{E}_{\Gamma_a} = m, \quad \dim \mathcal{E}_{\Gamma_b} = k .$$

### 3.1.3 The Cartan matrix and the bicolored Coxeter transformation

The Cartan matrix and the bicolored Coxeter transformation are constructed from the same blocks. More exactly, the matrix $\mathbf{B}$ and involutions $w_i$, where $i = 1, 2$, are constructed from the same blocks.

In the simply-laced case (i.e., for the symmetric Cartan matrix), we have

$$K = 2\mathbf{B}, \quad \text{where} \quad \mathbf{B} = \begin{pmatrix} I_m & D \\ D^t & I_k \end{pmatrix},$$

$$w_1 = \begin{pmatrix} -I_m & -2D \\ 0 & I_k \end{pmatrix}, \quad w_2 = \begin{pmatrix} I_m & 0 \\ -2D^t & -I_k \end{pmatrix}, \tag{3.2}$$

where the elements $d_{ij}$ that constitute matrix $D$ are given by the formula

$$d_{ij} = (a_i, b_j) = \begin{cases} -\dfrac{1}{2} & \text{if } |v(a_i) - v(b_j)| = 1 \,, \\ 0 & \text{if } |v(a_i) - v(b_j)| > 1 \,, \end{cases} \tag{3.3}$$

where $v(a_i)$ and $v(b_j)$ are vertices lying in the different sets of the bicolored partition, see §3.1.2.

In the multiply-laced case (i.e., for the symmetrizable and non-symmetric Cartan matrix $K$), we have

$$K = U\mathbf{B}, \quad \text{where} \quad K = \begin{pmatrix} 2I_m & 2D \\ 2F & 2I_k \end{pmatrix},$$

$$w_1 = \begin{pmatrix} -I_m & -2D \\ 0 & I_k \end{pmatrix}, \quad w_2 = \begin{pmatrix} I_m & 0 \\ -2F & -I_k \end{pmatrix} \tag{3.4}$$

with

$$d_{ij} = \frac{(a_i, b_j)}{(a_i, a_i)}, \qquad f_{pq} = \frac{(b_p, a_q)}{(b_p, b_p)},$$

where the $a_i$ and $b_j$ are simple roots in the root systems of the corresponding to $S_1$ and $S_2$ Kac-Moody Lie algebras [Kac80]. Let $U = (u_{ij})$ be the diagonal matrix (2.4). Then

$$u_{ii} = \frac{2}{(a_i, a_i)} = \frac{2}{\mathcal{B}(a_i)}.$$

We have

$$\mathbf{B} = \begin{pmatrix} (a_i, a_i) & \cdots & (a_i, b_j) \\ & \cdots & \\ (a_i, b_j) & \cdots & (b_j, b_j) \end{pmatrix},$$

and

$$K = U\mathbf{B} = \begin{pmatrix} 2 & \cdots & \dfrac{2(a_i, b_j)}{(a_i, a_i)} \\ & \cdots & \\ \dfrac{2(a_i, b_j)}{(b_j, b_j)} & \cdots & 2 \end{pmatrix}.$$

Dividing $U$ and $\mathbf{B}$ into blocks of size $m \times m$ and $k \times k$, we see that

$$U = \begin{pmatrix} 2U_1 & 0 \\ 0 & 2U_2 \end{pmatrix}, \quad \mathbf{B} = \begin{pmatrix} U_1^{-1} & A \\ A^t & U_2^{-1} \end{pmatrix}, \quad U_1 A = D, \quad U_2 A^t = F. \tag{3.5}$$

*Remark 3.1.* According to bicolored partition $S = S_1 \coprod S_2$ of the graph $\Gamma$ §3.1.1, and the corresponding partition of matrices (3.2), (3.4) into four blocks, we have also partition of any vector $v$ into two blocks:

$$v = \begin{pmatrix} v_x \\ v_y \end{pmatrix}.$$

The component $v_x$ (resp. $v_y$) of the vector $v$ is said to be $\mathbb{X}$-*component* (resp. $\mathbb{Y}$-*component*), see §3.3.1.

### 3.1.4 The dual graphs and dual forms

Every valued graph $\Gamma$ has a dual graph denoted by $\Gamma^\vee$. The dual graph is obtained by means of transposition $d_{ij} \leftrightarrow d_{ji}$. In other words, if $K$ is the Cartan matrix for $\Gamma$, then the Cartan matrix for $\Gamma^\vee$ is

$$K^\vee = K^t,$$

i.e.,

$$F^\vee = D^t, \qquad D^\vee = F^t. \tag{3.6}$$

Therefore,

$$F^\vee D^\vee = (FD)^t, \qquad D^\vee F^\vee = (DF)^t. \tag{3.7}$$

For any simply-laced graph, the Cartan matrix is symmetric and $F = D^t = F^\vee$. In this case the graph $\Gamma$ is dual to itself. Among extended Dynkin diagrams the following pairs of diagram are dual:

$$(\widetilde{B}_n, \widetilde{C}_n), \qquad (\widetilde{CD}_n, \widetilde{DD}_n), \qquad (\widetilde{G}_{21}, \widetilde{G}_{22}), \qquad (\widetilde{F}_{41}, \widetilde{F}_{42}).$$

Let dual Cartan matrices be factorized by means of the diagonal matrices $U$ and $U^\vee$:

$$K = U\mathbf{B}, \qquad K^\vee = U^\vee \mathbf{B}^\vee.$$

Then according to [Kac93, Ch.3, exs.3.1], we have

$$U^\vee = U^{-1},$$

see, e.g., matrices $U$ for dual diagrams $\widetilde{F}_{41}$, $\widetilde{F}_{42}$, eqs. (2.21), (2.23). Since $K^\vee = K^t = \mathbf{B}U = U^{-1}\mathbf{B}^\vee$, we see that dual Tits forms are related as follows:

$$\mathbf{B}^\vee = U\mathbf{B}U.$$

### 3.1.5 The eigenvalues of the Cartan matrix and the Coxeter transformation

There is a simple relation between eigenvalues of the Cartan matrix and the Coxeter transformation. Let vector $z \in \mathcal{E}_\Gamma$ be given in two-component form

$$z = \begin{pmatrix} x \\ y \end{pmatrix}, \qquad (3.8)$$

where $\dim \mathcal{E}_\Gamma = k + m = n$, $x \in \mathcal{E}_{\Gamma_a}$ and $y \in \mathcal{E}_{\Gamma_b}$. Consider the relation

$$\mathbf{C}z = \lambda z \quad \text{or} \quad w_2 z = \lambda w_1 z.$$

In the simply-laced case we deduce from (3.2) that

$$\mathbf{C}z = \lambda z \iff \begin{cases} \dfrac{\lambda+1}{2\lambda}x = -Dy \\ \dfrac{\lambda+1}{2}y = -D^t x \end{cases} \iff \mathbf{B}z = \dfrac{\lambda-1}{2}\begin{pmatrix} \frac{1}{\lambda}x \\ -y \end{pmatrix}. \quad (3.9)$$

In the multiply-laced case we deduce that

$$\mathbf{C}z = \lambda z \iff \begin{cases} \dfrac{\lambda+1}{2\lambda}x = -Dy \\ \dfrac{\lambda+1}{2}y = -Fx \end{cases} \iff \mathbf{B}z = \dfrac{\lambda-1}{2}\begin{pmatrix} \frac{1}{\lambda}x \\ -y \end{pmatrix}. \quad (3.10)$$

From (3.9) and (3.10) we have in the simply-laced and multiply-laced cases, respectively:

$$\begin{cases} DD^t x = \dfrac{(\lambda+1)^2}{4\lambda}x \\ D^t D y = \dfrac{(\lambda+1)^2}{4\lambda}y \end{cases} \qquad \begin{cases} DF x = \dfrac{(\lambda+1)^2}{4\lambda}x \\ FD y = \dfrac{(\lambda+1)^2}{4\lambda}y \end{cases} \qquad (3.11)$$

Similarly,

$$\mathbf{B}z = \gamma z \iff \begin{cases} (\gamma-1)x = Dy \\ (\gamma-1)y = D^t x \end{cases} \quad \text{resp.} \quad \begin{cases} (\gamma-1)x = Dy \\ (\gamma-1)y = Fx; \end{cases} \quad (3.12)$$

and from (3.12) we have

$$\begin{cases} DD^t x = (\gamma-1)^2 x \\ D^t D y = (\gamma-1)^2 y \end{cases} \quad \text{resp.} \quad \begin{cases} DF x = (\gamma-1)^2 x \\ FD y = (\gamma-1)^2 y. \end{cases}$$

By (2.4) we have

**Proposition 3.2 (On fixed and anti-fixed points).** *1) The eigenvalues $\lambda$ of the Coxeter transformation and the eigenvalues $\gamma$ of the matrix $\mathbf{B}$ of the Tits form are related as follows*

$$\dfrac{(\lambda+1)^2}{4\lambda} = (\gamma-1)^2.$$

*2) The kernel of the matrix* **B** *coincides with the kernel of the Cartan matrix* $K$ *and coincides with the space of fixed points of the Coxeter transformation*

$$\ker K = \ker \mathbf{B} = \{z \mid \mathbf{C}z = z\}.$$

*3) The space of fixed points of the matrix* **B** *coincides with the space of anti-fixed points of the Coxeter transformation*

$$\{z \mid \mathbf{B}z = z\} = \{z \mid \mathbf{C}z = -z\}.$$

For more information about fixed and anti-fixed points of the powers of the Coxeter transformation, see Appendix C.7.1.

## 3.2 An application of the Perron-Frobenius theorem

### 3.2.1 The pair of matrices $DD^t$ and $D^t D$ (resp. $DF$ and $FD$)

*Remark 3.3.* The matrices $DD^t$ and $D^t D$ have certain nice properties.

1) They give us complete information about the eigenvalues of Coxeter transformations and Cartan matrices. Our results hold for an arbitrary tree $\Gamma$.

2) The eigenvectors of Coxeter transformations are combinations of eigenvectors of the matrix $DD^t$ and eigenvectors of the matrix $D^t D$, see §3.3.1, relations (3.22), (3.23), (3.24).

3) They satisfy the Perron-Frobenius theorem.

More exactly, properties 1)–3) will be considered below in Proposition 3.4, Proposition 3.6, and in §3.2.2, §3.3.1.

**Proposition 3.4.** *1) The matrices* $DD^t$ *and* $D^t D$ *(resp.* $DF$ *and* $FD$*) have the same non-zero eigenvalues with equal multiplicities.*

*2) The eigenvalues* $\varphi_i$ *of the matrices* $DD^t$ *and* $D^t D$ *(resp.* $DF$ *and* $FD$*) are non-negative:*

$$\varphi_i \geq 0.$$

*3) The corresponding eigenvalues* $\lambda_{1,2}^{\varphi_i}$ *of the Coxeter transformations are*

$$\lambda_{1,2}^{\varphi_i} = 2\varphi_i - 1 \pm 2\sqrt{\varphi_i(\varphi_i - 1)}. \tag{3.13}$$

*The eigenvalues* $\lambda_{1,2}^{\varphi_i}$ *either lie on the unit circle or are real positive numbers. It the latter case* $\lambda_1^{\varphi_i}$ *and* $\lambda_2^{\varphi_i}$ *are mutually inverse:*

$$\lambda_1^{\varphi_i} \lambda_2^{\varphi_i} = 1.$$

*Proof.* 1) If $DD^t z = \mu z$, where $\mu \neq 0$, then $D^t z \neq 0$ and $D^t D(D^t z) = \mu(D^t z)$. We argue similarly for $D^t D$, $DF$, $FD$. The multiplicities of non-zero eigenvalues coincide since

$$x \neq 0, \; DD^t x = \mu x \implies D^t D(D^t x) = \mu(D^t x), \text{ and } D^t x \neq 0,$$

and

$$y \neq 0, \; D^t Dy = \mu y \implies DD^t(Dy) = \mu(Dy), \text{ and } Dy \neq 0.$$

*Remark 3.5.* The multiplicities of the zero eigenvalue are not equal. If $DD^t x = 0$ and $x$ is the eigenvector, $x \neq 0$, then it is possible that $D^t x = 0$ and $D^t x$ is not an eigenvector of $D^t D$, see Remark 3.7, 4) below.

2) The matrices $DD^t$ and $D^t D$ are symmetric and non-negative definite. For example,

$$\langle DD^t x, x \rangle = \langle D^t x, D^t x \rangle \geq 0.$$

So, if $DD^t x = \varphi_i x$, then

$$\langle DD^t x, x \rangle = \varphi_i \langle x, x \rangle$$

and

$$\varphi_i = \frac{\langle D^t x, D^t x \rangle}{\langle x, x \rangle} \geq 0.$$

In the multiply-laced case, we deduce from (3.5) that the matrix $DF$ is

$$DF = U_1 A U_2 A^t. \tag{3.14}$$

Let $\varphi$ be a non-zero eigenvalue for $DF = U_1 A U_2 A^t$ with eigenvector $x$:

$$U_1 A U_2 A^t x = \varphi x.$$

Since $U$ is a positive diagonal matrix, see §2.1.1, we have

$$(\sqrt{U_1} A U_2 A^t \sqrt{U_1})((\sqrt{U_1})^{-1} x) = \varphi(\sqrt{U_1})^{-1} x, \tag{3.15}$$

and $\varphi$ is also a non-zero eigenvalue with eigenvector $(\sqrt{U_1})^{-1} x$ for the matrix $\sqrt{U_1} A U_2 A^t \sqrt{U_1}$ which already is symmetric, so $\varphi \geq 0$.

3) From (3.13) if $0 \leq \varphi \leq 1$ we deduce that

$$|\lambda_{1,2}^{\varphi_i}|^2 = (2\varphi_i - 1)^2 + 4\varphi_i(1 - \varphi_i) = 1 .$$

If $\varphi_i > 1$, then

$$2\varphi_i - 1 > 2\sqrt{\varphi_i(\varphi_i - 1)} \implies \lambda_{1,2}^{\varphi_i} \geq 0 .$$

Thus,

$$\lambda_1^{\varphi_i} = 2\varphi_i - 1 + 2\sqrt{\varphi_i(\varphi_i - 1)} > 1,$$
$$\lambda_2^{\varphi_i} = 2\varphi_i - 1 - 2\sqrt{\varphi_i(\varphi_i - 1)} < 1. \quad \square \tag{3.16}$$

The pair of matrices $(A, B)$ is said to be a *PF-pair* if both matrices $A$ and $B$ satisfy conditions of the Perron-Frobenius theorem[1].

**Proposition 3.6.** *The matrix pair* $(DD^t, D^t D)$ *(resp.* $(DF, FD))$ *is a PF-pair, i.e.,*

1) $DD^t$ *and* $D^t D$ *(resp.* $DF$ *and* $FD$) *are non-negative;*
2) $DD^t$ *and* $D^t D$ *(resp.* $DF$ *and* $FD$) *are indecomposable.*

*Proof.* 1) Indeed, in the simply-laced case, the following relation holds

$$4(DD^t)_{ij} = 4 \sum_{p=1}^{k} (a_i, b_p)(b_p, a_j) =$$

$$\begin{cases} s_i \text{ the number of edges with a vertex } v_i & \text{if } i = j, \\ 1 & \text{if } |v_i - v_j| = 2, \\ 0 & \text{if } |v_i - v_j| > 2. \end{cases} \qquad (3.17)$$

In the multiply-laced case, we have

$$4(DF)_{ij} = 4 \sum_{p=1}^{k} \frac{(a_i, b_p)(b_p, a_j)}{(a_i, a_i)(b_p, b_p)} =$$

$$\begin{cases} 4 \sum_{p=1}^{k} \dfrac{(a_i, b_p)^2}{(a_i, a_i)(b_p, b_p)} = 4 \sum_{p=1}^{k} \cos^2\{a_i, b_p\} & \text{if } i = j, \\[2mm] 4 \dfrac{(a_i, b_p)(b_p, a_j)}{(a_i, a_i)(b_p, b_p)} & \text{if } |v_i - v_j| = 2, \\[2mm] 0 & \text{if } |v_i - v_j| > 2. \end{cases} \qquad (3.18)$$

2) Define the distance between two sets of vertices $A = \{a_i\}_{i \in I}$ and $B = \{b_j\}_{j \in J}$ to be

$$\min_{i,j} |a_i - b_j|.$$

If the matrix $DD^t$ is decomposable, then the set of vertices $\{v_1, \ldots, v_n\}$ can be partitioned into two subsets such that distance between these two subsets is $> 2$. This contradicts the assumption that $\Gamma$ is connected.    $\square$

*Remark 3.7.* 1) Eq. (3.17) is a particular case of eq. (3.18), since the angle between the adjacent simple roots $a_i$ and $b_j$ is $\dfrac{2\pi}{3}$, so $\cos\{a_i, b_j\} = -\dfrac{1}{2}$. Of course, the angles and the lengths of vectors are considered in the sense of the bilinear form $(\cdot, \cdot)$ from §2.1.1.

2) The case $|v_i - v_j| = 2$ from eq. (3.18) can be expressed in the following form:

---

[1] The Perron-Frobenius theorem is well known in the matrix theory, see §C.3 and [MM64], [Ga90].

$$4\frac{(a_i, b_p)(b_p, a_j)}{(a_i, a_i)(b_p, b_p)} = 4\frac{(a_i, b_p)(b_p, a_j)}{|a_i|^2|b_p|^2} =$$

$$4\frac{(a_i, b_p)}{|a_i||b_p|}\frac{(b_p, a_j)}{|a_j||b_p|}\frac{|a_j|}{|a_i|} = 4\frac{|a_j|}{|a_i|}\cos\{a_i, b_p\}\cos\{a_j, b_p\}.$$

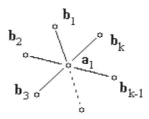

**Fig. 3.1.**    The star $*_{k+1}$ with $k$ rays

3) One can easily calculate the matrices $DD^t, D^tD, DF, FD$ by means of eq. (3.17) and eq. (3.18).

4) Consider a simple star $*_{k+1}$, Fig. 3.1. It is bipartite with respect to the following bicolored partition. One part of the graph consists of only one vertex $a_1$, i.e., $m = 1$, the other one consists of $k$ vertices $\{b_1, \ldots, b_k\}$, $n = k + 1$. According to (3.17) the $1 \times 1$ matrix $DD^t$ is

$$DD^t = k = n - 1,$$

and the $k \times k$ matrix $D^tD$ is

$$D^tD = \begin{pmatrix} 1\ 1\ 1 \ldots\ 1 \\ 1\ 1\ 1 \ldots\ 1 \\ 1\ 1\ 1 \ldots\ 1 \\ \ldots \\ 1\ 1\ 1 \ldots\ 1 \end{pmatrix}.$$

By Proposition 3.4, heading 2) the matrices $DD^t$ and $D^tD$ have only one non-zero eigenvalue $\varphi_1 = n - 1$. All the other eigenvalues of $D^tD$ are zeros and the characteristic polynomial of the $D^tD$ is

$$\varphi^{n-1}(\varphi - (n-1)).$$

### 3.2.2 The Perron-Frobenius theorem applied to $DD^t$ and $D^tD$ (resp. $DF$ and $FD$)

By Proposition 3.6 the pairs $(DD^t, D^tD)$ (resp. $(DF, FD)$) are PF-pairs, so we can apply the Perron-Frobenius theorem, see §C.3.

**Corollary 3.8.** *The matrices $DD^t$ and $D^tD$ (resp. $DF$ and $FD$) have a common simple (i.e., with multiplicity one) positive eigenvalue $\varphi_1$. This eigenvalue is the largest* (called dominant eigenvalue):

$$0 \leq \varphi_i \leq \varphi_1,$$

$$\varphi_1 = \begin{cases} \max\limits_{x \geq 0} \min\limits_{1 \leq i \leq m} \dfrac{(DD^t x)_i}{x_i} & \text{in the simply-laced case,} \\[3mm] \max\limits_{x \geq 0} \min\limits_{1 \leq i \leq m} \dfrac{(DF x)_i}{x_i} & \text{in the multiply-laced case.} \end{cases}$$

*There are positive eigenvectors $\mathbb{X}^{\varphi_1}$, $\mathbb{Y}^{\varphi_1}$ (i.e., non-zero vectors with non-negative coordinates) corresponding to the eigenvalue $\varphi_1$:*

$$\begin{aligned} DD^t\mathbb{X}^{\varphi_1} &= \varphi_1\mathbb{X}^{\varphi_1}, & D^tD\mathbb{Y}^{\varphi_1} &= \varphi_1\mathbb{Y}^{\varphi_1}, \\ DF\mathbb{X}^{\varphi_1} &= \varphi_1\mathbb{X}^{\varphi_1}, & FD\mathbb{Y}^{\varphi_1} &= \varphi_1\mathbb{Y}^{\varphi_1}. \end{aligned} \tag{3.19}$$

The matrices $DD^t$ (resp. $D^tD$) are symmetric and can be diagonalized in the some orthonormal basis of the eigenvectors from $\mathcal{E}_{\Gamma_a} = \mathbb{R}^h$ (resp. $\mathcal{E}_{\Gamma_b} = \mathbb{R}^k$). The Jordan normal forms of these matrices are shown in Fig. 3.2. The

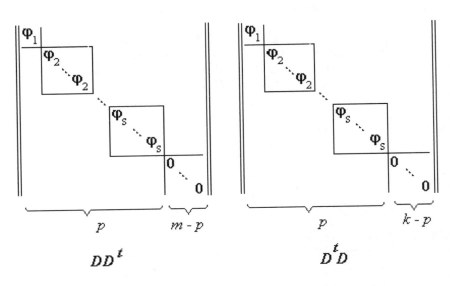

**Fig. 3.2.**      The Jordan normal forms of $DD^t$ and $D^tD$

normal forms of $DF$ and $FD$ are the same, since by (3.14), (3.15) the non-zero eigenvalues of the matrix $DF$ and symmetric matrix $\sqrt{U_1}AU_2A^t\sqrt{U_1}$ coincide. However, the normal bases (i.e., bases which consist of eigenvectors) for $DF$ and $FD$ are not necessarily orthonormal, since the eigenvectors of $DF$ are obtained from eigenvectors of $\sqrt{U_1}AU_2A^t\sqrt{U_1}$ by means of the matrix $\sqrt{U_1}$ which does not preserve orthogonality.

## 3.3 The basis of eigenvectors and a theorem on the Jordan form

### 3.3.1 An explicit construction of the eigenvectors

**Proposition 3.9.** *Let*

$$\mathbb{X}_1^{\varphi_1}, \mathbb{X}_1^{\varphi_2}, ..., \mathbb{X}_{t_2}^{\varphi_2}, ..., \mathbb{X}_1^{\varphi_s}, ..., \mathbb{X}_{t_s}^{\varphi_s}, \mathbb{X}_1^0, ..., \mathbb{X}_{m-p}^0 \qquad (3.20)$$

*be all the orthonormal eigenvectors for $DD^t$ and*

$$\mathbb{Y}_1^0, ..., \mathbb{Y}_{k-p}^0$$

*be all the orthonormal eigenvectors for $D^tD$ corresponding to the zero eigenvalue. Then*

$$D^t\mathbb{X}_1^{\varphi_1}, D^t\mathbb{X}_1^{\varphi_2}, ..., D^t\mathbb{X}_{t_2}^{\varphi_2}, ..., D^t\mathbb{X}_1^{\varphi_s}, ..., D^t\mathbb{X}_{t_s}^{\varphi_s}, \mathbb{Y}_1^0, ..., \mathbb{Y}_{k-p}^0 \qquad (3.21)$$

*is the set of all orthonormal eigenvectors for $DD^t$.*
*Bases for $DF, FD$ (not orthonormal) are similarly constructed.*

*Proof.* Indeed, if $\mathbb{X}^{\varphi_i}$ and $\mathbb{X}^{\varphi_j}$ are the eigenvectors corresponding to the $\varphi_i$ and $\varphi_j$, then the vectors $\mathbb{Y}^{\varphi_i} = D^t\mathbb{X}^{\varphi_i}$ and $\mathbb{Y}^{\varphi_j} = D^t\mathbb{X}^{\varphi_j}$ are eigenvectors of $D^tD$ and

$$\langle \mathbb{Y}^{\varphi_i}, \mathbb{Y}^{\varphi_j} \rangle = \langle \mathbb{X}^{\varphi_i}, DD^t\mathbb{X}^{\varphi_j} \rangle = \varphi_j \langle \mathbb{X}^{\varphi_i}, \mathbb{X}^{\varphi_j} \rangle = 0 . \square$$

Let us construct the eigenvectors for the Coxeter transformation. We set:
Case $\varphi_i \neq 0, 1$:

$$z_{r,\nu}^{\varphi_i} = \begin{pmatrix} \mathbb{X}_r^{\varphi_i} \\ -\dfrac{2}{\lambda_\nu^{\varphi_i}+1}D^t\mathbb{X}_r^{\varphi_i} \end{pmatrix}, \quad 1 \leq i \leq s, \ 1 \leq r \leq t_i, \ \nu = 1, 2 . \qquad (3.22)$$

Here $\lambda_{1,2}^{\varphi_i}$ is obtained by eq. (3.13).
Case $\varphi_i = 1$:

$$z_r^1 = \begin{pmatrix} \mathbb{X}_r^1 \\ -D^t\mathbb{X}_r^1 \end{pmatrix}, \quad \tilde{z}_r^1 = \frac{1}{4}\begin{pmatrix} \mathbb{X}_r^1 \\ D^t\mathbb{X}_r^1 \end{pmatrix}, \quad 1 \leq r \leq t_i. \qquad (3.23)$$

Case $\varphi_i = 0$:

$$z_{x_\eta}^0 = \begin{pmatrix} \mathbb{X}_\eta^0 \\ 0 \end{pmatrix}, \quad 1 \leq \eta \leq m-p, \qquad z_{y_\xi}^0 = \begin{pmatrix} 0 \\ \mathbb{Y}_\xi^0 \end{pmatrix}, \quad 1 \leq \xi \leq k-p. \quad (3.24)$$

**Proposition 3.10 ([SuSt75, SuSt78]).**

*1) The vectors (3.22), (3.23), (3.24) constitute a basis in $\mathcal{E}_\Gamma$ over $\mathbb{C}$.*

*2) The vectors (3.22) are eigenvectors of the Coxeter transformation corresponding to the eigenvalue $\lambda_{\varphi_i}$:*

$$\mathbf{C}z_{r,\nu}^{\varphi_i} = \lambda z_{r,\nu}^{\varphi_i}. \tag{3.25}$$

*The vectors (3.23) are eigenvectors and adjoint vectors of the Coxeter transformation corresponding to the eigenvalue 1:*

$$\mathbf{C}z_r^1 = z_r^1, \qquad \mathbf{C}\tilde{z}_r^1 = z_r^1 + \tilde{z}_r^1. \tag{3.26}$$

*The vectors (3.24) are eigenvectors of the Coxeter transformation corresponding to the eigenvalue $-1$:*

$$\mathbf{C}z_{x_\eta}^0 = -z_{x_\eta}^0, \qquad \mathbf{C}z_{y_\xi}^0 = -z_{y_\xi}^0. \tag{3.27}$$

*In other words, vectors (3.22), (3.23), (3.24) constitute an orthogonal basis which consists of eigenvectors and adjoint vectors of the Coxeter transformation. The number of the adjoint vectors $\tilde{z}_r^1$ is equal to the multiplicity of eigenvalue $\varphi = 1$.*

*Proof.* 1) The number of vectors (3.22), (3.23), (3.24) is $2p + (m-p) + (k-p) = n$ and it suffices to prove that these vectors are linearly independent. Let us write down the condition of linear dependence. It splits into two conditions: for the $\mathbb{X}$-component and for the $\mathbb{Y}$-component. The linear independence of vectors (3.22), (3.23), (3.24) follows from the linear independence of vectors (3.20), (3.21). $\quad\square$

2) To prove relation (3.25), the first relation from (3.26) and relation (3.27), it suffices to check (3.9). Let us check that

$$\mathbf{C}\tilde{z}_r^1 = z_r^1 + \tilde{z}_r^1.$$

We consider the multiply-laced case. Making use of (3.4) we see that

$$\mathbf{C} = \begin{pmatrix} 4DF - I_m & 2D \\ -2F & -I_k \end{pmatrix}.$$

Then

$$\mathbf{C}\tilde{z}_r^1 = \frac{1}{4} \begin{pmatrix} 4DF\mathbb{X}_r^1 - \mathbb{X}_r^1 + 2DF\mathbb{X}_r^1 \\ -2F\mathbb{X}_r^1 - F\mathbb{X}_r^1 \end{pmatrix} = \frac{1}{4} \begin{pmatrix} 5\mathbb{X}_r^1 \\ -3F\mathbb{X}_r^1 \end{pmatrix} = z_r^1 + \tilde{z}_r^1.$$

See Corollary 3.8, eq. (3.19). $\quad\square$

By heading 2) of Proposition 3.2 and eq. (3.23) from Proposition 3.9 we have

**Corollary 3.11.** *The Jordan normal form of the Coxeter transformation is diagonal if and only if the Tits form is nondegenerate[1]. $\quad\square$*

---

[1] For the Jordan canonical (normal) form, see, for example, [Ga90, Ch.VI] or [Pr94, Ch.III].

## 3.3.2 Monotonicity of the dominant eigenvalue

The following proposition is important for calculation of the number of $2 \times 2$ Jordan blocks in the Jordan normal form of the Coxeter transformation.

**Proposition 3.12 ([SuSt75, SuSt78]).** *Let us add an edge to some tree* $\Gamma$ *and let* $\hat{\Gamma}$ *be the new graph (§2.1.7). Then:*
*1) The dominant eigenvalue* $\varphi_1$ *may only grow:*

$$\varphi_1(\hat{\Gamma}) \geq \varphi_1(\Gamma) .$$

*2) Let* $\Gamma$ *be an extended Dynkin diagram, i.e.,* $\mathcal{B}$ *is non-negative definite. Then the spectra of* $DD^t(\hat{\Gamma})$ *and* $D^tD(\hat{\Gamma})$ *(resp.* $DF(\hat{\Gamma})$ *and* $FD(\hat{\Gamma})$*) do not contain 1, i.e.,*

$$\varphi_i(\hat{\Gamma}) \neq 1$$

*for all* $\varphi_i$ *are eigenvalues of* $DD^t(\hat{\Gamma})$.
*3) Let* $\mathcal{B}$ *be indefinite. Then*

$$\varphi_1(\hat{\Gamma}) > 1 .$$

*Proof.* 1) Adding an edge to the vertex $a_i$ we see, according to (3.17), that only one element of $DD^t$ changes: namely, $(DD^t)_{ii}$ changes from $\dfrac{s_i}{4}$ to $\dfrac{s_i + 1}{4}$. In the multiply-laced case, $(DF)_{ii}$ changes by $\cos^2(a_i, b_s)$, where $b_s$ is the new vertex incident with the vertex $a_i$. By Corollary 3.8 we have

$$\varphi_1(\hat{\Gamma}) = \max_{x \geq 0} \min_{1 \leq i \leq m} \frac{(DF(\hat{\Gamma})x)_i}{x_i} \geq$$

$$\min_{1 \leq i \leq m} \frac{(DF(\hat{\Gamma})\mathbb{X}_i^{\varphi_1})_i}{\mathbb{X}_i^{\varphi_1}} \geq \min_{1 \leq i \leq m} \frac{(DF\mathbb{X}_i^{\varphi_1})_i}{\mathbb{X}_i^{\varphi_1}} = \varphi_1 .$$

2) The characteristic polynomial of $DF(\hat{\Gamma})$ is

$$\det |DF(\hat{\Gamma}) - \mu I| = \det |DF - \mu I| + \cos^2\{a_i, b_s\} \det |A_i(\mu)| , \qquad (3.28)$$

where $A_i(\mu)$ is obtained by deleting the $i^{th}$ row and $i^{th}$ column from the matrix $DF - \mu I$. It corresponds to the operation *"Remove"* from the graph $\Gamma$ (§2.1.7), the graph obtained by removing vertex is $\overset{\vee}{\Gamma}$, i.e.,

$$\det |DF(\hat{\Gamma}) - \mu I| = \det |DF - \mu I| + \cos^2\{a_i, b_s\} \det |DF(\overset{\vee}{\Gamma}) - \mu I| . \quad (3.29)$$

According to §2.1.7 the quadratic form $\mathcal{B}(\overset{\vee}{\Gamma})$ is positive, $\overset{\vee}{\Gamma}$ is the Dynkin diagram, i.e., ker $\mathbf{B} = 0$. By Proposition 3.2 the Coxeter transformation for $\overset{\vee}{\Gamma}$

does not have eigenvalue 1. Then by (3.13) the corresponding matrix $DF(\overset{\vee}{\Gamma})$ does not have eigenvalue 1. Thus, in (3.29), $\mu = 1$ is a root of $\det|DF - \mu I|$ and is not a root of $\det|DF(\overset{\vee}{\Gamma}) - \mu I|$, and therefore $\mu = 1$ is not a root of $\det|DF(\overset{\wedge}{\Gamma}) - \mu I|$. The case $\cos^2\{a_i, b_s\} = 0$ is not possible since $\overset{\wedge}{\Gamma}$ is connected.

3) By 1) adding only one edge to an extended Dynkin diagram we get $\varphi_1(\overset{\wedge}{\Gamma}) \geq \varphi_1$. By 2) $\varphi_1(\overset{\wedge}{\Gamma}) > \varphi_1$. The form $\mathcal{B}$ becomes indefinite after we add some edges to the extended Dynkin diagram see §2.1.7, 3). So, $\varphi_1$ grows, and $\varphi_1 > 1$.  □

**Proposition 3.13.** *The common dominant eigenvalue of $DD^t$ and $D^tD$ (resp. $DF$ and $FD$) is equal to 1 if and only if $\Gamma$ is an extended Dynkin diagram.*

*Proof.* 1) Let $\Gamma$ be an extended Dynkin diagram, then $\mathcal{B}$ is non-negative definite. Since $\ker \mathbf{B} \neq 0$, we see by Proposition 3.2 that the eigenvalue $\varphi_1$ is the eigenvalue of the matrices $DF$ and $FD$. By (3.16) we have $\lambda_1^{\varphi_1} \geq 1$. Further, since the Weyl group preserves the quadratic form $\mathcal{B}$, we have

$$\mathcal{B}(z^{\varphi_1}) = \mathcal{B}(\mathbf{C}z^{\varphi_1}) = (\lambda_1^{\varphi_1})^2 \mathcal{B}(\mathbf{C}z^{\varphi_1}).$$

Therefore, either $\lambda_1^{\varphi_1} = 1$, i.e., $\varphi_1 = 1$, or $\mathcal{B}(z^{\varphi_1}) = 0$. We will show that in the latter case $\varphi_1 = 1$, too. By Proposition 3.9 the vectors $\mathbb{X}^{\varphi_1}$ and $F\mathbb{X}^{\varphi_1}$ have real coordinates. By (3.16) the eigenvalue $\lambda_1^{\varphi_1}$ is also real because $\varphi_1 \geq 1$. So, the vector $z^{\varphi_1}$ from (3.23) is real. Then from $\mathcal{B}(z^{\varphi_1}) = 0$ we have $z^{\varphi_1} \in \ker \mathbf{B}$ and again, $\lambda_1^{\varphi_1} = 1$.

2) Conversely, let $\lambda_1^{\varphi_1} = 1$. Then, by Proposition 3.2, $\ker \mathbf{B} \neq 0$, i.e., the form $\mathcal{B}$ is degenerate. Let us find whether $\mathcal{B}$ is non-negative definite or indefinite. By heading 3) of Proposition 3.12 if $\mathcal{B}$ is indefinite, then $\lambda_1^{\varphi_1} > 1$. Thus, $\mathcal{B}$ is non-negative definite and $\Gamma$ is an extended Dynkin diagram.  □

**Corollary 3.14.** *Let $\mathcal{B}$ be non-negative definite, i.e., let $\Gamma$ be an extended Dynkin diagram. Then*

*1) The kernel of the quadratic form $\mathcal{B}$ is one-dimensional.*

*2) The vector $z^{\varphi_1} \in \ker \mathbf{B}$ can be chosen so that all its coordinates are positive.*

*Proof.* 1) $\varphi_1$ is a simple eigenvalue of $DD^t$ and $D^tD$ (see Corollary 3.8).

2) $\mathbb{X}^{\varphi_1}$ is a positive vector (Corollary 3.8), $D^t\mathbb{X}^{\varphi_1}$ is a negative vector since $D^t$ is a nonpositive matrix (3.3). So, the vector $z^{\varphi_1}$ from (3.23) is positive.  □

The monotonicity of the dominant value $\varphi_1$ and the corresponding maximal eigenvalue $\lambda^{\varphi_1}$ of the Coxeter transformation is clearly demonstrated for the diagrams $T_{2,3,r}$, $T_{3,3,r}$, $T_{2,4,r}$, see Propositions 4.16, 4.17, 4.19 and Tables 4.4, 4.5, 4.6 in §4.4.

### 3.3.3 A theorem on the Jordan form

Now we can summarize.

**Theorem 3.15 ([SuSt75, SuSt78, St85]).** *1) The Jordan form of the Coxeter transformation is diagonal if and only if the Tits form is non-degenerate.*

*2) If $\mathcal{B}$ is non-negative definite ($\Gamma$ is an extended Dynkin diagram), then the Jordan form of the Coxeter transformation contains one $2 \times 2$ Jordan block. The remaining Jordan blocks are $1 \times 1$. All eigenvalues $\lambda_i$ lie on the unit circle.*

*3) If $\mathcal{B}$ is indefinite and degenerate, then the number of $2 \times 2$ Jordan blocks coincides with $\dim \ker \mathbf{B}$. The remaining Jordan blocks are $1 \times 1$. There is a simple maximal eigenvalue $\lambda_1^{\varphi_1}$ and a simple minimal eigenvalue $\lambda_2^{\varphi_1}$, and*

$$\lambda_1^{\varphi_1} > 1, \qquad \lambda_2^{\varphi_1} < 1.$$

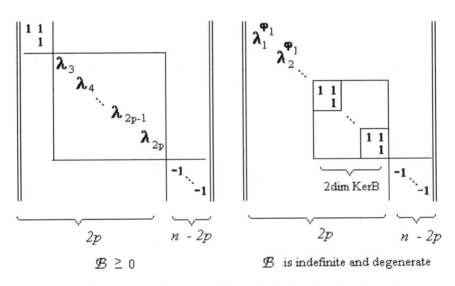

**Fig. 3.3.**     The Jordan normal form of the Coxeter transformation

*Example 3.16 (V. Kolmykov).* The example shows that there is a graph $\Gamma$ with indefinite and degenerate quadratic form $\mathcal{B}$ such that $\dim \ker \mathbf{B}$ is an arbitrarily number (see Fig. 3.4) and the Coxeter transformation has an arbitrarily number of $2 \times 2$ Jordan blocks. Consider $n$ copies of the extended Dynkin diagram $\widetilde{D}_4$ with centers $b_1, ..., b_n$ and marked vertices $a_1, ..., a_n$. We add new vertex $b_{n+1}$ and connect it with marked vertices $a_i$ for $i = 1, 2..., n$. The new graph is bipartite: one part consists of the vertices $b_1, ..., b_n, b_{n+1}$; the matrix $D^t D$ is of size $(n + 1) \times (n + 1)$ and has the form

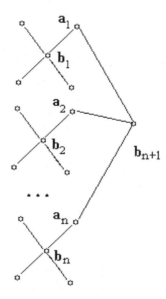

**Fig. 3.4.**    A graph $\Gamma$ such that $\dim \ker \mathbf{B}$ is an arbitrarily number

$$4D^t D = \begin{pmatrix} n & 1 & 1 & 1 & \dots & 1 & 1 \\ 1 & 4 & 0 & 0 & \dots & 0 & 0 \\ 1 & 0 & 4 & 0 & \dots & 0 & 0 \\ 1 & 0 & 0 & 4 & \dots & 0 & 0 \\ & & & \dots & & & \\ 1 & 0 & 0 & 0 & \dots & 4 & 0 \\ 1 & 0 & 0 & 0 & \dots & 0 & 4 \end{pmatrix}.$$

It is easy to show that

$$|4D^t D - \mu I| = (n - \mu)(4 - \mu)^n - n(4 - \mu)^{n-1}.$$

Thus, $\varphi_i = \dfrac{\mu_i}{4} = 1$ is of multiplicity $n - 1$.

In Proposition 4.11 of §4.3 we will see more examples of graphs $\Gamma(n)$ obtained by gluing $n$ copies of some graph $\Gamma$. By this proposition every eigenvalue $\lambda$ of $\Gamma$ is also an eigenvalue of $\Gamma(n)$ and the multiplicity of this eigenvalue for the graph $\Gamma(n)$ is $(n - 1) \times m$, where $m$ is the multiplicity of this eigenvalue in $\Gamma$.

**4**

# Eigenvalues, splitting formulas and diagrams $T_{p,q,r}$

> ... most of the fundamental results about simple
> Lie algebras, which were discovered by Killing are
> usually attributed to É. Cartan. This, despite the
> meticulousness with which Cartan noted his
> indebtedness to Killing. In Cartan's thesis there
> are 28 references to Lie and 60 to Killing!

A. J. Coleman, [Col89, p.447], 1989.

## 4.1 The eigenvalues of the affine Coxeter transformation are roots of unity

The Coxeter transformation corresponding to the extended Dynkin diagram, i.e., corresponding to affine Kac-Moody algebra is called the *affine Coxeter transformation*[1].

**Theorem 4.1 ([SuSt79], [St82a], [St85]).** *The eigenvalues of the affine Coxeter transformation are roots of unity. The proper eigenvalues are collected in Table* 4.1.

*Proof.* The eigenvalues for all cases of extended Dynkin diagrams are easily calculated by means of the generalized R. Steinberg theorem (Theorem 5.5) and Table 1.2. See Remark 4.2. □

*Remark 4.2.* According to generalized R. Steinberg's theorem (Theorem 5.5) orders (not always different) of eigenvalues (Table 4.1, col. 3) coincide with lengths of branches of the corresponding Dynkin diagram. If $g$ is the class

---

[1] For more details on the affine Weyl group and the affine Coxeter transformation, see Ch.6.

number (5.14) of the extended Dynkin diagram, the number of branches of the corresponding Dynkin diagram is $3 - g$. For $g = 0$, we have a simply-laced case of the extended Dynkin diagrams, i.e., $\widetilde{E}_6$, $\widetilde{E}_7$, $\widetilde{E}_8$, $\widetilde{D}_n(n \geq 4)$. In this case the number of branches is 3 . For $g = 1$, we have extended Dynkin diagrams $\widetilde{F}_{41}$, $\widetilde{F}_{42}$, $\widehat{CD}_n$, $\widetilde{CD}_n$, there exists two groups of eigenvalues, see (5.15). For $g = 2$, we have extended Dynkin diagrams $\widetilde{G}_{21}$, $\widetilde{G}_{22}$, $\widetilde{B}_n$, $\widetilde{C}_n$, $\widetilde{BC}_n$ and there exists only one group of eigenvalues. For $g = 3$, we have extended Dynkin diagrams $\widetilde{A}_{11}$, $\widetilde{A}_{12}$, in this case there is only one trivial eigenvalues 1.

*Remark 4.3.* Let $H$ be the hyperplane orthogonal to the adjoint vector $\tilde{z}^1$ from Proposition 3.10. The vector $\tilde{z}^1$ is responsible for a $2 \times 2$ block in the Jordan form of the affine Coxeter transformation. Due to the presence of a $2 \times 2$ block, the affine Coxeter transformation is of infinite order in the Weyl group. The restriction of the Coxeter transformation to the hyperplane $H$ is, however, of a finite order $h_a$. We call this number the *affine Coxeter number*. The affine Coxeter number $h_a$ is the least common multiple of orders of eigenvalues ($\neq 1$) of the affine Coxeter transformation (Table 4.1). We denote by $h$ the *Coxeter number* of a given Dynkin diagram. The value $h-1$ is the sum of coordinates of the highest root $\beta$ of the corresponding root system, see (1.2). The imaginary vector $z^1$ depicted in Fig. 2.6 coincides with the highest root $\beta$ extended to the vector with 1 at the additional vertex, the one that extends the Dynkin diagram to the extended Dynkin diagram. Thus,

$$h = n_1 + \cdots + n_l, \tag{4.1}$$

where $n_i$ are coordinates of the imaginary vector from Fig. 2.6. Similarly to (4.1), the *dual Coxeter number* is defined as

$$h^{\vee} = n_1^{\vee} + \cdots + n_l^{\vee}, \tag{4.2}$$

where $\{n_1^{\vee}, \ldots, n_l^{\vee}\}$ are coordinates of the imaginary vector $h^{\vee}$ of the dual Dynkin diagram, see Fig. 2.6. For values of the dual Coxeter numbers, see Table 4.2.

**Table 4.1.**    The eigenvalues of affine Coxeter transformations

| Diagram | Eigenvalues $\lambda$ | Orders of eigenvalues[1] | Affine Coxeter number $h_a$ [2] |
|---|---|---|---|
| $\widetilde{E}_6$ | $\lambda^1_{1,2} = 1,$ <br> $\lambda^2_{1,2} = \lambda^3_{1,2} = e^{\pm 2\pi i/3},$ <br> $\lambda_7 = -1$ | <br> $3,3$ <br> $2$ | $6$ |
| $\widetilde{E}_7$ | $\lambda^1_{1,2} = 1,$ <br> $\lambda^2_{1,2} = e^{\pm 2\pi i/3},$ <br> $\lambda^3_{1,2} = e^{\pm \pi i/2},$ <br> $\lambda_{7,8} = -1$ | <br> $3$ <br> $4$ <br> $2$ | $12$ |
| $\widetilde{E}_8$ | $\lambda^1_{1,2} = 1,$ <br> $\lambda^2_{1,2} = e^{\pm 2\pi i/3},$ <br> $\lambda^3_{1,2} = e^{\pm 2\pi i/5}, \quad \lambda^3_{1,2} = e^{\pm 4\pi i/5},$ <br> $\lambda_9 = -1$ | <br> $3$ <br> $5$ <br> $2$ | $30$ |
| $\widetilde{D}_n$ | $\lambda^1_{1,2} = 1,$ <br> $\lambda^2_s = e^{2s\pi i/(n-2)}, s = 1,2,...,n-3,$ <br> $\lambda_n = \lambda_{n+1} = -1$ | <br> $n-2$ <br> $2,2$ | $n-2$ for $n = 2k$; <br> $2(n-2)$ for $n = 2k+1$ |
| $\widetilde{G}_{21}, \widetilde{G}_{22}$ | $\lambda^1_{1,2} = 1,$ <br> $\lambda_3 = -1$ | <br> $2$ | $2$ |
| $\widetilde{F}_{41}, \widetilde{F}_{42}$ | $\lambda^1_{1,2} = 1,$ <br> $\lambda^2_{1,2} = e^{2\pi i/3},$ <br> $\lambda_3 = -1$ | <br> $3$ <br> $2$ | $6$ |
| $\widetilde{A}_{11}, \widetilde{A}_{12}$ | $\lambda^1_{1,2} = 1$ | $1$ | $1$ |
| $\widetilde{B}_n, \widetilde{C}_n$ <br> $\widetilde{BC}_n$ | $\lambda^1_{1,2} = 1,$ <br> $\lambda^2_s = e^{2s\pi i/n}, s = 1,2,...,n-1$ | <br> $n$ | $n$ |
| $\widetilde{CD}_n$ <br> $\widetilde{DD}_n$ | $\lambda^1_{1,2} = 1,$ <br> $\lambda^2_s = e^{2s\pi i/(n-1)}, s = 1,2,...,n-2,$ <br> $\lambda_{n+1} = -1$ | <br> $n-1$ <br> $2$ | $n-1$ for $n = 2k-1$; <br> $2(n-1)$ for $n = 2k$ |

[1] See Remark 4.2.
[2] See Remark 4.3.

**Table 4.2.**    The Coxeter numbers and affine Coxeter numbers

| Extended Dynkin Diagram | Notation in context of twisted affine Lie algebra [1] | Affine Coxeter number $h_a$ | Coxeter number $h$ | Dual [2] Coxeter number $h^\vee$ |
|---|---|---|---|---|
| $\widetilde{E}_6$ | $E_6^{(1)}$ | 6 | 12 | 12 |
| $\widetilde{E}_7$ | $E_7^{(1)}$ | 12 | 18 | 18 |
| $\widetilde{E}_8$ | $E_8^{(1)}$ | 30 | 30 | 30 |
| $\widetilde{D}_n$ | $D_n^{(1)}$ | $n-2$ for $n=2k$; $2(n-2)$ for $n=2k+1$ | $2(n-1)$ | $2(n-1)$ |
| $\widetilde{A}_{11}$ | $A_2^{(2)}$ | 1 | 3 | 3 |
| $\widetilde{A}_{12}$ | $A_1^{(1)}$ | 1 | 2 | 2 |
| $\widetilde{BC}_n$ | $A_{2n}^{(2)}$ | $n$ | $2n+1$ | $2n+1$ |
| $\widetilde{G}_{22}=\widetilde{G}_{21}^\vee$ | $G_2^{(1)}$ | 2 | 6 | $\boxed{4}$ |
| $\widetilde{G}_{21}=\widetilde{G}_{22}^\vee$ | $D_4^{(3)}$ | 2 | $\boxed{4}$ | 6 |
| $\widetilde{F}_{42}=\widetilde{F}_{41}^\vee$ | $F_4^{(1)}$ | 6 | 12 | $\boxed{9}$ |
| $\widetilde{F}_{41}=\widetilde{F}_{42}^\vee$ | $E_6^{(2)}$ | 6 | $\boxed{9}$ | 12 |
| $\widetilde{C}_n=\widetilde{B}_n^\vee$ | $C_n^{(1)}$ | $n$ | $2n$ | $\boxed{n+1}$ |
| $\widetilde{B}_n=\widetilde{C}_n^\vee$ | $D_{n+1}^{(2)}$ | $n$ | $\boxed{n+1}$ | $2n$ |
| $\widetilde{CD}_n=\widetilde{DD}_n^\vee$ | $B_n^{(1)}$ | $n-1$ for $n=2k-1$; $2(n-1)$ for $n=2k$ | $2n$ | $\boxed{2n-1}$ |
| $\widetilde{DD}_n=\widetilde{CD}_n^\vee$ | $A_{2n-1}^{(2)}$ | $n-1$ for $n=2k-1$; $2(n-1)$ for $n=2k$ | $\boxed{2n-1}$ | $2n$ |

*Remark 4.4.* For the first time, the notation of *twisted affine Lie algebras* from Table 4.2, col. 2 appeared in [Kac69] in the description of finite order automorphisms; see also [Kac80], [GorOnVi94, p.123], [OnVi90], [Kac93]. The upper index $r$ in the notation of twisted affine Lie algebras has an invariant sense: it is the order of the diagram automorphism $\mu$ of $\mathfrak{g}$, where $\mathfrak{g}$ is a complex simple finite dimensional Lie algebra of type $X_N = A_{2l}, A_{2l-1}, D_{l+1}, E_6, D_4$, [Kac93, Th.8.3].

---

[1] See Remark 4.4.

[2] For emphasis and to distinguish $h$ from $h^\vee$, we put the value $\min(h, h^\vee)$ in a box if $h \neq h^\vee$.

The affine Lie algebra associated to a generalized Cartan matrix of type $X_l^{(1)}$ is called a *non-twisted affine Lie algebra*, [Kac93, Ch.7].

The affine Lie algebras associated to a generalized Cartan matrix of type $X_l^{(2)}$ and $X_l^{(3)}$ are called *twisted affine Lie algebras*. [Kac93, Ch.8].

The corresponding $\mathbb{Z}/r\mathbb{Z}$-gradings of $\mathfrak{g} = \mathfrak{g}(X_N)$ are (here $\bar{i} \in \mathbb{Z}/r\mathbb{Z}$):

$$\mathfrak{g} = \mathfrak{g}_{\bar{0}} + \mathfrak{g}_{\bar{1}} \qquad \text{for } r = 2,$$
$$\mathfrak{g} = \mathfrak{g}_{\bar{0}} + \mathfrak{g}_{\bar{1}} + \mathfrak{g}_{\bar{2}} \quad \text{for } r = 3,$$

see [Kac93, 8.3.1, 8.3.2], [GorOnVi94, Ch.3, §3]

**Proposition 4.5. ([Kac93, exs.6.3])** *Let $A$ be a Cartan matrix of type $X_N^{(r)}$ from Table 4.2, col. 2, let $l = \text{rank of } A$, and let $h$ be the Coxeter number. Let $\Delta$ be the finite root system of type $X_N$. Then*

$$rlh = |\Delta|.$$

*Proof.* For $r = 1$, we have H. S. M. Coxeter's proposition (1.1), §1.1. We consider only the remaining cases:

$$A_2^{(2)}, \quad A_{2n}^{(2)}(l \geq 2), \quad A_{2n-1}^{(2)}(l \geq 3), \quad D_{n+1}^{(2)}(l \geq 2), \quad E_6^{(2)}, \quad D_4^{(3)}.$$

(see Table 4.2, col. 5).

(a) $A_2^{(2)}$, $r = 2$; rank $l = 1$; $h = 3$; $X_N = A_2$, $|\Delta(A_2)| = 6$, see [Bo, Tab.I].

(b) $A_{2n}^{(2)}$, $r = 2$; rank $l = n$; $h = 2n+1$, $X_N = A_{2n}$, $|\Delta(A_{2n})| = 2n(2n+1)$, see [Bo, Tab.I].

(c) $A_{2n-1}^{(2)}$, $r = 2$; rank $l = n$; $h = 2n - 1$; $X_N = A_{2n-1}$, $|\Delta(A_{2n-1})| = 2n(2n-1)$, see [Bo, Tab.I].

(d) $D_{n+1}^{(2)}$, $r = 2$; rank $l = n$; $h = n + 1$, $X_N = D_{n+1}$, $|\Delta(D_{n+1})| = 2n(n+1)$, see [Bo, Tab.IV].

(e) $E_6^{(2)}$, $r = 2$; rank $l = 4$; $h = 9$, $X_N = E_6$, $|\Delta(E_6)| = 72$, see [Bo, Tab.V].

(f) $D_4^{(3)}$, $r = 3$; rank $l = 2$; $h = 4$, $X_N = D_4$, $|\Delta(D_4)| = 24$, see [Bo, Tab.IV].

Cases (a)-(f) are collected in Table 4.3. □

## 4.2 Bibliographical notes on the spectrum of the Coxeter transformation

The eigenvalues of affine Coxeter transformations were also calculated by S. Berman, Y.S. Lee and R. Moody in [BLM89]. Theorem 3.15 was also proved by N. A'Campo [A'C76] and R. Howlett [How82].

Natural difficulties in the study of Cartan matrices and Coxeter transformations for the graphs containing cycles are connected with the following two facts:

**Table 4.3.**    The Kac relation $rlh = |\Delta|$ for $r = 2, 3$, Proposition 4.5.

| Extended Dynkin Diagram | Index $r$ | Rank of $A$ | Coxeter number $h$ | Root System $X_N$ | $|\Delta| = |\Delta(X_N)|$ |
|---|---|---|---|---|---|
| $A_2^{(2)}$ | 2 | 1 | 3 | $A_2$ | 6 |
| $A_{2n}^{(2)}$ | 2 | $n$ | $2n+1$ | $A_{2n}$ | $2n(2n+1)$ |
| $A_{2n-1}^{(2)}$ | 2 | $n$ | $2n-1$ | $A_{2n-1}$ | $2n(2n-1)$ |
| $D_{n+1}^{(2)}$ | 2 | $n$ | $2n+1$ | $D_{n+1}$ | $2n(2n+1)$ |
| $E_6^{(2)}$ | 2 | 4 | 9 | $E_6$ | 72 |
| $D_4^{(3)}$ | 3 | 2 | 4 | $D_4$ | 24 |

1) these graphs have non-symmetrizable Cartan matrices,

2) in general, there are several conjugacy classes of the Coxeter transformation.

We distinguish several works related to the Coxeter transformation for the graphs with cycles: C. M. Ringel [Rin94], A. J. Coleman [Col89], Shi Jian-yi [Shi00], Menshikh and Subbotin [MeSu82], [Men85], Boldt and Takane [BT97].

For generalized Cartan matrices (see §2.1.1), i.e., for graphs with cycles, C. M. Ringel [Rin94] showed that the *spectral radius* $\rho(\mathbf{C})$ of the Coxeter transformation lies outside the unit circle, and is an eigenvalue of multiplicity one. This result generalizes Theorem 3.15, 3), proved in [SuSt75, SuSt78, St85] only for trees. The spectral radius $\rho(\mathbf{C})$ is used by V. Dlab and C. M. Ringel to determine the Gelfand-Kirillov dimension of the preprojective algebras, [DR81].

A. J. Coleman [Col89] computed characteristic polynomials for the Coxeter transformation for all extended Dynkin diagrams, including the case with cycles $\widetilde{A}_n$. He baptized these polynomials *Killing polynomials*, (see the epigraph to this chapter). Coleman also shows that $\widetilde{A}_n$ has $\dfrac{n}{2}$ spectral conjugacy classes.

V. Dlab and P. Lakatos [DL03] gave a number of upper bounds of the spectral radius $\rho(\mathbf{C})$. For an arbitrary tree:

$$\rho(\mathbf{C}) < 4d - 6,$$

where $d$ is the maximal branching degree. For a wild tree with two branching vertices which are not neighbors:

$$\rho(\mathbf{C}) < d.$$

P. Lakatos [Lak99a] showed that the Coxeter polynomial of a wild star has exactly two real roots and one irreducible non-cyclotomic factor.

**Definition 4.6.** The valued trees $\Gamma_1$ and $\Gamma_2$ are called *quasi-cospectral* if they have the same non-cyclotomic irreducible factor of their Coxeter polynomials.

*Example 4.7.* The wild star $*_{m,\dots,m}$ (consisting of $r$ arms of length $m$) and the wild star $*_{m-1,1\dots,1}$ (consisting of $r$ arms of length 1 and one arm of the length $m-1$) are quasi-cospectral. For this and similar results concerning quasi-cospectral trees, see [Lak99b, Prop 2.5].

V. V. Menshikh and V. F. Subbotin in [MeSu82], and V. V. Menshikh in [Men85] established a connection between an orientation $\Omega$ of the graph $\Gamma$ and spectral classes of conjugacy of the Coxeter transformation. For any orientation $\Omega$ of a given graph $\Gamma$ containing several cycles, they consider an invariant $R_\Omega$ equal to the number of arrows directed in a clockwise direction. For any graph $\Gamma$ containing disjoint cycles, they show that $R_{\Omega_1} = R_{\Omega_2}$ if and only if orientations $\Omega_1$ and $\Omega_2$ can be obtained from each other by applying a sink-admissible or a source-admissible sequence of reflections $\sigma_i$, see §2.2.6, i.e.,

$$R_{\Omega_1} = R_{\Omega_2} \quad \Longleftrightarrow \quad \Omega_1 = \sigma_{i_n}\dots\sigma_{i_2}\sigma_{i_1}(\Omega_2)$$

for any sink-admissible or source-admissible sequence $i_1, i_2, \dots, i_n$. Menshikh and Subbotin also showed that two Coxeter transformations $\mathbf{C}_{\Omega_1}$ and $\mathbf{C}_{\Omega_2}$ are conjugate if and only if $R_{\Omega_1} = R_{\Omega_2}$. The number $R_\Omega$ is called the *index* of the conjugacy class of the Coxeter transformation. Menshikh and Subbotin also calculated the characteristic polynomial of the Coxeter transformation for every class equivalent to $\Omega$ for the extended Dynkin diagram $\widetilde{A}_n$; this polynomial is

$$\det|\mathbf{C} - \lambda I| = \lambda^{n+1} - \lambda^{n-k+1} - \lambda^k + 1, \tag{4.3}$$

where $k = R_\Omega$ is the index of the conjuagacy class of the Coxeter transformation.

Shi Jian-yi [Shi00] considers conjugacy relation on Coxeter transformations for the case where $\Gamma$ is just a cycle. Different equivalence relations on Coxeter transformations are considered in more difficult cases. In [Shi00], Shi also obtained an explicit formula (4.3).

A graph $\Gamma$ is called *unicyclic* if it contains precisely one cycle of type $\widetilde{A}_n$. Boldt and Takane showed in [Bol96], [BT97] how the characteristic polynomial of the Coxeter transformation for the unicyclic graph $\Gamma$ can be reduced to the characteristic polynomial of the cycle. They also arrive at the explicit formula (4.3).

## 4.3 Splitting and gluing formulas for the characteristic polynomial

The purpose of this section is to prove the Subbotin-Sumin *splitting along the edge* formula for the characteristic polynomial of the Coxeter transformation [SuSum82], to prove its generalization for the multiply-laced case and to get some of its corollaries that will be used in the following sections.

Let us consider a characteristic polynomial of the Coxeter transformation

$$\mathcal{X}(\Gamma, \lambda) = \det |\Gamma - \lambda I|$$

for the graph $\Gamma$ with a *splitting edge*. Recall that an edge $l$ is said to be *splitting* if by deleting it we split the graph $\Gamma$ into two graphs $\Gamma_1$ and $\Gamma_2$, see Fig. 4.1.

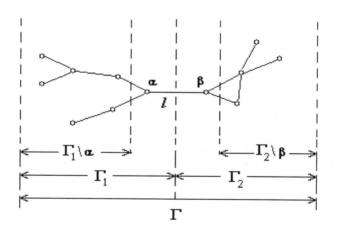

**Fig. 4.1.**     A split graph $\Gamma$

**Proposition 4.8 ([SuSum82]).** *For a given graph $\Gamma$ with a splitting edge $l$, we have*

$$\mathcal{X}(\Gamma, \lambda) = \mathcal{X}(\Gamma_1, \lambda)\mathcal{X}(\Gamma_2, \lambda) - \lambda\mathcal{X}(\Gamma_1\backslash\alpha, \lambda)\mathcal{X}(\Gamma_2\backslash\beta, \lambda), \qquad (4.4)$$

*where $\alpha$ and $\beta$ are the endpoints of the deleted edge $l$.*

*Proof.* We select a basis in $\mathcal{E}_\Gamma$ such that

$$\begin{aligned}
&\mathcal{E}_\Gamma = \mathcal{E}_{\Gamma_1} \cup \mathcal{E}_{\Gamma_2}, \\
&\mathcal{E}_{\Gamma_1} = \{a_1, ..., a_m\}, \quad a_m = \alpha, \quad \dim \mathcal{E}_{\Gamma_1} = m, \\
&\mathcal{E}_{\Gamma_2} = \{b_1, ..., b_k\}, \quad b_1 = \beta, \quad \dim \mathcal{E}_{\Gamma_2} = k,
\end{aligned}$$

and a Coxeter transformation $\mathbf{C}$ such that

$$\mathbf{C} = w_{\Gamma_1} w_{\Gamma_2}, \quad w_{\Gamma_1} = \sigma_{a_m}...\sigma_{a_1}, \quad w_{\Gamma_2} = \sigma_{b_1}...\sigma_{b_k}.$$

An ordering of reflections $\sigma_{a_i}$ (resp. $\sigma_{b_j}$) in $w_{\Gamma_1}$ (resp. $w_{\Gamma_2}$) is selected so that $\sigma_{a_m}$ (resp. $\sigma_{b_1}$) is the last to be executed. Then the reflections $\sigma_{a_i}$ ($i \neq m$) (resp. $\sigma_{b_j}$ ($j \neq 1$)) do not act on the coordinates $b_j$ (resp. $a_i$). Only $\sigma_{a_m}$ acts on the coordinate $b_1$ and $\sigma_{b_1}$ acts on the coordinate $a_m$. We have

$$w_{\Gamma_1} = \sigma_{a_m}...\sigma_{a_1} = \begin{pmatrix} \mathbf{C}_{\Gamma_1} & \delta_{a_m b_1} \\ 0 & I \end{pmatrix} \begin{matrix} \}m \\ \}k \end{matrix},$$

$$w_{\Gamma_2} = \sigma_{b_1}...\sigma_{b_k} = \begin{pmatrix} I & 0 \\ \delta_{b_1 a_m} & \mathbf{C}_{\Gamma_2} \end{pmatrix} \begin{matrix} \}m \\ \}k \end{matrix},$$

(4.5)

where $\mathbf{C}_{\Gamma_1}$ (resp. $\mathbf{C}_{\Gamma_2}$) is the Coxeter transformation for the graph $\Gamma_1$ (resp. $\Gamma_2$) and $\delta_{a_m b_1}$ (resp. $\delta_{b_1 a_m}$) means the matrix with zeros everywhere except 1 in the $(a_m b_1)$th (resp. $(b_1 a_m)$th) slot.

Then we have

$$\mathbf{C} = w_{\Gamma_1} w_{\Gamma_2} = \begin{pmatrix} \mathbf{C}_{\Gamma_1} + \delta_{a_m a_m} & \delta_{a_m b_1} \mathbf{C}_{\Gamma_2} \\ \delta_{b_1 a_m} & \mathbf{C}_{\Gamma_2} \end{pmatrix},$$

$$\det |\mathbf{C} - \lambda I| = \begin{vmatrix} \mathbf{C}_{\Gamma_1} + \delta_{a_m a_m} - \lambda I & \delta_{a_m b_1} \mathbf{C}_{\Gamma_2} \\ \delta_{b_1 a_m} & \mathbf{C}_{\Gamma_2} - \lambda I \end{vmatrix}.$$

(4.6)

Here, $\delta_{a_m b_1} \mathbf{C}_{\Gamma_2}$ is the $b_1$st line (the first line in $\mathbf{C}_{\Gamma_2}$ for the selected basis) of the matrix $\mathbf{C}_{\Gamma_2}$ and we subtract now this line from the line $a_m$. Then

$$\det |\mathbf{C} - \lambda I| = \begin{vmatrix} \mathbf{C}_{\Gamma_1} - \lambda I & \lambda \delta_{a_m b_1} \\ \delta_{b_1 a_m} & \mathbf{C}_{\Gamma_2} - \lambda I \end{vmatrix}$$

or

$$\det |\mathbf{C} - \lambda I| = \begin{vmatrix} \mathbf{C}_{\Gamma_1} - \lambda I & \begin{matrix} 0\,0\,...\,0 \\ 0\,0\,...\,0 \\ ... \\ 0\,0\,...\,0 \\ \lambda\,0\,...\,0 \end{matrix} \\ \begin{matrix} 0\,...\,0\,1 \\ 0\,...\,0\,0 \\ ... \\ 0\,...\,0\,0 \\ 0\,...\,0\,0 \end{matrix} & \mathbf{C}_{\Gamma_2} - \lambda I \end{vmatrix}.$$

(4.7)

Let us expand the determinant $\det|\mathbf{C} - \lambda I|$ in eq. (4.7) with respect to the minors corresponding to the line $b_1$ (the line containing $\lambda$ in the column $b_1$, which is the next after $a_m$). We see that

$$\det|\mathbf{C} - \lambda I| = \det|\mathbf{C}_{\Gamma_1} - \lambda I|\det|\mathbf{C}_{\Gamma_2} - \lambda I| + (-1)^{a_m + (a_m + 1)}\lambda R, \quad (4.8)$$

where $R$ is the determinant shown on the Fig. 4.2. Expanding the determinant $R$ we get the Subbotin-Sumin formula (4.4).      □

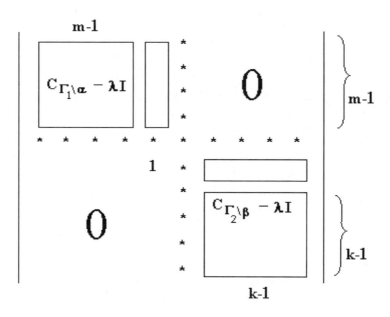

**Fig. 4.2.**     The asterisks mark the deleted line $a_m$ and column $b_1$

Now we will extend the Subbotin-Sumin formula to the multiply-laced case where the endpoints $\alpha$ and $\beta$ correspond to simple roots of different lengths. In this case, eq. (4.4) is modified to eq. (4.9).

**Proposition 4.9.** *For a given graph $\Gamma$ with a splitting weighted edge $l$ corresponding to roots with different lengths, we have*

$$\mathcal{X}(\Gamma, \lambda) = \mathcal{X}(\Gamma_1, \lambda)\mathcal{X}(\Gamma_2, \lambda) - \rho\lambda\mathcal{X}(\Gamma_1\backslash\alpha, \lambda)\mathcal{X}(\Gamma_2\backslash\beta, \lambda), \quad (4.9)$$

*where $\alpha$ and $\beta$ are the endpoints of the deleted edge $l$, and $\rho$ is the following factor:*

$$\rho = k_{\alpha\beta}k_{\beta\alpha},$$

*where $k_{ij}$ is an element of the Cartan matrix (2.4).*

*Proof.* Eq. (4.5) is modified to the following

$$w_{\Gamma_1} = \begin{pmatrix} \mathbf{C}_{\Gamma_1} & k_{a_m b_1} \delta_{a_m b_1} \\ 0 & I \end{pmatrix}, \qquad w_{\Gamma_2} = \begin{pmatrix} I & 0 \\ k_{b_1 a_m} \delta_{b_1 a_m} & \mathbf{C}_{\Gamma_2} \end{pmatrix},$$

and eq. (4.6) is modified as follows

$$\mathbf{C} = \begin{pmatrix} \mathbf{C}_{\Gamma_1} + k_{a_m b_1} k_{b_1 a_m} \delta_{a_m a_m} & k_{a_m b_1} \delta_{a_m b_1} \mathbf{C}_{\Gamma_2} \\ k_{b_1 a_m} \delta_{b_1 a_m} & \mathbf{C}_{\Gamma_2} \end{pmatrix}.$$

Multiply the line $b_1$ by $k_{a_m b_1}$ and subtract this product from the line $a_m$, then we obtain

$$\det |\mathbf{C} - \lambda I| = \begin{vmatrix} \mathbf{C}_{\Gamma_1} + k_{a_m b_1} k_{b_1 a_m} \delta_{a_m a_m} - \lambda I & k_{a_m b_1} \delta_{a_m b_1} \mathbf{C}_{\Gamma_2} \\ k_{b_1 a_m} \delta_{b_1 a_m} & \mathbf{C}_{\Gamma_2} - \lambda I \end{vmatrix} =$$

$$\begin{vmatrix} \mathbf{C}_{\Gamma_1} - \lambda I & k_{a_m b_1} \lambda \delta_{a_m b_1} \\ k_{b_1 a_m} \delta_{b_1 a_m} & \mathbf{C}_{\Gamma_2} - \lambda I \end{vmatrix} = \begin{vmatrix} \mathbf{C}_{\Gamma_1} - \lambda I & \rho \lambda \delta_{a_m b_1} \\ \delta_{b_1 a_m} & \mathbf{C}_{\Gamma_2} - \lambda I \end{vmatrix}$$

(4.10)

Further, as in Proposition 4.8, we obtain (4.9). $\square$

**Corollary 4.10.** *Let $\Gamma_2$ in Proposition 4.9 be a component containing a single point. Then, the following formula holds*

$$\mathcal{X}(\Gamma, \lambda) = -(\lambda + 1)\mathcal{X}(\Gamma_1, \lambda) - \rho \lambda \mathcal{X}(\Gamma_1 \backslash \alpha, \lambda), \tag{4.11}$$

*Proof.* Relation (4.5) is modified to the following

$$w_{\Gamma_1} = \left. \begin{pmatrix} & & 0 \\ & & 0 \\ \mathbf{C}_{\Gamma_1} & & \cdots \\ & & 0 \\ & & \rho \\ 0 & & 1 \end{pmatrix} \begin{matrix} \Big\} m \\ \\ \Big\} 1 \end{matrix} \right.,$$

$$w_{\Gamma_2} = \begin{pmatrix} I & 0 \\ 0\,0\ldots 1 & -1 \end{pmatrix} \begin{matrix} \} m \\ \} 1 \end{matrix},$$

and (4.10) is modified to

$$\mathbf{C} = \begin{pmatrix} & & & 0 \\ & & & 0 \\ & \mathbf{C}_{\Gamma_1} + \rho\delta_{a_m a_m} & & \cdots \\ & & & 0 \\ & & & -\rho \\ & 0\;0\ldots 1 & & -1 \end{pmatrix},$$

$$\det|\mathbf{C} - \lambda I| = \begin{vmatrix} & & 0 \\ & & 0 \\ \mathbf{C}_{\Gamma_1} + \rho\delta_{a_m a_m} - \lambda I & & \cdots \\ & & 0 \\ & & -\rho \\ 0\;0\ldots 1 & & -1-\lambda \end{vmatrix} = \begin{vmatrix} & & 0 \\ & & 0 \\ \mathbf{C}_{\Gamma_1} - \lambda I & & \cdots \\ & & 0 \\ & & \rho\lambda \\ 0\;0\ldots 1 & -1-\lambda \end{vmatrix}$$

Further, as in Proposition 4.8, we obtain (4.11). □

The next proposition follows from eq. (4.4) and allows one to calculate the spectrum of the graph $\Gamma(n)$ obtained by gluing $n$ copies of the graph $\Gamma$.

**Proposition 4.11 ([SuSum82], [KMSS83], [KMSS83a]).** *Let $*_n$ be a star with $n$ rays coming from a vertex. Let $\Gamma(n)$ be the graph obtained from $*_n$ by gluing $n$ copies of the graph $\Gamma$ to the endpoints of its rays . Then*

$$\mathcal{X}(\Gamma(n), \lambda) = \mathcal{X}(\Gamma, \lambda)^{n-1}\varphi_{n-1}(\lambda),$$

*where*

$$\varphi_n(\lambda) = \mathcal{X}(\Gamma + \beta, \lambda) - n\lambda\mathcal{X}(\Gamma\backslash\alpha, \lambda).$$

*Proof.* By splitting along the edge $l$ (4.4) we get, for $n = 2$, (Fig. 4.3)

$$\mathcal{X}(\Gamma(2), \lambda) = \mathcal{X}(\Gamma, \lambda)\mathcal{X}(\Gamma + \beta, \lambda) - \lambda\mathcal{X}(\Gamma\backslash\alpha, \lambda)\mathcal{X}(\Gamma, \lambda) =$$
$$\mathcal{X}(\Gamma, \lambda)(\mathcal{X}(\Gamma + \beta, \lambda) - \lambda\mathcal{X}(\Gamma\backslash\alpha, \lambda)) = \mathcal{X}(\Gamma, \lambda)\varphi_1(\lambda).$$

Let the proposition be true for $n = r$ and

$$\mathcal{X}(\Gamma(r), \lambda) = \mathcal{X}(\Gamma, \lambda)^{r-1}\varphi_{r-1}(\lambda).$$

Then, for $n = r + 1$, we have

$$\mathcal{X}(\Gamma(r + 1), \lambda) = \mathcal{X}(\Gamma, \lambda)\mathcal{X}(\Gamma(r), \lambda) - \lambda\mathcal{X}(\Gamma\backslash\alpha, \lambda)(\mathcal{X}(\Gamma, \lambda))^r =$$
$$\mathcal{X}(\Gamma, \lambda)\mathcal{X}(\Gamma, \lambda)^{r-1}\varphi_{r-1}(\lambda) - \lambda\mathcal{X}(\Gamma\backslash\alpha, \lambda)\mathcal{X}(\Gamma, \lambda)^r =$$
$$\mathcal{X}(\Gamma, \lambda)^r(\varphi_{r-1}(\lambda) - \lambda\mathcal{X}(\Gamma\backslash\alpha, \lambda)) = \mathcal{X}(\Gamma, \lambda)^r\varphi_r(\lambda). \quad \square$$

The following proposition is due to V. Kolmykov. For the case $\lambda = 1$, it is formulated in [Kac93, Ch.4, exs.4.12].

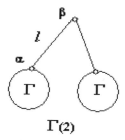

**Fig. 4.3.**    Splitting along the edge $l$ of the graph $\Gamma(2)$.
Here, the graph $\Gamma(2)$ is obtained by gluing two copies of the graph $\Gamma$.


**Proposition 4.12.** *If the spectrum of the Coxeter transformations for graphs* $\Gamma_1$ *and* $\Gamma_2$ *contains the eigenvalue* $\lambda$, *then this eigenvalue is also the eigenvalue of the Coxeter transformation for the graph* $\Gamma$ *obtained by gluing as described in Proposition 4.11.*

*Proof.* By splitting along the edge $l$ (4.4) we get

$$\mathcal{X}(\Gamma_1 + \beta + \Gamma_2, \lambda) =$$
$$\mathcal{X}(\Gamma_1, \lambda)\mathcal{X}(\Gamma_2 + \beta, \lambda) - \lambda\mathcal{X}(\Gamma\backslash\alpha, \lambda)\mathcal{X}(\Gamma_2, \lambda),$$

see Fig. 4.4. If $\mathcal{X}(\Gamma_1, \lambda)|_{\lambda=\lambda_0} = 0$ and $\mathcal{X}(\Gamma_2, \lambda)|_{\lambda=\lambda_0} = 0$, then also $\mathcal{X}(\Gamma, \lambda)|_{\lambda=\lambda_0} = 0$.    □

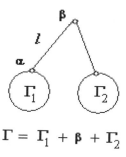

$$\Gamma = \Gamma_1 + \beta + \Gamma_2$$

**Fig. 4.4.**    Splitting along the edge $l$, $\Gamma_1 \neq \Gamma_2$


Let $\mathcal{X}(A_n)$ be the characteristic polynomial of the Coxeter transformation for the Dynkin diagram $A_n$. Then, from the recurrence formula (4.11), we have

$$
\begin{aligned}
\mathcal{X}(A_1) &= -(\lambda + 1),\\
\mathcal{X}(A_2) &= \lambda^2 + \lambda + 1,\\
\mathcal{X}(A_3) &= -(\lambda^3 + \lambda^2 + \lambda + 1),\\
\mathcal{X}(A_4) &= \lambda^4 + \lambda^3 + \lambda^2 + \lambda + 1,
\end{aligned}
\tag{4.12}
$$

$$
\cdots
$$

$$
\mathcal{X}(A_n) = -(\lambda + 1)\mathcal{X}(A_{n-1}) - \lambda\mathcal{X}(A_{n-2}), \qquad n > 2.
$$

*Remark 4.13.* Since the characteristic polynomial can be chosen up to a factor, we prefer to consider the polynomial

$$
\mathcal{X}_n = (-1)^n \mathcal{X}(A_n)
$$

as the characteristic polynomial of the Coxeter transformation of $A_n$, thus the leading coefficient is positive. Pay attention that the sign $(-1)^n$ for diagrams with $n$ vertices should be taken into account only in recurrent calculations using eq. (4.11) (cf. with calculation for $\widetilde{D}_n$ and $\widetilde{D}_4$ below).

*Remark 4.14.* 1) J. S. Frame in [Fr51, p.784] obtained that

$$
\mathcal{X}(A_{m+n}) = \mathcal{X}(A_m)\mathcal{X}(A_n) - \lambda\mathcal{X}(A_{m-1})\mathcal{X}(A_{n-1}),
\tag{4.13}
$$

which easily follows from eq. (4.4). Another *Frame formula* will be proved in Proposition 5.2.

2) S. M. Gussein-Zade in [Gu76] obtained the formula

$$
\mathcal{X}(A_n) = -(\lambda + 1)\mathcal{X}_{n-1} - \lambda\mathcal{X}_{n-2},
$$

for the characteristic polynomial of the classical monodromy, see (4.12).

From (4.12) for $n > 0$, we see that $\mathcal{X}_n$ is the *cyclotomic polynomial* whose roots are the primitive $(n+1)$th roots of unity:

$$
\mathcal{X}_n = \sum_{i=1}^{n} \lambda^i = \frac{\lambda^{n+1} - 1}{\lambda - 1}.
\tag{4.14}
$$

## 4.4 Formulas of the characteristic polynomials for the diagrams $T_{p,q,r}$

We give here explicit formulas of characteristic polynomials of the Coxeter transformations for the three classes of diagrams: $T_{2,3,r}$, $T_{3,3,r}$, and $T_{2,4,r}$, see §2.1.8, Fig. 2.1.

## 4.4.1 The diagrams $T_{2,3,r}$

The case $T_{2,3,r}$, where $r \geq 2$, includes diagrams $D_5$, $E_6$, $E_7$, $E_8$, $\widetilde{E}_8$, $E_{10}$. Since $D_5 = T_{2,3,2} = E_5$ and $\widetilde{E}_8 = T_{2,3,6} = E_9$, we call these diagrams the $E_n$-series, where $n = r + 3$. In Table 1.1 and Table 1.2 we see that characteristic polynomials of $E_6$, $E_7$, $E_8$, $\widetilde{E}_8$ constitute a series, and this series can be continued, see (4.15) and Table 4.4.

*Remark 4.15.* C. McMullen observed in [McM02] that the spectral radius $\rho(\mathbf{C})$ of the Coxeter transformation for all graphs with indefinite Tits form attains its minimum when the diagram is $T_{2,3,7} = E_{10}$, see Fig. 4.5, [Hir02], [McM02], and §2.1.8 about *hyperbolic Dynkin diagrams*. McMullen showed that $\rho(\mathbf{C})$ is the so-called *Lehmer's number*,

$$\lambda_{Lehmer} \approx 1.176281... \; , \; \text{see Table 4.4.}$$

For details and definitions, see §C.2.

*Lehmer's number* is a root of the polynomial

$$1 + x - x^3 - x^4 - x^5 - x^6 - x^7 + x^9 + x^{10},$$

see [Hir02], [McM02].

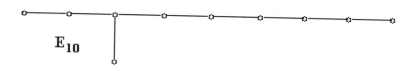

**Fig. 4.5.**    The spectral radius attains its minimum on the graph $E_{10}$

**Proposition 4.16.** *The characteristic polynomials of Coxeter transformations for the diagrams $T_{2,3,n}$ are as follows:*

$$\mathcal{X}(T_{2,3,n-3}) = \mathcal{X}(E_n) = \lambda^n + \lambda^{n-1} - \sum_{i=3}^{n-3} \lambda^i + \lambda + 1, \qquad (4.15)$$

*where $\mathcal{X}(E_n)$ is given up to sign $(-1)^n$, see Remark 4.13. In other words,*

$$\mathcal{X}(T_{2,3,n-3}) = \mathcal{X}(E_n) = \frac{\lambda^{n-2}(\lambda^3 - \lambda - 1) + (\lambda^3 + \lambda^2 - 1)}{\lambda - 1}. \qquad (4.16)$$

The spectral radius $\rho(T_{2,3,n-3})$ converges to the maximal root $\rho_{max}$ of the equation

$$\lambda^3 - \lambda - 1 = 0, \tag{4.17}$$

*and*

$$\rho_{max} = \sqrt[3]{\frac{1}{2} + \sqrt{\frac{23}{108}}} + \sqrt[3]{\frac{1}{2} - \sqrt{\frac{23}{108}}} \approx 1.324717... \; . \tag{4.18}$$

*Proof.* Use induction and (4.11). We have

$$(-1)^{n+2}\mathcal{X}(E_{n+2}) = -(\lambda + 1)(-1)^{n+1}\mathcal{X}(E_{n+1}) - \lambda(-1)^n\mathcal{X}(E_n) =$$
$$(-1)^{n+2}[(\lambda + 1)\mathcal{X}(E_{n+1}) - \lambda\mathcal{X}(E_n)].$$

Thus,

$$\mathcal{X}(E_{n+2}) = (\lambda + 1)\mathcal{X}(E_{n+1}) - \lambda\mathcal{X}(E_n) = \lambda(\mathcal{X}(E_{n+1} - \mathcal{X}(E_n)) + \mathcal{X}(E_{n+1}).$$

By the induction hypothesis, (4.15) yields

$$\lambda(\mathcal{X}(E_{n+1} - \mathcal{X}(E_n)) = \lambda^{n+2} - \lambda^n - \lambda^{n-1},$$

and

$$\mathcal{X}(E_{n+2}) = \lambda^{n+2} - \lambda^n - \lambda^{n-1} + (\lambda^{n+1} + \lambda^n - \sum_{i=3}^{n-2}\lambda^i + \lambda + 1) =$$

$$\lambda^{n+2} + \lambda^{n+1} - \sum_{i=3}^{n-1}\lambda^i + \lambda + 1,$$

and hence (4.15) is proved. Further,

$$\sum_{i=3}^{n-3}\lambda^i = \lambda^3\sum_{i=0}^{n-6}\lambda^i = \lambda^3\frac{\lambda^{n-5} - 1}{\lambda - 1} = \frac{\lambda^{n-2} - \lambda^3}{\lambda - 1},$$

and

$$\mathcal{X}(E_n) = \lambda^n + \lambda^{n-1} - \frac{\lambda^{n-2} - \lambda^3}{\lambda - 1} + \lambda + 1 =$$
$$\frac{\lambda^{n+1} - \lambda^{n-1} - \lambda^{n-2} + \lambda^3 + \lambda^2 - 1}{\lambda - 1} =$$
$$\frac{\lambda^{n-2}(\lambda^3 - \lambda - 1) + (\lambda^3 + \lambda^2 - 1)}{\lambda - 1}.$$

This proves (4.16). For $\lambda > 1$, the characteristic polynomial $\mathcal{X}(E_n)$ converges to

$$\frac{\lambda^{n-2}(\lambda^3 - \lambda - 1)}{\lambda - 1}.$$

We will prove[1] in §4.4.4 that the spectral radius $\rho_{max}$ converges to the maximal root of equation (4.17). For the results of calculations related to the Proposition 4.16, see Table 4.4. □

---
[1] See Proposition 4.22, Corollary 4.24, Table 4.7.

**Table 4.4.** The characteristic polynomials for $T_{2,3,r}$

| $T_{2,3,r}$ and its well-known name | Characteristic polynomial | Maximal eigenvalue outside the unit circle |
|---|---|---|
| $(2,3,2)$ $D_5$ | $\lambda^5 + \lambda^4 + \lambda + 1$ | – |
| $(2,3,3)$ $E_6$ | $\lambda^6 + \lambda^5 - \lambda^3 + \lambda + 1$ | – |
| $(2,3,4)$ $E_7$ | $\lambda^7 + \lambda^6 - \lambda^4 - \lambda^3 + \lambda + 1$ | – |
| $(2,3,5)$ $E_8$ | $\lambda^8 + \lambda^7 - \lambda^5 - \lambda^4 - \lambda^3 + \lambda + 1$ | – |
| $(2,3,6)$ $\widetilde{E}_8$ | $\lambda^9 + \lambda^8 - \lambda^6 - \lambda^5 - \lambda^4 - \lambda^3 + \lambda + 1$ | – |
| $(2,3,7)$ $E_{10}$ | $\lambda^{10} + \lambda^9 - \lambda^7 - \lambda^6 - \lambda^5 - \lambda^4 - \lambda^3 + \lambda + 1$ | 1.176281... |
| $(2,3,8)$ $E_{11}$ | $\lambda^{11} + \lambda^{10} - \lambda^8 - \lambda^7 - \lambda^6 - \lambda^5 - \lambda^4 - \lambda^3 + \lambda + 1$ | 1.230391... |
| $(2,3,9)$ $E_{12}$ | $\lambda^{12} + \lambda^{11} - \lambda^9 - \lambda^8 - \lambda^7 - \lambda^6 - \lambda^5 - \lambda^4 - \lambda^3 + \lambda + 1$ | 1.261231... |
| $(2,3,10)$ $E_{13}$ | $\lambda^{13} + \lambda^{12} - \lambda^{10} - \lambda^9 - \lambda^8 - \lambda^7 - \lambda^6 - \lambda^5 - \lambda^4 - \lambda^3 + \lambda + 1$ | 1.280638... |
| ... | ... | ... |
| $(2,3,n)$ $E_n$ | $\dfrac{\lambda^{n-2}(\lambda^3 - \lambda - 1) + (\lambda^3 + \lambda^2 - 1)}{\lambda - 1}$ | 1.324717... as $n \to \infty$ |

## 4.4.2 The diagrams $T_{3,3,r}$

The case $T_{3,3,r}$, where $r \geq 2$, includes diagrams $E_6$ and $\widetilde{E}_6$, and so we call these diagrams the $E_{6,n}$-*series*, where $n = r - 2$. Thus, $E_6 = T_{3,3,2} = E_{6,0}$ and $\widetilde{E}_6 = T_{3,3,3} = E_{6,1}$. In Table 1.1 and Table 1.2 we see that the characteristic polynomials of $E_6, \widetilde{E}_6$ constitute a series, and this series can be continued, see (4.19) and Table 4.5.

**Proposition 4.17.** *The characteristic polynomials of Coxeter transformations for the diagrams $T_{3,3,n}$ with $n \geq 3$ are calculated as follows:*

$$\mathcal{X}(T_{3,3,n}) = \mathcal{X}(E_{6,n-2}) = \lambda^{n+4} + \lambda^{n+3} - 2\lambda^{n+1} - 3\sum_{i=4}^{n} \lambda^i - 2\lambda^3 + \lambda + 1, \quad (4.19)$$

*where $\mathcal{X}(E_{6,n})$ is given up to sign $(-1)^n$, see Remark 4.13. In other words,*

$$\mathcal{X}(T_{3,3,n}) = \mathcal{X}(E_{6,n-2}) =$$
$$\frac{\lambda^{n+1}(\lambda^4 - \lambda^2 - 2\lambda - 1) + (\lambda^4 + 2\lambda^3 + \lambda^2 - 1)}{\lambda - 1} =$$
$$\frac{(\lambda^2 + \lambda + 1)[\lambda^{n+1}(\lambda^2 - \lambda - 1) + (\lambda^2 + \lambda - 1)]}{\lambda - 1}.$$

*The spectral radius $\rho(T_{3,3,n})$ converges to the maximal root $\rho_{max}$ of the equation*

$$\lambda^2 - \lambda - 1 = 0, \quad (4.20)$$

*and*

$$\rho_{max} = \frac{\sqrt{5} + 1}{2} \approx 1.618034... .$$

*Remark 4.18.* For $n = 3$, the sum $\sum_{i=4}^{n} \lambda^i$ in (4.19) disappears and we have

$$\mathcal{X}(T_{3,3,3}) = \mathcal{X}(E_{6,1}) = \mathcal{X}(\widetilde{E}_6) =$$
$$\lambda^7 + \lambda^6 - 2\lambda^4 - 2\lambda^3 + \lambda + 1 = (\lambda^3 - 1)^2(\lambda + 1).$$

*Proof.* As above in (4.15), we use induction and (4.11). So, by (4.11) we have

$$\mathcal{X}(E_{6,n+2}) = (\lambda + 1)\mathcal{X}(E_{6,n+1}) - \lambda\mathcal{X}(E_{6,n}) =$$
$$\lambda(\mathcal{X}(E_{6,n+1}) - \mathcal{X}(E_{6,n})) + \mathcal{X}(E_{6,n+1}). \quad (4.21)$$

By the induction hypothesis and (4.19) we have

$$\lambda(\mathcal{X}(E_{6,n+1}) - \mathcal{X}(E_{6,n})) = \lambda^{n+8} - \lambda^{n+6} - 2\lambda^{n+5} - \lambda^{n+4},$$

and

**Table 4.5.**    The characteristic polynomials for $T_{3,3,r}$

| $T_{3,3,r}$ and its well-known name | Characteristic polynomial | Maximal eigenvalue outside the unit circle |
|---|---|---|
| $(3,3,2)$ $E_6$ | $\lambda^6 + \lambda^5 - \lambda^3 + \lambda + 1$ | – |
| $(3,3,3)$ $\widetilde{E}_6$ | $\lambda^7 + \lambda^6 - 2\lambda^4 - 2\lambda^3 + \lambda + 1$ | – |
| $(3,3,4)$ $E_{6,2}$ | $\lambda^8 + \lambda^7 - 2\lambda^5 - 3\lambda^4 - 2\lambda^3 + \lambda + 1$ | 1.401268... |
| $(3,3,5)$ $E_{6,3}$ | $\lambda^9 + \lambda^8 - 2\lambda^6 - 3\lambda^5 - 3\lambda^4 - 2\lambda^3 + \lambda + 1$ | 1.506136... |
| $(3,3,6)$ $E_{6,4}$ | $\lambda^{10} + \lambda^9 - 2\lambda^7 - 3\lambda^6 - 3\lambda^5 - 3\lambda^4 - 2\lambda^3 + \lambda + 1$ | 1.556030... |
| $(3,3,7)$ $E_{6,5}$ | $\lambda^{11} + \lambda^{10} - 2\lambda^8 - 3\lambda^7 - 3\lambda^6 - 3\lambda^5 - 3\lambda^4 - 2\lambda^3 + \lambda + 1$ | 1.582347... |
| $(3,3,8)$ $E_{6,6}$ | $\lambda^{12} + \lambda^{11} - 2\lambda^9 - 3\lambda^8 - 3\lambda^7$ $-3\lambda^6 - 3\lambda^5 - 3\lambda^4 - 2\lambda^3 + \lambda + 1$ | 1.597005... |
| $\ldots$ | $\ldots$ | $\ldots$ |
| $(3,3,n)$ $E_{6,n-2}$ | $\dfrac{(\lambda^2 + \lambda + 1)[\lambda^{n+1}(\lambda^2 - \lambda - 1) + (\lambda^2 + \lambda - 1)]}{\lambda - 1}$ | 1.618034... as $n \to \infty$ |

$$\mathcal{X}(E_{6,n+2}) =$$
$$\lambda^{n+8} - \lambda^{n+6} - 2\lambda^{n+5} - \lambda^{n+4} +$$
$$(\lambda^{n+7} + \lambda^{n+6} - 2\lambda^{n+4} - 3\sum_{i=4}^{n+3} \lambda^i - 2\lambda^3 + \lambda + 1) =$$
$$\lambda^{n+8} + \lambda^{n+7} - 2\lambda^{n+5} - 3\sum_{i=4}^{n+4} \lambda^i - 2\lambda^3 + \lambda + 1;$$

this proves (4.19). Further,

$$\mathcal{X}(T_{3,3,n}) = \lambda^{n+4} + \lambda^{n+3} - 2\lambda^{n+1} - 3\sum_{i=4}^{n} \lambda^i - 2\lambda^3 + \lambda + 1 =$$
$$\lambda^{n+4} + \lambda^{n+3} + \lambda^{n+1} - 3\sum_{i=3}^{n+1} \lambda^i + \lambda^3 + \lambda + 1 =$$

$$\lambda^{n+1}(\lambda^3 + \lambda^2 + 1) - 3\lambda^3 \frac{\lambda^{n-1} - 1}{\lambda - 1} + \lambda^3 + \lambda + 1,$$

i.e.,

$$\mathcal{X}(T_{3,3,n}) = \frac{\lambda^{n+1}(\lambda^4 - \lambda^2 + \lambda - 1) - 3\lambda^{n+2} + 3\lambda^3 + (\lambda^4 - \lambda^3 + \lambda^2 - 1)}{\lambda - 1} =$$

$$\frac{\lambda^{n+1}(\lambda^4 - \lambda^2 - 2\lambda - 1) + (\lambda^4 + 2\lambda^3 + \lambda^2 - 1)}{\lambda - 1}.$$

For $\lambda > 1$, the characteristic polynomial $\mathcal{X}(T_{3,3,n})$ converges to

$$\frac{\lambda^{n+1}(\lambda^4 - \lambda^2 - 2\lambda - 1)}{\lambda - 1} = \frac{\lambda^{n+1}(\lambda^2 - \lambda - 1)(\lambda^2 + \lambda + 1)}{\lambda - 1}.$$

We will prove[1] in §4.4.4 that the spectral radius $\rho_{max}$ converges to the maximal root of the equation (4.20). For the results of calculations related to the Proposition 4.17, see Table 4.5. □

### 4.4.3 The diagrams $T_{2,4,r}$

The case $T_{2,4,r}$, where $r \geq 2$, includes diagrams $D_6, E_7, \widetilde{E}_7$ and we call these diagrams the $E_{7,n}$-*series*, where $n = r - 3$. Thus, $D_6 = T_{2,4,2} = E_{7,-1}$, $E_7 = T_{2,4,3} = E_{7,0}$, $\widetilde{E}_7 = T_{2,4,4} = E_{7,1}$. In Table 1.1 and Table 1.2 we see that characteristic polynomials of $E_7, \widetilde{E}_7$ constitute a series, and this series can be continued, see (4.22) and Table 4.6.

---

[1] See Proposition 4.22, Corollary 4.24, Table 4.7.

**Proposition 4.19.** *The characteristic polynomials of Coxeter transformations for diagrams $T_{2,4,n}$, where $n \geq 3$, are calculated as follows:*

$$\mathcal{X}(T_{2,4,n}) = \mathcal{X}(E_{7,n-3}) = \lambda^{n+4} + \lambda^{n+3} - \lambda^{n+1} - 2\sum_{i=4}^{n} \lambda^i - \lambda^3 + \lambda + 1, \quad (4.22)$$

*where $\mathcal{X}(E_{7,n})$ is given up to sign $(-1)^n$, see Remark 4.13. In other words*

$$\mathcal{X}(T_{2,4,n}) = \mathcal{X}(E_{7,n-3}) = \frac{\lambda^{n+1}(\lambda^4 - \lambda^2 - \lambda - 1) + (\lambda^4 + \lambda^3 + \lambda - 1)}{\lambda - 1} =$$

$$\frac{(\lambda + 1)(\lambda^{n+1}(\lambda^3 - \lambda^2 - 1) + (\lambda^3 + \lambda - 1))}{\lambda - 1}.$$

*The spectral radius $\rho(T_{2,4,n})$ converges to the maximal root $\rho_{max}$ of the equation*

$$\lambda^3 - \lambda^2 - 1 = 0, \quad (4.23)$$

*and*

$$\rho_{max} = \frac{1}{3} + \sqrt[3]{\frac{58}{108} + \sqrt{\frac{31}{108}}} + \sqrt[3]{\frac{58}{108} - \sqrt{\frac{31}{108}}} \approx 1.465571... \; .$$

*Remark 4.20.* For $n = 3$, the sum $\sum_{i=4}^{n} \lambda^i$ disappears from (4.22) and we have

$$\mathcal{X}(T_{2,4,3}) = \mathcal{X}(E_{7,0}) = \mathcal{X}(E_7) =$$
$$\lambda^7 + \lambda^6 - \lambda^4 - \lambda^3 + \lambda + 1 = (\lambda^4 - 1)(\lambda^3 - 1)(\lambda + 1).$$

*Proof.* Use induction and (4.11). As above in (4.21), we have

$$\mathcal{X}(E_{7,n+2}) =$$
$$(\lambda + 1)\mathcal{X}(E_{7,n+1}) - \lambda\mathcal{X}(E_{7,n}) =$$
$$\lambda(\mathcal{X}(E_{7,n+1}) - \mathcal{X}(E_{7,n})) + \mathcal{X}(E_{7,n+1}).$$

By the induction hypothesis and (4.22) we have

$$\lambda(\mathcal{X}(E_{7,n+1}) - \mathcal{X}(E_{7,n})) = \lambda^{n+9} - \lambda^{n+7} - \lambda^{n+6} - \lambda^{n+5},$$

and we see that

$$\mathcal{X}(E_{7,n+2}) =$$
$$\lambda^{n+9} - \lambda^{n+7} - \lambda^{n+6} - \lambda^{n+5}$$
$$+ (\lambda^{n+8} + \lambda^{n+7} - \lambda^{n+5} - 2\sum_{i=4}^{n+4} \lambda^i - \lambda^3 + \lambda + 1) =$$
$$\lambda^{n+9} + \lambda^{n+8} - \lambda^{n+6} - 2\sum_{i=4}^{n+5} \lambda^i - \lambda^3 + \lambda + 1.$$

this proves (4.22). Further,

$$\mathcal{X}(T_{2,4,n}) = \lambda^{n+4} + \lambda^{n+3} - \lambda^{n+1} - 2\sum_{i=4}^{n} \lambda^i - \lambda^3 + \lambda + 1 =$$

$$(\lambda^{n+4} + \lambda^{n+3} + \lambda^{n+1}) - 2\sum_{i=3}^{n+1} \lambda^i + (\lambda^3 + \lambda + 1) =$$

$$\lambda^{n+1}(\lambda^3 + \lambda^2 + 1) - 2\frac{\lambda^3(\lambda^{n-1} - 1)}{\lambda - 1} + (\lambda^3 + \lambda + 1) =$$

$$\frac{\lambda^{n+1}(\lambda^4 - \lambda^2 + \lambda - 1) - 2\lambda^{n+2} + 2\lambda^3 + (\lambda^4 - \lambda^3 + \lambda^2 - 1)}{\lambda - 1} =$$

$$\frac{\lambda^{n+1}(\lambda^4 - \lambda^2 - \lambda - 1) + (\lambda^4 + \lambda^3 + \lambda^2 - 1)}{\lambda - 1} =$$

$$\frac{\lambda^{n+1}(\lambda + 1)(\lambda^3 - \lambda^2 - 1) + (\lambda + 1)(\lambda^3 + \lambda - 1)}{\lambda - 1}.$$

For $\lambda > 1$, the characteristic polynomial $\mathcal{X}(T_{2,4,n})$ converges to

$$\frac{\lambda^{n+1}(\lambda + 1)(\lambda^3 - \lambda^2 - 1)}{\lambda - 1}.$$

We will prove[1] in §4.4.4 that the spectral radius $\rho_{max}$ converges to the maximal root of the equation (4.23). For the results of calculations related to the Proposition 4.19, see Table 4.6. □

*Remark 4.21 (Bibliographical references).* 1) The maximal eigenvalues (Tables 4.4, 4.5, 4.6) lying outside the unit circle are obtained by means of the online service "Polynomial Roots Solver" of "EngineersToolbox"[2].

2) Some convenient rules for the calculating coefficients of the Coxeter transformation for trees are obtained in [KMSS03], e.g., the characteristic polynomial is reciprocal[3], the free coefficient and the coefficient of $\lambda$ are equal to 1, and the coefficient of $\lambda^2$ vanishes if and only if the tree is a *trefoil* (i.e., has exactly one node of branching degree 3). These rules are easily checked by Tables 4.4, 4.5, 4.6.

3) After the first version of this text was put to arXiv I have discovered the work [Lak99b, Cor. 2.9] of P. Lakatos who obtained results on the convergence of the spectral radii $\rho_{max}$ similar to Propositions 4.19, 4.17, 4.16.

---

[1] See Proposition 4.22, Corollary 4.24, Table 4.7.

[2] See http://www.engineerstoolbox.com.

[3] The polynomial $P$ of degree $n$ is said to be *reciprocal* if $P(z) = z^n P(1/z)$.

**Table 4.6.**    The characteristic polynomials for $T_{2,4,r}$

| $T_{2,4,r}$ and its well-known name | Characteristic polynomial | Maximal eigenvalue outside the unit circle |
|---|---|---|
| $(2,4,2)$ $D_6$ | $\lambda^6 + \lambda^5 + \lambda + 1$ | – |
| $(2,4,3)$ $E_7$ | $\lambda^7 + \lambda^6 - \lambda^4 - \lambda^3 + \lambda + 1$ | – |
| $(2,4,4)$ $\widetilde{E}_7$ | $\lambda^8 + \lambda^7 - \lambda^5 - 2\lambda^4 - \lambda^3 + \lambda + 1$ | – |
| $(2,4,5)$ $E_{7,2}$ | $\lambda^9 + \lambda^8 - \lambda^6 - 2\lambda^5 - 2\lambda^4 - \lambda^3 + \lambda + 1$ | 1.280638... |
| $(2,4,6)$ $E_{7,3}$ | $\lambda^{10} + \lambda^9 - \lambda^7 - 2\lambda^6 - 2\lambda^5 - 2\lambda^4 - \lambda^3 + \lambda + 1$ | 1.360000... |
| $(2,4,7)$ $E_{7,4}$ | $\lambda^{11} + \lambda^{10} - \lambda^8 - 2\lambda^7 - 2\lambda^6 - 2\lambda^5 - 2\lambda^4 - \lambda^3 + \lambda + 1$ | 1.401268... |
| $(2,4,8)$ $E_{7,5}$ | $\lambda^{12} + \lambda^{11} - \lambda^9 - 2\lambda^8 - 2\lambda^7$ $-2\lambda^6 - 2\lambda^5 - 2\lambda^4 - \lambda^3 + \lambda + 1$ | 1.425005... |
| $\ldots$ | $\ldots$ | $\ldots$ |
| $(2,4,n)$ $E_n$ | $\dfrac{\lambda^{n+1}(\lambda^4 - \lambda^2 - \lambda - 1) + (\lambda^4 + \lambda^3 + \lambda - 1)}{\lambda - 1}$ | 1.465571... as $n \to \infty$ |

4) Thanks are due to C. McMullen who kindly informed me about the work by J. F. McKee, P. Rowlinson, C. J. Smyth [MRS99]. In this work authors also obtained certain results on the convergence of the spectral radii. In [MS05] McKee and Smyth introduced notions of a *Salem graph* and a *Pisot graph*. For simplicity, let $\Gamma$ be a star-like tree. Denote it by $T_{a_1,a_2,...,a_r}$, where $a_1, a_2, \ldots, a_n$ are the lengths of arms of the star $G$. This graph is called a *Salem graph* if the characteristic polynomial of the corresponding Coxeter transformation has only one eigenvalue $\lambda > 1$ and only $-\lambda$ lies in $(-\infty, -1)$. It is shown in [MRS99, Cor. 9] that the spectral radius of a Salem graph is a *Salem number*[1]. A *Pisot graph* is defined as a member of the family of graphs obtained by letting the number of the edges of one arm tend to infinity. These Pisot graphs represent, in fact, a sequence of Salem numbers tending to the Pisot number, [MS05, Th.1][2].

### 4.4.4 Convergence of the sequence of eigenvalues

In this section we give a substantiation (Proposition 4.22) of the fact that the maximal eigenvalues (i.e., spectral radii) of the characteristic polynomials for $T_{p,q,r}$ converge to a real number which is the maximal value of a known polynomial (see, Propositions 4.16, 4.17, 4.19). We will give this substantiation in a generic form. Similarly, we have a dual fact (Proposition 4.25): the minimal eigenvalues of the characteristic polynomials for $T_{p,q,r}$ converges to a real number which is the minimal value of a known polynomial.

**Proposition 4.22.** *Let $f(\lambda)$ and $g(\lambda)$ be some polynomials in $\lambda$, and let*

$$P_n(\lambda) = \lambda^n f(\lambda) + g(\lambda)$$

*be a sequence of polynomials in $\lambda$. Let $z_n \in \mathbb{R}$ be a root of $P_n(\lambda)$, lying in the interval $(1, 2)$:*

$$P_n(z_n) = 0, \quad z_n \in (1, 2), \quad n = 1, 2, \ldots \tag{4.24}$$

*Suppose that the sequence $\{z_n\}, n = 1, 2, \ldots$ is non-decreasing:*

$$1 < z_1 \leq z_2 \leq \cdots \leq z_n \leq \ldots \ . \tag{4.25}$$

*Then the sequence $z_n$ converges to a real number $z_0$ which is a root of $f(\lambda)$:*

$$f(z_0) = 0.$$

*Proof.* According to (4.24), the sequence (4.25) is non-decreasing and bounded from above. Therefore, there exists a real number $z_0$ such that

$$\lim_{k \to \infty} z_k = z_0. \tag{4.26}$$

---

[1] For a definition of Salem numbers, see §C.2
[2] For a definition of Pisot numbers, see §C.2.

Let us estimate $|f(z_0)|$ as follows:

$$|f(z_0)| < |f(z_n)| + |f(z_n) - f(z_0)|.$$

By (4.26) we obtain

$$|f(z_0)| < |f(z_n)| + \varepsilon_n , \quad \text{where } \varepsilon_n = |f(z_n) - f(z_0)| \to 0 \quad \text{as } n \to \infty.$$
$$(4.27)$$

By (4.24), we have

$$f(z_n) = -\frac{g(z_n)}{z_n^n}.$$

Let

$$\delta_n = \left| \frac{g(z_n)}{z_n^n} \right| \quad \text{for each } n = 1, 2, \dots$$

Since the function $g(z)$ is uniformly bounded on the interval $(1, 2)$, it follows that $\delta_n \to 0$. By (4.27) we have

$$|f(z_0)| < \delta_n + \varepsilon_n \to 0 \quad \text{as } n \to \infty.$$

Thus, $f(z_0) = 0$. $\square$

**Corollary 4.23.** *Proposition 4.22 holds also for the following non-polynomial functions $P_n$:*
   *1) For*

$$P_n(\lambda) = \lambda^{n+k} f(\lambda) + g(\lambda),$$

*where $k \in \mathbb{Z}$ is independent of $n$.*
   *2) For*

$$P_n(\lambda) = D(\lambda)(\lambda^n f(\lambda) + g(\lambda)),$$

*where $D(\lambda)$ is a rational function independent of $n$ and without roots on the interval $(1, 2)$.*


**Corollary 4.24.** *The maximal values of characteristic polynomials of the Coxeter transformation for diagrams $T_{p,q,r}$ satisfy the conditions of Proposition 4.22, see Table 4.7.*

   *Proof.* Let $\varphi_{1,r} = \varphi_1(T_{p,q,r})$ be the dominant value of the matrix $DD^t$ (see §3.3.2) for the diagram $T_{p,q,r}$. According to Proposition 3.12 the dominant value $\varphi_{1,r}$ may only grow, i.e.,

$$\varphi_{1,r} \leq \varphi_{1,r+1}.$$

Therefore, the corresponding eigenvalue $\lambda_1^{max}$ also only grows. Indeed, by (3.13) we get

**Table 4.7.**    The diagrams $T_{p,q,r}$ and characteristic polynomials $\lambda^n f(\lambda) + g(\lambda)$

| Diagram | $f(\lambda)$ | $g(\lambda)$ | $D(\lambda)$ | $P_n(\lambda)$ |
|---------|--------------|--------------|--------------|----------------|
| $T(2,3,r)$ | $\lambda^3 - \lambda - 1$ | $\lambda^3 + \lambda^2 - 1$ | $\dfrac{1}{\lambda - 1}$ | $\lambda^{n-2} f(\lambda) + g(\lambda)$ |
| $T(3,3,r)$ | $\lambda^2 - \lambda - 1$ | $\lambda^2 + \lambda - 1$ | $\dfrac{\lambda^2 + \lambda + 1}{\lambda - 1}$ | $\lambda^{n+1} f(\lambda) + g(\lambda)$ |
| $T(2,4,r)$ | $\lambda^3 - \lambda^2 - 1$ | $\lambda^3 + \lambda - 1$ | $\dfrac{\lambda + 1}{\lambda - 1}$ | $\lambda^{n+1} f(\lambda) + g(\lambda)$ |

$$\lambda_{1,r}^{max} = 2\varphi_{1,r} - 1 + 2\sqrt{\varphi_{1,r}(\varphi_{1,r} - 1)} \leq$$
$$2\varphi_{1,r+1} - 1 + 2\sqrt{\varphi_{1,r+1}(\varphi_{1,r+1} - 1)} = \lambda_{1,r+1}^{max}. \tag{4.28}$$

In addition, from Proposition 3.12 we deduce that $\varphi_{1,r} > 1$, therefore the corresponding eigenvalue $\lambda_{1,r}$ of the Coxeter transformation is also $> 1$.

It remains to show that every $\lambda_{1,r} < 2$. This is clear, because $f(\lambda) > 0$ and $g(\lambda) > 0$ for every $\lambda > 2$, see Table 4.7. $\square$

**Proposition 4.25.** *Let $f(\lambda)$ and $g(\lambda)$ be some polynomials in $\lambda$, and let*

$$P_n(\lambda) = \lambda^n f(\lambda) + g(\lambda), \quad \text{where } n \in \mathbb{Z}_+.$$

*Let $z_n \in \mathbb{R}$ be a root of $P_n(\lambda)$, lying in the interval $\left(\dfrac{1}{2}, 1\right)$:*

$$P_n(z_n) = 0, \quad z_n \in \left(\frac{1}{2}, 1\right). \tag{4.29}$$

*Suppose that the sequence $\{z_n\}_{n \in \mathbb{Z}_+}$ is non-increasing:*

$$1 > z_1 \geq z_2 \geq \cdots \geq z_n \geq \cdots .$$

*Then the sequence $z_n$ converges to a real number $z_0$ which is a root of $g(\lambda)$:*

$$g(z_0) = 0.$$

*Proof.* As in Proposition 4.22, there is a real number $z_0$ such that $z_n \to z_0$ as $n \to \infty$. Since $g(z_n) \to g(z_0)$, we obtain

$$|g(z_0)| < |g(z_n)| + \varepsilon_n , \qquad \text{where } \varepsilon_n = |g(z_n) - g(z_0)|.$$

By (4.29), we have

$$g(z_n) = -z_n^n f(z_n).$$

Let

$$\delta_n = |z_n^n f(z_n)| \quad \text{for each } n = 1, 2, \ldots$$

Since the function $f(z_n)$ is uniformly bounded on the interval $\left(\frac{1}{2}, 1\right)$, it follows that $\delta_n \to 0$. Thus, we have

$$|g(z_0)| < \delta_n + \varepsilon_n \to 0 \quad \text{as } n \to \infty.$$

Thus, $g(z_0) = 0$. $\square$

**Corollary 4.26.** *The minimal values of characteristic polynomials of the Coxeter transformation for diagrams $T_{p,q,r}$ satisfy the conditions of Proposition 4.22, see Table 4.7.*

*Proof.* The minimal value $\lambda_1^{min}$ and the maximal value $\lambda_1^{max}$ are reciprocal:

$$\lambda_1^{min} = \frac{1}{\lambda_1^{max}}.$$

Therefore, by (4.28), we see that the sequence of eigenvalues $\lambda_{1,r}^{min}$ is non-increasing:

$$\lambda_{1,r}^{min} \geq \lambda_{1,r+1}^{min}.$$

Since the maximal eigenvalue $\lambda_{1,r}^{max} > 1$, then $\lambda_{1,r}^{min} < 1$.

It remains to show that $\lambda_{1,r}^{min} > \frac{1}{2}$ for every $r \in \mathbb{Z}$. But this is true since $f(\lambda) < 0$ and $g(\lambda) < 0$ for every $\lambda < \frac{1}{2}$, see Table 4.7. $\square$

# 5

## R. Steinberg's theorem, B. Kostant's construction

### 5.1 R. Steinberg's theorem and a $(p, q, r)$ mystery

R. Steinberg in [Stb85, p.591, $(*)$] observed a property of affine Coxeter transformations (i.e., transformations corresponding to extended Dynkin diagrams), which plays the main role in his derivation of the McKay correspondence. Let $(p, q, r)$ be the same as in Table A.1, Table A.2 and relations (A.2), (A.6).

**Theorem 5.1 ([Stb85]).** *The affine Coxeter transformation for the extended Dynkin diagram $\tilde{\Gamma}$ has the same eigenvalues as the product of three Coxeter transformations of types $A_n$, where $n = p - 1$, $q - 1$, and $r - 1$, corresponding to the branches of the Dynkin diagram $\Gamma$.*

Essentially, R. Steinberg observed that the orders of eigenvalues of the affine Coxeter transformation corresponding to the *extended* Dynkin diagram $\tilde{\Gamma}$ and given in Table 4.1 coincide with the lengths of branches of the Dynkin diagram $\Gamma$.

Now, we give the proof of R. Steinberg's theorem for the simply-laced extended Dynkin diagram by using the Subbotin-Sumin splitting formula (4.4).

In §5.3.2, we generalize this theorem with some modifications to the multiply-laced case, see Theorem 5.5, Table 1.2.

Let $T_{p,q,r}$ be a connected graph with three branches of lengths $p, q, r$. Let $\mathcal{X}$ be the characteristic polynomial of the Coxeter transformation. Split the graph along the edge $\{\alpha\beta\}$, Fig. 5.1. Then, by the Subbotin-Sumin formula (4.4) we have

$$
\begin{aligned}
\mathcal{X} =&\mathcal{X}(A_{p+q-1})\mathcal{X}(A_{r-1}) - \lambda\mathcal{X}(A_{p-1})\mathcal{X}(A_{q-1})\mathcal{X}(A_{r-2}) = \\
&(-1)^{p+q+r}(\mathcal{X}_{p+q-1}\mathcal{X}_{r-1} - \lambda\mathcal{X}_{p-1}\mathcal{X}_{q-1}\mathcal{X}_{r-2}).
\end{aligned}
\tag{5.1}
$$

By (4.13), we have, up to a factor $(-1)^{p+q+r}$,

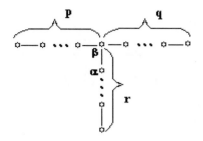

**Fig. 5.1.**    Splitting the graph $T_{p,q,r}$ along the edge $\{\alpha\beta\}$

$$\mathcal{X} = \frac{(\lambda^{p+q}-1)(\lambda^r-1)}{(\lambda-1)^2} - \frac{(\lambda^p-1)(\lambda^q-1)(\lambda^r-\lambda)}{(\lambda-1)^3}. \qquad (5.2)$$

For $p = q$ (e.g., $E_6$, $\widetilde{E}_6$, $\widetilde{E}_7$, and $D_n$ for $n \geq 4$), we have, up to a factor $(-1)^{2p+r}$,

$$\mathcal{X} = \frac{(\lambda^{2p}-1)(\lambda^r-1)}{(\lambda-1)^2} - \frac{(\lambda^p-1)^2(\lambda^r-\lambda)}{(\lambda-1)^3} =$$
$$\frac{(\lambda^p-1)}{(\lambda-1)^3}((\lambda^p+1)(\lambda^r-1)(\lambda-1) - (\lambda^p-1)(\lambda^r-\lambda)). \qquad (5.3)$$

For $q = 2p$ (e.g., $E_7$, $\widetilde{E}_8$), we have, up to a factor $(-1)^{3p+r}$,

$$\mathcal{X} = \frac{(\lambda^{3p}-1)(\lambda^r-1)}{(\lambda-1)^2} - \frac{(\lambda^{2p}-1)(\lambda^p-1)(\lambda^r-\lambda)}{(\lambda-1)^3} =$$
$$\frac{(\lambda^p-1)}{(\lambda-1)^3}((\lambda^{2p}+\lambda^p+1)(\lambda^r-1)(\lambda-1) - (\lambda^{2p}-1)(\lambda^r-\lambda)). \qquad (5.4)$$

1) Case $p = q = 3$, $r = 3$ ($\widetilde{E}_6$). From eq. (5.3) we have

$$\mathcal{X} = \frac{(\lambda^3-1)}{(\lambda-1)^3}[(\lambda^3+1)(\lambda^3-1)(\lambda-1) - (\lambda^3-1)(\lambda^3-\lambda)] =$$
$$\frac{(\lambda^3-1)^2}{(\lambda-1)^3}[(\lambda^3+1)(\lambda-1) - (\lambda^3-\lambda)] =$$
$$\frac{(\lambda^3-1)^2}{(\lambda-1)^3}(\lambda^4 - 2\lambda^3 + 2\lambda - 1) =$$
$$\frac{(\lambda^3-1)^2(\lambda-1)^2(\lambda^2-1)}{(\lambda-1)^3} = (\lambda-1)^2\mathcal{X}_2^2\mathcal{X}_1.$$

The polynomials $\mathcal{X}_1$ and $\mathcal{X}_2^2$ have, respectively, eigenvalues of orders $2, 3, 3$ which are equal to the lengths of branches of $E_6$.

2) Case $p = q = 4$, $r = 2$ ($\widetilde{E}_7$). Here, from eq. (5.3) we have

$$
\begin{aligned}
\mathcal{X} &= \frac{(\lambda^4 - 1)}{(\lambda - 1)^3}[(\lambda^4 + 1)(\lambda^2 - 1)(\lambda - 1) - (\lambda^4 - 1)(\lambda^2 - \lambda)] = \\
&\frac{(\lambda^4 - 1)(\lambda^2 - 1)(\lambda - 1)}{(\lambda - 1)^3}[(\lambda^4 + 1) - \lambda(\lambda^2 + 1)] = \\
&\frac{(\lambda^4 - 1)(\lambda^3 - 1)(\lambda^2 - 1)(\lambda - 1)^2}{(\lambda - 1)^3} = (\lambda - 1)^2 \mathcal{X}_3 \mathcal{X}_2 \mathcal{X}_1.
\end{aligned}
$$

The polynomials $\mathcal{X}_1, \mathcal{X}_2$ and $\mathcal{X}_3$ have, respectively, eigenvalues of orders $2, 3, 4$ which are equal to the lengths of branches of $E_7$.

3) Case $p = 3$, $q = 6$, $r = 2$ ($\widetilde{E}_8$). From eq. (5.4), we have

$$
\begin{aligned}
\mathcal{X} &= \frac{(\lambda^3 - 1)}{(\lambda - 1)^3}[(\lambda^6 + \lambda^3 + 1)(\lambda^2 - 1)(\lambda - 1) - (\lambda^6 - 1)(\lambda^2 - \lambda)] = \\
&\frac{(\lambda^3 - 1)(\lambda^2 - 1)(\lambda - 1)}{(\lambda - 1)^3}[(\lambda^6 + \lambda^3 + 1) - \lambda(\lambda^4 + \lambda^2 + 1)] = \\
&\frac{(\lambda^5 - 1)(\lambda^3 - 1)(\lambda^2 - 1)(\lambda - 1)^2}{(\lambda - 1)^3} = (\lambda - 1)^2 \mathcal{X}_4 \mathcal{X}_2 \mathcal{X}_1.
\end{aligned}
$$

The polynomials $\mathcal{X}_1, \mathcal{X}_2$ and $\mathcal{X}_4$ have, respectively, eigenvalues of orders $2, 3, 5$ which are equal to the lengths of branches of $E_8$.

4) Case $p = q = 2$, $r = 2, 3, \dots$ ($D_{r+2}$). From eq. (5.3) we have

$$
\begin{aligned}
(-1)^r \mathcal{X}(D_{r+2}) &= \\
\frac{(\lambda^2 - 1)}{(\lambda - 1)^3}&[(\lambda^2 + 1)(\lambda^r - 1)(\lambda - 1) - (\lambda^2 - 1)(\lambda^r - \lambda)] = \\
\frac{(\lambda^2 - 1)(\lambda - 1)}{(\lambda - 1)^3}&[(\lambda^2 + 1)(\lambda^r - 1) - (\lambda + 1)(\lambda^r - \lambda)] = \\
\frac{(\lambda^2 - 1)(\lambda - 1)^2(\lambda^{r+1} + 1)}{(\lambda - 1)^3} &= (\lambda + 1)(\lambda^{r+1} + 1),
\end{aligned}
\qquad (5.5)
$$

and, up to a sign,

$$
\mathcal{X} = (\lambda + 1)(\lambda^{r+1} + 1).
$$

Pay attention that, as in (5.5), the sign $(-1)^r$ should be taken into account only in recurrent calculations (e.g., for $\widetilde{D}_n$ and $\widetilde{D}_4$ below), see Remark 4.13.

5) Case $\widetilde{D}_{r+2}$ (diagram contains $r + 3$ points).

Here, from eq. (4.4) we have

$$
\mathcal{X}(\widetilde{D}_{r+2}) = \mathcal{X}(D_r)\mathcal{X}(A_3) - \lambda\mathcal{X}(D_{r-1})\mathcal{X}(A_1)^2. \qquad (5.6)
$$

From eq. (5.6) and eq. (5.5) we have

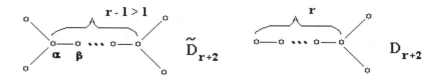

**Fig. 5.2.**    The diagrams $\widetilde{D}_{r+2}$ and $D_{r+2}$

$$(-1)^{r-1}\mathcal{X}(\widetilde{D}_{r+2}) =$$
$$(\lambda+1)(\lambda^{r-1}+1)(\lambda^3+\lambda^2+\lambda+1) - \lambda(\lambda+1)(\lambda^{r-2}+1)(\lambda+1)^2 =$$
$$(\lambda+1)^2[(\lambda^{r-1}+1)(\lambda^2+1) - (\lambda+1)(\lambda^{r-1}+\lambda)] =$$
$$(\lambda+1)^2[(\lambda^{r+1}+\lambda^2+\lambda^{r-1}+1) - (\lambda^r+\lambda^{r-1}+\lambda^2+\lambda)] =$$
$$(\lambda+1)^2[(\lambda^{r+1}-\lambda^r) - (\lambda-1)] = (\lambda+1)^2(\lambda^r-1)(\lambda-1) =$$
$$(\lambda-1)^2\mathcal{X}_1^2\mathcal{X}_{r-1},$$

and so

$$\mathcal{X} = (\lambda-1)^2\mathcal{X}_1^2\mathcal{X}_{r-1}. \tag{5.7}$$

The polynomials $\mathcal{X}_1^2$ and $\mathcal{X}_{r-1}$ have, respectively, eigenvalues of orders $2, 2, r$ which are equal to the lengths of branches of $D_{r+2}$.

6) Case $\widetilde{D}_4$. Here, by (5.5) we have

$$\mathcal{X}(\widetilde{D}_4) = \mathcal{X}(D_4)\mathcal{X}(A_1) - \lambda\mathcal{X}(A_1)^3 = -(\lambda+1)^2(\lambda^3+1) + \lambda(\lambda+1)^3 =$$
$$- (\lambda+1)^2(\lambda^3+1-\lambda^2-\lambda) = -(\lambda+1)^2(\lambda^2-1)(\lambda-1) =$$
$$- (\lambda-1)^2(\lambda+1)^3,$$

and so

$$\mathcal{X} = (\lambda-1)^2\mathcal{X}_1^3.$$

Polynomial $\mathcal{X}_1^3$ have eigenvalues of orders $2, 2, 2$ which are equal to the lengths of branches of $D_4$.

In (4.13), we saw that the simple Frame formula is a particular case of the Subbotin-Sumin formula (4.4). Now we will show another *Frame formula* from [Fr51, p.785] used by H. S. M. Coxeter in [Cox51].

**Proposition 5.2 ([Fr51]).** *Let* $T_{p,q,r}$ *be a connected graph with three branches of lengths* $p, q, r$, *where the lengths include the branch point, see Fig. 2.1. Then*

$$\mathcal{X}(T_{p,q,r}) = \mathcal{X}(A_{p+q+r-2}) - \lambda^2\mathcal{X}(A_{p-2})\mathcal{X}(A_{q-2})\mathcal{X}(A_{r-2}). \tag{5.8}$$

*Proof.* By (4.4) we have

$$\mathcal{X}(T_{p,q,r}) = \mathcal{X}(A_{p+q-1})\mathcal{X}(A_{r-1}) - \lambda\mathcal{X}(A_{p-1})\mathcal{X}(A_{q-1})\mathcal{X}(A_{r-2}).$$

This is equivalent to

$$\mathcal{X}(T_{p,q,r}) = \mathcal{X}(A_{p+q-1})\mathcal{X}(A_{r-1}) - \lambda\mathcal{X}(A_{p+q-2})\mathcal{X}(A_{r-2})$$
$$+ \lambda\mathcal{X}(A_{p+q-2})\mathcal{X}(A_{r-2}) - \lambda\mathcal{X}(A_{p-1})\mathcal{X}(A_{q-1})\mathcal{X}(A_{r-2}).$$

By the first Frame formula (4.13) we have

$$\mathcal{X}(A_{p+q+r-2}) = \mathcal{X}(A_{p+q-1})\mathcal{X}(A_{r-1}) - \lambda\mathcal{X}(A_{p+q-2})\mathcal{X}(A_{r-2}),$$

and

$$\mathcal{X}(T_{p,q,r}) =$$
$$\mathcal{X}(A_{p+q+r-2}) + \lambda\mathcal{X}(A_{p+q-2})\mathcal{X}(A_{r-2}) - \lambda\mathcal{X}(A_{p-1})\mathcal{X}(A_{q-1})\mathcal{X}(A_{r-2}) =$$
$$\mathcal{X}(A_{p+q+r-2}) + \lambda\mathcal{X}(A_{r-2})[\mathcal{X}(A_{p+q-2}) - \mathcal{X}(A_{p-1})\mathcal{X}(A_{q-1})].$$
$$(5.9)$$

Again, by (4.13) we have

$$\mathcal{X}(A_{p+q-2}) - \mathcal{X}(A_{p-1})\mathcal{X}(A_{q-1}) = -\lambda\mathcal{X}(A_{p-2})\mathcal{X}(A_{q-2}),$$

and from eq. (5.9) we deduce

$$\mathcal{X}(T_{p,q,r}) = \mathcal{X}(A_{p+q+r-2}) - \lambda^2\mathcal{X}(A_{r-2})\mathcal{X}(A_{p-2})\mathcal{X}(A_{q-2}). \quad \square$$

## 5.2 The characteristic polynomials for the Dynkin diagrams

In order to calculate characteristic polynomials of the Coxeter transformations for the Dynkin diagrams, we use the Subbotin-Sumin formula (4.4), the generalized Subbotin-Sumin formula (4.9), its specialization (5.2) for arbitrary $(p, q, r)$-trees, and particular cases of (5.2): formula (5.3) for $p = q$, and (5.4) for $p = 2q$.

The results of calculations of this section concerning Dynkin diagrams are collected in Table 1.1.

1) Case $p = q = 3$, $r = 2$ ($E_6$). From eq. (5.3) we have

$$\mathcal{X} = \frac{(\lambda^3 - 1)}{(\lambda - 1)^3}[(\lambda^3 + 1)(\lambda^2 - 1)(\lambda - 1) - (\lambda^3 - 1)(\lambda^2 - \lambda)] =$$
$$\frac{(\lambda^3 - 1)}{(\lambda - 1)^3}(\lambda - 1)^2[(\lambda^3 + 1)(\lambda + 1) - \lambda(\lambda^2 + \lambda + 1)] =$$
$$\frac{(\lambda^3 - 1)}{(\lambda - 1)}(\lambda^4 - \lambda^2 + 1) =$$
$$(\lambda^2 + \lambda + 1)(\lambda^4 - \lambda^2 + 1) = \lambda^6 + \lambda^5 - \lambda^3 + \lambda + 1.$$

In another form, we have

$$\mathcal{X} = \frac{(\lambda^3 - 1)}{(\lambda - 1)}(\lambda^4 - \lambda^2 + 1) = \frac{(\lambda^6 + 1)}{(\lambda^2 + 1)}\frac{(\lambda^3 - 1)}{(\lambda - 1)}.$$

2) Case $p = 2$, $q = 4$, $r = 3$ ($E_7$). From eq. (5.4) we have, up to a sign,

$$\mathcal{X} = \frac{(\lambda^2 - 1)}{(\lambda - 1)^3}[(\lambda^4 + \lambda^2 + 1)(\lambda^3 - 1)(\lambda - 1) - (\lambda^4 - 1)(\lambda^3 - \lambda)] =$$

$$\frac{(\lambda^2 - 1)(\lambda - 1)^2}{(\lambda - 1)^3}[(\lambda^4 + \lambda^2 + 1)(\lambda^2 + \lambda + 1) -$$

$$(\lambda^3 + \lambda^2 + \lambda + 1)\lambda(\lambda + 1)] =$$

$$(\lambda + 1)[(\lambda^4 + \lambda^2 + 1)(\lambda^2 + \lambda + 1) -$$

$$(\lambda^3 + \lambda^2 + \lambda + 1)(\lambda^2 + \lambda)] =$$

$$(\lambda + 1)(\lambda^6 - \lambda^3 + 1) = \lambda^7 + \lambda^6 - \lambda^4 - \lambda^3 + \lambda + 1.$$

In another form, we have

$$\mathcal{X} = -(\lambda + 1)(\lambda^6 - \lambda^3 + 1) = -\frac{(\lambda + 1)(\lambda^9 + 1)}{(\lambda^3 + 1)}.$$

3) Case $p = 3$, $q = 2$, $r = 5$ ($E_8$). From eq. (5.2) we have

$$\mathcal{X} = \frac{(\lambda^5 - 1)^2}{(\lambda - 1)^2} - \frac{(\lambda^3 - 1)(\lambda^2 - 1)(\lambda^5 - \lambda)}{(\lambda - 1)^3} =$$

$$(\lambda^4 + \lambda^3 + \lambda^2 + \lambda + 1)^2 -$$

$$(\lambda^2 + \lambda + 1)(\lambda + 1)\lambda(\lambda^3 + \lambda^2 + \lambda + 1) =$$

$$\lambda^8 + \lambda^7 - \lambda^5 - \lambda^4 - \lambda^3 + \lambda + 1.$$

Since

$$(\lambda^8 + \lambda^7 - \lambda^5 - \lambda^4 - \lambda^3 + \lambda + 1)(\lambda^2 - \lambda + 1) = \lambda^{10} - \lambda^5 + 1,$$

we have

$$\mathcal{X} = \frac{\lambda^{10} - \lambda^5 + 1}{\lambda^2 - \lambda + 1} = \frac{(\lambda^{15} + 1)(\lambda + 1)}{(\lambda^5 + 1)(\lambda^3 + 1)}.$$

4) For the case $D_n$, see (5.5),

$$\mathcal{X}(D_n) = (-1)^n(\lambda + 1)(\lambda^{n-1} + 1),$$
$$\mathcal{X} = (\lambda + 1)(\lambda^{n-1} + 1). \tag{5.10}$$

5) Case $F_4$. We use eq. (4.9), splitting the diagram $F_4$ along the weighted edge into two diagrams $\Gamma_1 = A_2$ and $\Gamma_2 = A_2$. Here, $\rho = 2$.

$$\mathcal{X} = \mathcal{X}_2^2 - 2\lambda\mathcal{X}_1^2 = (\lambda^2 + \lambda + 1)^2 - 2\lambda(\lambda + 1)^2 = \lambda^4 - \lambda^2 + 1.$$

In another form, we have

$$\mathcal{X} = \lambda^4 - \lambda^2 + 1 = \frac{\lambda^6 + 1}{\lambda^2 + 1}.$$

6) Case $G_2$. A direct calculation of the Coxeter transformation gives

$$\mathbf{C} = \begin{pmatrix} -1 & 3 \\ 0 & 1 \end{pmatrix} \begin{pmatrix} 1 & 0 \\ 1 & -1 \end{pmatrix} = \begin{pmatrix} 2 & -3 \\ 1 & -1 \end{pmatrix},$$

and

$$\mathcal{X} = \lambda^2 - \lambda + 1 = \frac{\lambda^3 + 1}{\lambda + 1}.$$

7) The dual cases $B_n$ and $C_n = B_n^\vee$. Since the spectra of the Coxeter transformations of the dual graphs coincide, we need consider only the case $B_n$.

a) Consider $B_2$. A direct calculation of the Coxeter transformation gives

$$\mathbf{C} = \begin{pmatrix} -1 & 2 \\ 0 & 1 \end{pmatrix} \begin{pmatrix} 1 & 0 \\ 1 & -1 \end{pmatrix} = \begin{pmatrix} 1 & -2 \\ 1 & -1 \end{pmatrix},$$

and

$$\mathcal{X}(B_2) = \lambda^2 + 1. \tag{5.11}$$

b) Consider $B_3$. We use eq. (4.9), splitting the diagram $B_3$ along the weighted edge

$$\begin{aligned}
\mathcal{X}(B_3) =& \mathcal{X}(A_2)\mathcal{X}(A_1) - 2\lambda\mathcal{X}(A_1) = -(\mathcal{X}_2\mathcal{X}_1 - 2\lambda\mathcal{X}_1) = \\
& - \mathcal{X}_1(\mathcal{X}_2 - 2\lambda) = -(\lambda+1)(\lambda^2 + \lambda + 1 - 2\lambda) = \\
& - (\lambda+1)(\lambda^2 - \lambda + 1) = -(\lambda^3 + 1).
\end{aligned} \tag{5.12}$$

c) Case $B_n$ for $n \geq 4$.

$$\mathcal{X}(B_n) =$$
$$\mathcal{X}(A_{n-3})\mathcal{X}(B_3) - \lambda\mathcal{X}(A_{n-4})\mathcal{X}(B_2) = \mathcal{X}_{n-3}\mathcal{X}(B_3) - \lambda\mathcal{X}_{n-4}\mathcal{X}(B_2),$$

i.e.,

$$\begin{aligned}
(-1)^n\mathcal{X}(B_n) =& \\
\frac{\lambda^{n-2} - 1}{\lambda - 1}(\lambda^3 + 1) &- \lambda\frac{\lambda^{n-3} - 1}{\lambda - 1}(\lambda^2 + 1) = \\
\frac{(\lambda^{n-2} - 1)(\lambda^3 + 1) &- (\lambda^{n-3} - 1)(\lambda^3 + \lambda)}{\lambda - 1} = \\
\frac{\lambda^{n+1} - \lambda^n + \lambda - 1}{\lambda - 1} &= \lambda^n + 1,
\end{aligned}$$

and

$$\mathcal{X}(B_n) = (-1)^n(\lambda^n + 1). \tag{5.13}$$

By (5.11), (5.12) and (5.13), we see, for the case $B_n$, that, up to a sign,

$$\mathcal{X}(B_n) = \lambda^n + 1.$$

## 5.3 A generalization of R. Steinberg's theorem

We will show now that R. Steinberg's theorem 5.1 can be extended to the multiply-laced case.

### 5.3.1 The folded Dynkin diagrams and branch points

**Definition 5.3.** A vertex is said to be a *branch point* of the Dynkin diagram in one of the following cases:

(a) if it is the common endpoint of three edges ($E_6, E_7, E_8, D_n$);

(b) if it is the common endpoint of two edges, one of which is non-weighted and one is weighted ($B_n, C_n, F_4$). Such a vertex is said to be a *non-homogeneous point*.

(c) By definition let both points of $G_2$ be branch points.

*Remark 5.4.* Every multiply-laced Dynkin diagram (and also every extended Dynkin diagram) can be obtained by a so-called *folding* operation from a simply-laced diagrams, see Fig. 5.3, such that the branch point is transformed into the *non-homogeneous point*, see, e.g., I. Satake, [Sat60, p.109], J. Tits, [Ti66, p.39], P. Slodowy, [Sl80, Appendices I and III], or more recent works [FSS96], [Mohr04]. This fact was our motivation for considering *non-homogeneous points* from (b) as *branch points*.

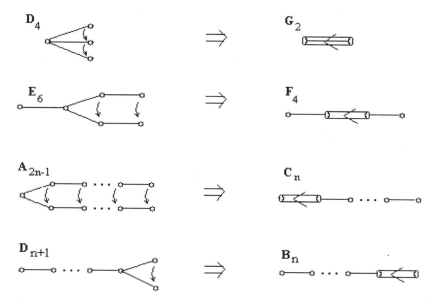

**Fig. 5.3.**     The folding operation applied to Dynkin diagrams

Let us divide all extended Dynkin diagrams (trees) into 4 classes and let $g$ be the number of the class:

Class $g = 0$ contains all simply-laced diagrams.

Class $g = 1$ contains multiply-laced diagrams which has only one weighted edge of type (1,2) or (2,1).

Class $g = 2$ contains multiply-laced diagrams with either two weighted edges or one weighted edge of type (1,3) or (3,1).

Class $g = 3$ contains multiply-laced diagrams with two vertices.

Thus,

$$
\begin{aligned}
& g = 0 \text{ for } \widetilde{E}_6, \widetilde{E}_7, \widetilde{E}_8, \widetilde{D}_n, \\
& g = 1 \text{ for } \widetilde{F}_{41}, \widetilde{F}_{42}, \widetilde{CD}_n, \widetilde{DD}_n, \\
& g = 2 \text{ for } \widetilde{G}_{12}, \widetilde{G}_{22}, \widetilde{B}_n, \widetilde{C}_n, \widetilde{BC}_n, \\
& g = 3 \text{ for } \widetilde{A}_{11}, \widetilde{A}_{12}.
\end{aligned}
\tag{5.14}
$$

### 5.3.2 R. Steinberg's theorem for the multiply-laced case

Now, we can generalize R. Steinberg's theorem for the multiply-laced case.

**Theorem 5.5.** *The affine Coxeter transformation with the extended Dynkin diagram $\widetilde{\Gamma}$ of class $g$ (see (5.14)) has the same eigenvalues as the product of $(3 - g)$ multipliers, every one of which is the Coxeter transformation of type $A_i$, where $i \in \{p - 1, q - 1, r - 1\}$. In other words,*

$$
\begin{aligned}
& \text{For } g = 0, \text{ the product } \mathcal{X}_{p-1}\mathcal{X}_{q-1}\mathcal{X}_{r-1} \text{ is taken.} \\
& \text{For } g = 1, \text{ the product } \mathcal{X}_{p-1}\mathcal{X}_{q-1} \text{ is taken.} \\
& \text{For } g = 2, \text{ the product consists only of } \mathcal{X}_{p-1}. \\
& \text{For } g = 3, \text{ the product is trivial } (= 1).
\end{aligned}
\tag{5.15}
$$

*The remaining two eigenvalues of the affine Coxeter transformation are both equal to 1, see Table 1.2.*

*Proof.* For $g = 0$, we have a simply-laced case considered in R. Steinberg's theorem (Theorem 5.1).

Let $g = 1$.

1) Cases $\widetilde{F}_{41}, \widetilde{F}_{42}$. Since the spectra of the characteristic polynomials of the Coxeter transformations of the dual graphs coincide, we consider $\widetilde{F}_{41}$. We use eq. (4.9), splitting the diagram $\widetilde{F}_{41}$ along the weighted edge. We have

$$
\begin{aligned}
\mathcal{X}(\widetilde{F}_{41}) &= -\left(\mathcal{X}_3\mathcal{X}_2 - 2\lambda\mathcal{X}_2\mathcal{X}_1\right) = -\mathcal{X}_2(\mathcal{X}_3 - 2\lambda\mathcal{X}_1) = \\
&\quad - \mathcal{X}_2(\lambda^3 + \lambda^2 + \lambda + 1 - 2\lambda(\lambda + 1)) = \\
&\quad - \mathcal{X}_2(\lambda^3 - \lambda^2 - \lambda + 1) = \\
&\quad - \mathcal{X}_2(\lambda^2 - 1)(\lambda - 1) = -(\lambda - 1)^2\mathcal{X}_2\mathcal{X}_1,
\end{aligned}
$$

and, up to a sign, we have

$$\mathcal{X} = (\lambda - 1)^2 \mathcal{X}_2 \mathcal{X}_1. \tag{5.16}$$

The polynomials $\mathcal{X}_1$ and $\mathcal{X}_2$ have, respectively, eigenvalues of orders 2 and 3 which are equal to the lengths of the branches $A_1$ and $A_2$ of $F_4$ without a non-homogeneous branch point.

2) Cases $\widetilde{CD}_n$, $\widetilde{DD}_n$. These diagrams are obtained as extensions of $B_n$, see [Bo, Tab.II]. By (4.9) and by splitting the diagram $\widetilde{CD}_n$ along the weighted edge, we have

$$\mathcal{X}(\widetilde{CD}_n) = \mathcal{X}(D_n)\mathcal{X}(A_1) - 2\lambda\mathcal{X}(D_{n-1}).$$

By (5.10)

$$\mathcal{X}(D_n) = (-1)^n(\lambda + 1)(\lambda^{n-1} + 1).$$

Thus,

$$\begin{aligned}
(-1)^{n-1}\mathcal{X}(\widetilde{CD}_n) =&(\lambda + 1)^2(\lambda^{n-1} + 1) - 2\lambda(\lambda + 1)(\lambda^{n-2} + 1) = \\
&(\lambda + 1)(\lambda^n + \lambda + \lambda^{n-1} + 1 - 2\lambda^{n-1} - 2\lambda) = \\
&(\lambda + 1)(\lambda^n - \lambda^{n-1} - \lambda + 1) = \\
&(\lambda + 1)(\lambda - 1)(\lambda^{n-1} - 1) = (\lambda - 1)^2\mathcal{X}_1\mathcal{X}_{n-2},
\end{aligned}$$

and, up to a sign, we have

$$\mathcal{X} = (\lambda - 1)^2\mathcal{X}_1\mathcal{X}_{n-2}. \tag{5.17}$$

The polynomials $\mathcal{X}_1$ and $\mathcal{X}_{n-2}$ have, respectively, eigenvalues of orders 2 and $n - 1$ which are equal to the lengths of the branches $A_1$ and $A_{n-2}$ of $B_n$ without a non-homogeneous branch point.

Let $g = 2$.

3) Cases $\widetilde{G}_{12}$, $\widetilde{G}_{22}$. Since the spectra of the Coxeter transformations of the dual graphs coincide, we consider $\widetilde{G}_{12}$. By (4.9), splitting the diagram $\widetilde{G}_{12}$ along the weighted edge we have

$$\begin{aligned}
\mathcal{X}(\widetilde{G}_{12}) =& - (\mathcal{X}_2\mathcal{X}_1 - 3\lambda\mathcal{X}_1) = -\mathcal{X}_1(\mathcal{X}_2 - 3\lambda) = \\
& - (\lambda + 1)(\lambda^2 + \lambda + 1 - 3\lambda) = \\
& - (\lambda - 1)^2(\lambda + 1) = -(\lambda - 1)^2\mathcal{X}_1,
\end{aligned}$$

and, up to a sign, we have

$$\mathcal{X} = (\lambda - 1)^2\mathcal{X}_1. \tag{5.18}$$

Polynomial $\mathcal{X}_1$ has the single eigenvalue of order 2, it corresponds to the length of the single branch $A_1$ of $G_2$ without a non-homogeneous branch point.

4) Cases $\widetilde{B}_n$, $\widetilde{C}_n$, $\widetilde{BC}_n$. The characteristic polynomials of the Coxeter transformations of these diagrams coincide. Consider $\widetilde{B}_n$. By (4.9), splitting the diagram $\widetilde{B}_n$ along the weighted edge we have

$$\mathcal{X}(\widetilde{B}_n) = \mathcal{X}(B_n)\mathcal{X}(A_1) - 2\lambda\mathcal{X}(B_{n-1}), \text{ i.e.,}$$
$$(-1)^{n-1}\mathcal{X}(\widetilde{B}_n) = (\lambda^n + 1)(\lambda + 1) - 2\lambda(\lambda^{n-1} + 1) =$$
$$\lambda^{n+1} + \lambda^n + \lambda + 1 - 2\lambda^{n-1} - 2\lambda =$$
$$\lambda^{n+1} - \lambda^n - \lambda + 1 = (\lambda - 1)(\lambda^n - 1),$$

and, up to a sign, we have

$$\mathcal{X} = (\lambda - 1)^2 \mathcal{X}_{n-1}. \tag{5.19}$$

All eigenvalues of the polynomial $\mathcal{X}_{n-1}$ are of order $n$, it corresponds to the length of the single branch $A_{n-1}$ of $B_n$ without a non-homogeneous branch point.

Let $g = 3$.

5) Cases $\widetilde{A}_{11}$, $\widetilde{A}_{12}$. Direct calculation of the Coxeter transformation gives

$$\text{for } \widetilde{A}_{11} : \mathbf{C} = \begin{pmatrix} -1 & 4 \\ 0 & 1 \end{pmatrix} \begin{pmatrix} 1 & 0 \\ 1 & -1 \end{pmatrix} = \begin{pmatrix} 3 & -4 \\ 1 & -1 \end{pmatrix},$$

$$\text{for } \widetilde{A}_{12} : \mathbf{C} = \begin{pmatrix} -1 & 2 \\ 0 & 1 \end{pmatrix} \begin{pmatrix} 1 & 0 \\ 2 & -1 \end{pmatrix} = \begin{pmatrix} 3 & -2 \\ 2 & -1 \end{pmatrix},$$

and in both cases we have

$$\mathcal{X} = (\lambda - 3)(\lambda + 1) + 4 = (\lambda - 1)^2. \quad \square$$

## 5.4 The Kostant generating function and Poincaré series

### 5.4.1 The generating function

Let $\text{Sym}(\mathbb{C}^2)$ be the symmetric algebra on $\mathbb{C}^2$, in other words, $\text{Sym}(\mathbb{C}^2) = \mathbb{C}[x_1, x_2]$, see (2.28). The symmetric algebra $\text{Sym}(\mathbb{C}^2)$ is a graded $\mathbb{C}$-algebra, see (2.29):

$$\text{Sym}(\mathbb{C}^2) = \bigoplus_{n=0}^{\infty} \text{Sym}^n(\mathbb{C}^2).$$

For $n \in \mathbb{Z}_+$, let $\pi_n$ be[1] the representation of $SU(2)$ in $\text{Sym}^n(\mathbb{C}^2)$ induced by its action on $\mathbb{C}^2$. The set $\{\pi_n \mid n \in \mathbb{Z}_+\}$ is the set of all irreducible

---

[1] We use the following notation: $\mathbb{Z}_+ = \{0, 1, 2, \dots\}$.

representations of $SU(2)$, see, e.g., [Zhe73, §37]. Let $G$ be any finite subgroup of $SU(2)$, see §A.1. B. Kostant in [Kos84] considered the question:

*How does $\pi_n | G$ decompose for any $n \in \mathbb{N}$?*

The answer — the decomposition $\pi_n | G$ — is as follows:

$$\pi_n | G = \sum_{i=0}^{r} m_i(n) \rho_i, \tag{5.20}$$

where $\rho_i$ are irreducible representations of $G$, considered in the context of the *McKay correspondence*, see §A.4. Thus, the decomposition (5.20) reduces the question to the following one:

*What are the multiplicities $m_i(n)$ equal to?*

B. Kostant in [Kos84] obtained the multiplicities $m_i(n)$ by means the orbit structure of the Coxeter transformation on the highest root of the corresponding Lie algebra. For further details concerning this orbit structure and the multiplicities $m_i(n)$, see §5.5.

Note, that multiplicities $m_i(n)$ in (5.20) are calculated as follows:

$$m_i(n) = \langle \pi_n | G, \rho_i \rangle, \tag{5.21}$$

(for the definition of the inner product $\langle \cdot, \cdot \rangle$, see (5.33) ).

*Remark 5.6.* For further considerations, we extend the relation for multiplicity (5.21) to the cases of *restricted representations* $\rho_i^{\downarrow} := \rho_i \downarrow_H^G$ and *induced representations* $\rho_i^{\uparrow} := \rho_i \uparrow_H^G$, where $H$ is any subgroup of $G$ (see §A.5.1):

$$m_i^{\downarrow}(n) = \langle \pi_n | H, \rho_i^{\downarrow} \rangle, \quad m_i^{\uparrow}(n) = \langle \pi_n | G, \rho_i^{\uparrow} \rangle. \tag{5.22}$$

We do not have any decomposition like (5.20) for restricted representations $\rho_i^{\downarrow}$ or for induced representations $\rho_i^{\uparrow}$. Nevertheless, we may use multiplicities $m_i^{\downarrow}(n)$ and $m_i^{\uparrow}(n)$ for the generalization of the Kostant generating functions, see below (5.27), (5.28).

*Remark 5.7.* 1) Let $GL_k(V)$ be the group of invertible linear transformations of the $k$-vector space $V$. A representation $\rho : G \longrightarrow GL_k(V)$ defines a $k$-linear action $G$ on $V$ by the rule

$$gv = \rho(g)v.$$

The pair $(V, \rho)$ is called a *G-module*. The case where $\rho(g) = \mathrm{Id}_V$ is called the *trivial representation* in $V$. In this case

$$gv = v \text{ for all } g \in V. \tag{5.23}$$

In (5.20), the trivial representation $\rho_0$ corresponds to a particular vertex (see [McK80]), which extends the Dynkin diagram to the extended Dynkin diagram.

2) Let $\rho_0(H)$ (resp. $\rho_0(G)$) be the trivial representation of a subgroup $H \subset G$ (resp. of group $G$). The trivial representation $\rho_0(H) : H \longrightarrow GL_k(V)$ coincides with the *restricted representation* $\rho_0 \downarrow_H^G : G \longrightarrow GL_k(V)$, and the trivial representation $\rho_0(G) : G \longrightarrow GL_k(V)$ coincides with the *induced representation* $\rho_0 \uparrow_H^G : H \longrightarrow GL_k(V)$.

Since there is a one-to-one correspondence between the $\rho_i$ and the vertices of the Dynkin diagram, we can define (see [Kos84, p.211]) the vectors $v_n$, where $n \in \mathbb{Z}_+$, as follows:

$$v_n = \sum_{i=0}^{r} m_i(n)\alpha_i, \quad \text{where } \pi_n|G = \sum_{i=0}^{r} m_i(n)\rho_i, \tag{5.24}$$

where $\alpha_i$ are simple roots of the corresponding extended Dynkin diagram. Similarly, for the multiply-laced case, we define vectors $v_n$ to be:

$$v_n = \sum_{i=0}^{r} m_i^\uparrow(n)\alpha_i \quad \text{or} \quad v_n = \sum_{i=0}^{r} m_i^\downarrow(n)\alpha_i, \tag{5.25}$$

where the multiplicities $m_i^\uparrow(n)$ and $m_i^\downarrow(n)$ are defined by (5.22). The vector $v_n$ belongs to the root lattice generated by simple roots. Following B. Kostant, we define the generating function $P_G(t)$ for cases (5.24) and (5.25) as follows:

$$P_G(t) = \begin{pmatrix} [P_G(t)]_0 \\ [P_G(t)]_1 \\ \cdots \\ [P_G(t)]_r \end{pmatrix} := \sum_{n=0}^{\infty} v_n t^n, \tag{5.26}$$

Let us introduce $P_{G\uparrow}(t)$ (resp. $P_{G\downarrow}(t)$) for the case of induced representations $\rho_i \uparrow_H^G$ (resp. restricted representation $\rho_i \downarrow_H^G$), playing the same role as the Kostant generating functions for the representation $\rho_i$.

$$P_{G\uparrow}(t) = \begin{pmatrix} [P_{G\uparrow}(t)]_0 \\ \cdots \\ [P_{G\uparrow}(t)]_r \end{pmatrix} := \sum_{n=0}^{\infty} v_n t^n, \quad \text{where } v_n = \sum_{i=0}^{r} m_i^\uparrow(n)\alpha_i,$$

$$\tag{5.27}$$

$$P_{G\downarrow}(t) = \begin{pmatrix} [P_{G\downarrow}(t)]_0 \\ \cdots \\ [P_{G\downarrow}(t)]_r \end{pmatrix} := \sum_{n=0}^{\infty} v_n t^n, \quad \text{where } v_n = \sum_{i=0}^{r} m_i^\downarrow(n)\alpha_i,$$

The components of vectors $P_G(t)$, $P_{G\uparrow}(t)$, $P_{G\downarrow}(t)$ are the following series

$$[P_G(t)]_i = \sum_{n=0}^{\infty} m_i(n)t^n,$$

$$[P_{G\uparrow}(t)]_i = \sum_{n=0}^{\infty} m_i^{\uparrow}(n)t^n, \qquad (5.28)$$

$$[P_{G\downarrow}(t)]_i = \sum_{n=0}^{\infty} m_i^{\downarrow}(n)t^n,$$

where $i = 0, 1, \ldots, r$. In particular, for $i = 0$, we have

$$[P_G(t)]_0 = \sum_{n=0}^{\infty} m_0(n)t^n,$$

$$[P_{G\uparrow}(t)]_0 = \sum_{n=0}^{\infty} m_0^{\uparrow}(n)t^n,$$

$$[P_{G\downarrow}(t)]_0 = \sum_{n=0}^{\infty} m_0^{\downarrow}(n)t^n,$$

where $m_0(n)$ is the multiplicity of the trivial representation $\rho_0$ (see Remark 5.7) in $\mathrm{Sym}^n(\mathbb{C}^2)$. By §2.3 the *algebra of invariants* $R^G$ is a subalgebra of the *symmetric algebra* $\mathrm{Sym}(\mathbb{C}^2)$. Thanks to (5.23), we see that $R^G$ coincides with $\mathrm{Sym}(\mathbb{C}^2)$, and $[P_G(t)]_0$ is the Poincaré series of the algebra of invariants $\mathrm{Sym}(\mathbb{C}^2)^G$, i.e.,

$$[P_G(t)]_0 = P(\mathrm{Sym}(\mathbb{C}^2)^G, t). \qquad (5.29)$$

(see [Kos84, p.221, Rem.3.2]).

The series $[P_{G\uparrow}(t)]_0$ (resp. $[P_{G\downarrow}(t)]_0$) are said to be the *generalized Poincaré series for induced representations (resp. for restricted representations)*. Let

$$\widetilde{P}_G(t) = \begin{cases} P_G(t), & \text{for } v_n = \sum_{i=0}^{r} m_i(n)\alpha_i, \\ P_{G\uparrow}(t), & \text{for } v_n = \sum_{i=0}^{r} m_i^{\uparrow}(n)\alpha_i, \\ P_{G\downarrow}(t), & \text{for } v_n = \sum_{i=0}^{r} m_i^{\downarrow}(n)\alpha_i, \end{cases}$$

$$(5.30)$$

$$[\widetilde{P}_G(t)]_0 = \begin{cases} [P_G(t)]_0, & \text{for } v_n = \sum_{i=0}^{r} m_i(n)\alpha_i, \\ [P_{G\uparrow}(t)]_0, & \text{for } v_n = \sum_{i=0}^{r} m_i^{\uparrow}(n)\alpha_i, \\ [P_{G\downarrow}(t)]_0, & \text{for } v_n = \sum_{i=0}^{r} m_i^{\downarrow}(n)\alpha_i, \end{cases}$$

The series $\widetilde{P}_G(t)$ are said to be the *generalized Kostant generating functions*, and the series $[\widetilde{P}_G(t)]_0$ are said to be *generalized Poincaré series*.

The following theorem gives a remarkable formula for calculating the Poincaré series for the binary polyhedral groups. The theorem is known in different forms. B. Kostant in [Kos84] proves it in the context of the Coxeter number $h$.

**Theorem 5.8 (Kostant, Knörrer, Gonzalez-Sprinberg, Verdier).** *The Poincaré series $[P_G(t)]_0$ can be calculated as the following rational function:*

$$[P_G(t)]_0 = \frac{1 + t^h}{(1 - t^a)(1 - t^b)}, \qquad (5.31)$$

*where*

$$b = h + 2 - a, \ \ and \ ab = 2|G|. \qquad (5.32)$$

For a proof, see Theorem 1.4 and Theorem 1.8 from [Kos84], [Kn85, p.185], [GV83, p.428]. We call the numbers $a$ and $b$ the *Kostant numbers*. They can be easily calculated, see Table 5.1, compare also with Table A.2 and Table A.3. Note, that $a = 2d$, where $d$ is the maximal coordinate of the *nil-root* vector from the kernel of the Tits form, see §2.2.1 and Fig. 2.6.

### 5.4.2 The characters and the McKay operator

Let $\chi_1, \chi_2, \ldots \chi_r$ be all irreducible $\mathbb{C}$-characters of a finite group $G$ corresponding to irreducible representations $\rho_1, \rho_2, \ldots, \rho_r$, so that $\chi_1$ corresponds to the trivial representation, i.e., $\chi_1(g) = 1$ for all $g \in G$. The set of all characters of $G$ constitute the *character algebra* $C(G)$ since the set $C(G)$ is also a vector space over $\mathbb{C}$. An *hermitian inner product* $\langle \cdot, \cdot \rangle$ on $C(G)$ is defined as follows. For characters $\alpha, \beta \in C(G)$, let

$$\langle \alpha, \beta \rangle = \frac{1}{|G|} \sum_{g \in G} \alpha(g)\overline{\beta(g)}. \qquad (5.33)$$

Sometimes, we will write *inner product* $\langle \rho_i, \rho_j \rangle$ of the representations meaning the inner product of the corresponding characters $\langle \chi_{\rho_i}, \chi_{\rho_j} \rangle$. Let $z_{ij}^k = \langle \chi_i \chi_j, \chi_k \rangle$, where $\chi_i \chi_j$ corresponds to the representation $\rho_i \otimes \rho_j$. It is known that $z_{ij}^k$ is the multiplicity of the representation $\rho_k$ in $\rho_i \otimes \rho_j$ and $z_{ij}^k = z_{ji}^k$. The numbers $z_{ij}^k$ are integers and are called the *structure constants*, see, e.g., [Kar92, p.765].

For every $i \in \{1, \ldots, r\}$, there exists some $\overset{\wedge}{i} \in \{1, \ldots, r\}$ such that

$$\chi_{\overset{\wedge}{i}}(g) = \overline{\chi_i(g)} \text{ for all } g \in G.$$

The character $\chi_{\overset{\wedge}{i}}$ corresponds to the *contragredient representation* $\rho_{\overset{\wedge}{i}}$ determined from the relation

**Table 5.1.**    The binary polyhedral groups (BPG) and the Kostant numbers $a$, $b$

| Dynkin diagram | Order of group | BPG | Coxeter number | $a$ | $b$ |
|---|---|---|---|---|---|
| $A_{n-1}$ | $n$ | $\mathbb{Z}/n\mathbb{Z}$ | $n$ | 2 | $n$ |
| $D_{n+2}$ | $4n$ | $\mathcal{D}_n$ | $2n+2$ | 4 | $2n$ |
| $E_6$ | 24 | $\mathcal{T}$ | 12 | 6 | 8 |
| $E_7$ | 48 | $\mathcal{O}$ | 18 | 8 | 12 |
| $E_8$ | 120 | $\mathcal{J}$ | 30 | 12 | 20 |

$$\rho_{\hat{i}}(g) = (\rho_i(g)^t)^{-1}. \tag{5.34}$$

We have

$$\langle \chi_i \chi_j, \chi_k \rangle = \langle \chi_i, \chi_{\hat{j}} \chi_k \rangle \tag{5.35}$$

since

$$\langle \chi_i \chi_j, \chi_k \rangle = \frac{1}{|G|} \sum_{g \in G} \chi_i(g) \chi_j(g) \overline{\chi_k(g)} =$$

$$\frac{1}{|G|} \sum_{g \in G} \chi_i(g) (\overline{\overline{\chi_j(g)} \chi_k(g)}) = \langle \chi_i, \chi_{\hat{j}} \chi_k \rangle.$$

*Remark 5.9.* The group $SU(2)$ is the set of all unitary unimodular $2 \times 2$ matrices $u$, i.e.,

$$u^* = u^{-1} \ (unitarity) \,,$$
$$\det(u) = 1 \ (unimodularity) \,,$$

where

$$(u^*)_{ij} = \overline{u_{ji}},$$

where the complex conjugate of $u_{ij} = c + di$ is $c - di$. The matrices $u \in SU(2)$ have the following form:

$$u = \begin{pmatrix} a & b \\ -\bar{b} & \bar{a} \end{pmatrix}, \text{ and } u^* = \begin{pmatrix} \bar{a} & -b \\ \bar{b} & a \end{pmatrix}, \text{ where } a\bar{a} + b\bar{b} = 1, \tag{5.36}$$

see, e.g., [Ha89, Ch.9, §6]. The mutually inverse matrices $u$ and $u^{-1}$ are

$$u = \begin{pmatrix} a & b \\ c & d \end{pmatrix}, \text{ and } u^{-1} = \begin{pmatrix} d & -b \\ -c & a \end{pmatrix}, \text{ where } ad - bc = 1. \tag{5.37}$$

Set

$$s = \begin{pmatrix} 0 & -1 \\ 1 & 0 \end{pmatrix}, \text{ then } s^{-1} = s^3 = \begin{pmatrix} 0 & 1 \\ -1 & 0 \end{pmatrix}.$$

For any $u \in SU(2)$, we have

$$sus^{-1} = (u^{-1})^t = (u^t)^{-1}. \tag{5.38}$$

The element $s$ is called the *Weyl element*.  □

According to (5.34) and (5.38) we see that every finite dimensional representation of the group of $SU(2)$ is equivalent to its contragredient representation, see [Zhe73, §37, Rem.3]. Thus by (5.35), for representations $\rho_i$ of any finite subgroup $G \subset SU(2)$, we have

$$\langle \chi_i \chi_j, \chi_k \rangle = \langle \chi_i, \chi_j \chi_k \rangle. \tag{5.39}$$

The matrix of multiplicities $A := A(G)$ from eq. (A.13) was introduced by J. McKay in [McK80]; it plays the central role in the *McKay correspondence*, see §A.4. We call this matrix — or the corresponding operator — the *McKay matrix* or the *McKay operator*.

Similarly, let $\widetilde{A}$ and $\widetilde{A}^\vee$ be matrices of multiplicities (A.14), (A.16). These matrices were introduced by P. Slodowy [Sl80] by analogy with the McKay matrix for the multiply-laced case, see §A.5.1. We call these matrices the *Slodowy operators*.

The following result of B. Kostant [Kos84], which holds for the McKay operator holds also for the Slodowy operators.

**Proposition 5.10.** *If B is either the McKay operator A or one of the Slodowy operator $\widetilde{A}$ or $\widetilde{A}^\vee$, then*

$$B v_n = v_{n-1} + v_{n+1}. \tag{5.40}$$

*Proof.* From now on

$$
\rho_i = \begin{cases} \rho_i & \text{for } B = A, \\ \rho_i^{\downarrow} & \text{for } B = \widetilde{A}, \\ \rho_i^{\uparrow} & \text{for } B = \widetilde{A}^{\vee}, \end{cases} \qquad m_i(n) = \begin{cases} m_i(n) & \text{for } B = A, \\ m_i^{\downarrow}(n) & \text{for } B = \widetilde{A}, \\ m_i^{\uparrow}(n) & \text{for } B = \widetilde{A}^{\vee}. \end{cases}
$$

By (5.24), (5.25), and by definition of the McKay operator (A.13) and by definition of the Slodowy operator (A.14) and (A.16), we have

$$
Bv_n = B \begin{pmatrix} m_0(n) \\ \cdots \\ m_r(n) \end{pmatrix} = \begin{pmatrix} \sum a_{0i} m_i(n) \\ \cdots \\ \sum a_{ri} m_i(n) \end{pmatrix} = \begin{pmatrix} \sum a_{0i} \langle \rho_i, \pi_n \rangle \\ \cdots \\ \sum a_{ri} \langle \rho_i, \pi_n \rangle \end{pmatrix}. \tag{5.41}
$$

By (A.13), (A.14) and (A.16) we have

$$
\sum_{i=1}^{r} a_{0i} \langle \rho_i, \pi_n \rangle = \langle \sum_{i=1}^{r} a_{0i} \rho_i, \pi_n \rangle = \langle \rho \otimes \rho_i, \pi_n \rangle,
$$

and from eq. (5.41) we obtain

$$
Bv_n = \begin{pmatrix} \langle \rho \otimes \rho_0, \pi_n \rangle \\ \cdots \\ \langle \rho \otimes \rho_r, \pi_n \rangle \end{pmatrix}, \tag{5.42}
$$

where $\rho$ is the irreducible two-dimensional representation which coincides with the representation $\pi_1$ in $\mathrm{Sym}^2(\mathbb{C}^2)$ from §5.4.1. Thus,

$$
Bv_n = \begin{pmatrix} \langle \pi_1 \otimes \rho_0, \pi_n \rangle \\ \cdots \\ \langle \pi_1 \otimes \rho_r, \pi_n \rangle \end{pmatrix}. \tag{5.43}
$$

From eq. (5.39) we obtain

$$
Bv_n = \begin{pmatrix} \langle \rho_0, \pi_1 \otimes \pi_n \rangle \\ \cdots \\ \langle \rho_r, \pi_1 \otimes \pi_n \rangle \end{pmatrix}. \tag{5.44}
$$

By Clebsch-Gordan formula we have

$$
\pi_1 \otimes \pi_n = \pi_{n-1} \oplus \pi_{n+1}, \tag{5.45}
$$

where $\pi_{-1}$ is the zero representation, see [Sp77, exs.3.2.4] or [Ha89, Ch.5, §6,§7]. From eq. (5.44) and eq. (5.45) we have (5.40). $\square$

For the following corollary, see [Kos84, p.222] and also [Sp87, §4.1].

**Corollary 5.11.** *Let $x = \widetilde{P}_G(t)$ be given by (5.30). Then*

$$tBx = (1 + t^2)x - v_0, \tag{5.46}$$

*where $B$ is either the McKay operator $A$ or one of the Slodowy operators $\widetilde{A}$, $\widetilde{A}^\vee$.*

*Proof.* From (5.40) we obtain

$$Bx = \sum_{n=0}^{\infty} Bv_n t^n = \sum_{n=0}^{\infty} (v_{n-1} + v_{n+1}) t^n = \sum_{n=0}^{\infty} v_{n-1} t^n + \sum_{n=0}^{\infty} v_{n+1} t^n =$$

$$t \sum_{n=1}^{\infty} v_{n-1} t^{n-1} + t^{-1} \sum_{n=0}^{\infty} v_{n+1} t^{n+1} = tx + t^{-1} \left( \sum_{n=0}^{\infty} v_n t^n - v_0 \right) =$$

$$tx + t^{-1}x - t^{-1}v_0.\square$$

### 5.4.3 The Poincaré series and W. Ebeling's theorem

W. Ebeling in [Ebl02] makes use of the Kostant relation (5.40) and deduces a new remarkable fact about the Poincaré series, a fact that shows that the Poincaré series of a binary polyhedral group (see (5.31)) is the quotient of two polynomials: the characteristic polynomial of the Coxeter transformation and the characteristic polynomial of the corresponding affine Coxeter transformation, see [Ebl02, Th.2].

We show W. Ebeling's theorem also for the multiply-laced case, i.e., for the generalized Poincaré series, see (5.30).

**Theorem 5.12 (generalized W.Ebeling's theorem [Ebl02]).** *Let $G$ be a binary polyhedral group and let $[\widetilde{P}_G(t)]_0$ be the generalized Poincaré series (5.30). Then*

$$[\widetilde{P}_G(t)]_0 = \frac{\det M_0(t)}{\det M(t)},$$

*where*

$$\det M(t) = \det |t^2 I - \mathbf{C}_a|, \qquad \det M_0(t) = \det |t^2 I - \mathbf{C}|,$$

$\mathbf{C}$ *is the Coxeter transformation and* $\mathbf{C}_a$ *is the corresponding affine Coxeter transformation.*

*Proof.* By (5.46) we have

$$[(1 + t^2)I - tB]x = v_0,$$

where $x$ is the vector $\widetilde{P}_G(t)$ and by Cramer's rule the first coordinate $\widetilde{P}_G(t)$ is

$$[\widetilde{P}_G(t)]_0 = \frac{\det M_0(t)}{\det M(t)},$$

**Table 5.2.**    The characteristic polynomials $\mathcal{X}$, $\tilde{\mathcal{X}}$ and the Poincaré series

| Dynkin diagram | Coxeter transformation $\mathcal{X}$ | Affine Coxeter transformation $\tilde{\mathcal{X}}$ | Quotient $p(\lambda) = \dfrac{\mathcal{X}}{\tilde{\mathcal{X}}}$ |
|---|---|---|---|
| $D_4$ | $(\lambda + 1)(\lambda^3 + 1)$ | $(\lambda - 1)^2(\lambda + 1)^3$ | $\dfrac{\lambda^3 + 1}{(\lambda^2 - 1)^2}$ |
| $D_{n+1}$ | $(\lambda + 1)(\lambda^n + 1)$ | $(\lambda^{n-1} - 1)(\lambda - 1)(\lambda + 1)^2$ | $\dfrac{\lambda^n + 1}{(\lambda^{n-1} - 1)(\lambda^2 - 1)}$ |
| $E_6$ | $\dfrac{(\lambda^6 + 1)}{(\lambda^2 + 1)}\dfrac{(\lambda^3 - 1)}{(\lambda - 1)}$ | $(\lambda^3 - 1)^2(\lambda + 1)$ | $\dfrac{\lambda^6 + 1}{(\lambda^4 - 1)(\lambda^3 - 1)}$ |
| $E_7$ | $\dfrac{(\lambda + 1)(\lambda^9 + 1)}{(\lambda^3 + 1)}$ | $(\lambda^4 - 1)(\lambda^3 - 1)(\lambda + 1)$ | $\dfrac{\lambda^9 + 1}{(\lambda^4 - 1)(\lambda^6 - 1)}$ |
| $E_8$ | $\dfrac{(\lambda^{15} + 1)(\lambda + 1)}{(\lambda^5 + 1)(\lambda^3 + 1)}$ | $(\lambda^5 - 1)(\lambda^3 - 1)(\lambda + 1)$ | $\dfrac{\lambda^{15} + 1}{(\lambda^{10} - 1)(\lambda^6 - 1)}$ |
| $B_n$ | $\lambda^n + 1$ | $(\lambda^{n-1} - 1)(\lambda^2 - 1)$ | $\dfrac{\lambda^n + 1}{(\lambda^{n-1} - 1)(\lambda^2 - 1)}$ |
| $C_n$ | $\lambda^n + 1$ | $(\lambda^n - 1)(\lambda - 1)$ | $\dfrac{\lambda^n + 1}{(\lambda^n - 1)(\lambda - 1)}$ |
| $F_4$ | $\dfrac{\lambda^6 + 1}{\lambda^2 + 1}$ | $(\lambda^2 - 1)(\lambda^3 - 1)$ | $\dfrac{\lambda^6 + 1}{(\lambda^4 - 1)(\lambda^3 - 1)}$ |
| $G_2$ | $\dfrac{\lambda^3 + 1}{\lambda + 1}$ | $(\lambda - 1)^2(\lambda + 1)$ | $\dfrac{\lambda^3 + 1}{(\lambda^2 - 1)^2}$ |
| $A_n$ | $\dfrac{\lambda^{n+1} - 1}{\lambda - 1}$ | $(\lambda^{n-k+1} - 1)(\lambda^k - 1)$ | $\dfrac{\lambda^{n+1} - 1}{(\lambda - 1)(\lambda^{n-k+1} - 1)(\lambda^k - 1)}$ |
| $A_{2n-1}$ | $\dfrac{\lambda^{2n} - 1}{\lambda - 1}$ | $(\lambda^n - 1)^2$ for $k = n$ | $\dfrac{\lambda^n + 1}{(\lambda^n - 1)(\lambda - 1)}$ |

where
$$\det M(t) = \det \left((1+t^2)I - tB\right),\qquad(5.47)$$
and $M_0(t)$ is the matrix obtained by replacing the first column of $M(t)$ by $v_0 = (1,0,...,0)^t$. The vector $v_0$ corresponds to the trivial representation $\pi_0$, and by the McKay correspondence, $v_0$ corresponds to the particular vertex which extends the Dynkin diagram to the extended Dynkin diagram. Therefore, if $\det M(t)$ corresponds to the affine Coxeter transformation, and
$$\det M(t) = \det |t^2 I - \mathbf{C}_a|,\qquad(5.48)$$
then $\det M_0(t)$ corresponds to the Coxeter transformation, and
$$\det M_0(t) = \det |t^2 I - \mathbf{C}|.\qquad(5.49)$$
So, it suffices to prove (5.48), i.e.,
$$\det[(1+t^2)I - tB] = \det |t^2 I - \mathbf{C}_a|.\qquad(5.50)$$
If $B$ is the McKay operator $A$ given by (A.13), then
$$B = 2I - K = \begin{pmatrix} 2I & 0 \\ 0 & 2I \end{pmatrix} - \begin{pmatrix} 2I & 2D \\ 2D^t & 2I \end{pmatrix} = \begin{pmatrix} 0 & -2D \\ -2D^t & 0 \end{pmatrix},\qquad(5.51)$$
where $K$ is a symmetric Cartan matrix (3.2). If $B$ is the Slodowy operator $\widetilde{A}$ or $\widetilde{A}^\vee$ given by (A.14), (A.16), then
$$B = 2I - K = \begin{pmatrix} 2I & 0 \\ 0 & 2I \end{pmatrix} - \begin{pmatrix} 2I & 2D \\ 2F & 2I \end{pmatrix} = \begin{pmatrix} 0 & -2D \\ -2F & 0 \end{pmatrix},\qquad(5.52)$$
where $K$ is the symmetrizable Cartan matrix (3.4). Thus, in the generic case
$$M(t) = (1+t^2)I - tB = \begin{pmatrix} 1+t^2 & 2tD \\ 2tF & 1+t^2 \end{pmatrix}.\qquad(5.53)$$
Assuming $t \neq 0$ we deduce from (5.53) that
$$M(t)\begin{pmatrix} x \\ y \end{pmatrix} = 0 \Longleftrightarrow \begin{cases} (1+t^2)x = -2tDy, \\ 2tFx = -(1+t^2)y. \end{cases}$$
$$\Longleftrightarrow \begin{cases} \dfrac{(1+t^2)^2}{4t^2}x = FDy, \\[2mm] \dfrac{(1+t^2)^2}{4t^2}y = DFy. \end{cases}\qquad(5.54)$$
According to (3.11), Proposition 3.4 and Proposition 3.10 we see that $t^2$ is an eigenvalue of the affine Coxeter transformation $\mathbf{C}_a$, i.e., (5.50) together with (5.48) are proved. $\square$

For the results of calculations using W. Ebeling's theorem, see Table 5.2.

*Remark 5.13.* 1) The characteristic polynomials $\mathcal{X}$ for the Coxeter transformation and $\tilde{\mathcal{X}}$ for the affine Coxeter transformation in Table 5.2 are taken from Tables 1.1 and 1.2. Pay attention to the fact that the affine Dynkin diagram for $B_n$ is $\widetilde{CD}_n$, ([Bo, Tab.2]), and the affine Dynkin diagram for $C_n$ is $\widetilde{C}_n$, ([Bo, Tab.3]), see Fig. 2.6.

2) The characteristic polynomial $\mathcal{X}$ for the affine Coxeter transformation of $A_n$ depends on the *index of the conjugacy class* $k$ of the Coxeter transformation, see (4.3). In the case of $A_n$ (for every $k = 1, 2, ..., n$) the quotient $p(\lambda) = \dfrac{\mathcal{X}}{\tilde{\mathcal{X}}}$ contains three factors in the denominator, and its form is different from (5.31), see Table 5.2.

For the case $A_{2n-1}$ and $k = n$, we have

$$p(\lambda) = \frac{\lambda^{2n} - 1}{(\lambda - 1)(\lambda^{2n-k} - 1)(\lambda^k - 1)} =$$
$$\frac{\lambda^{2n} - 1}{(\lambda - 1)(\lambda^n - 1)(\lambda^n - 1)} = \frac{\lambda^n + 1}{(\lambda^n - 1)(\lambda - 1)} \tag{5.55}$$

and $p(\lambda)$ again is of the form (5.31), see Table 5.2.

3) The quotients $p(\lambda)$ coincide for the following pairs:

$$D_4 \text{ and } G_2, \qquad E_6 \text{ and } F_4,$$
$$D_{n+1} \text{ and } B_n(n \geq 4), \qquad A_{2n-1} \text{ and } C_n.$$

Note that the second elements of the pairs are obtained by *folding* operation from the first ones, see Remark 5.4.

## 5.5 The orbit structure of the Coxeter transformation

### 5.5.1 The Kostant generating functions and polynomials $z(t)_i$

Let $\mathfrak{g}$ be a simple complex Lie algebra of type $A, D$ or $E$, $\tilde{\mathfrak{g}}$ be the affine Kac-Moody Lie algebra associated to $\mathfrak{g}$, and $\mathfrak{h} \subseteq \tilde{\mathfrak{h}}$ be, respectively, Cartan subalgebras of $\mathfrak{g} \subseteq \tilde{\mathfrak{g}}$. Let $\mathfrak{h}^\vee$ (resp. $\tilde{\mathfrak{h}}^\vee$) be the dual space to $\mathfrak{h}$ (resp. $\tilde{\mathfrak{h}}$) and $\alpha_i \in \mathfrak{h}^\vee, i = 1, \ldots, l$ be an ordered set of simple positive roots. Here, we follow B. Kostant's description [Kos84] of the orbit structure of the Coxeter transformation $\mathbf{C}$ on the highest root in the root system of $\mathfrak{g}$. We consider a bipartite graph and a bicolored Coxeter transformation from §3.1.1, §3.1.2. Let $\beta$ be the highest root of $(\mathfrak{h}, \mathfrak{g})$, see §4.1. Then

$$w_2\beta = \beta \quad \text{or} \quad w_1\beta = \beta.$$

In the second case we just swap $w_1$ and $w_2$, i.e., we always have

$$w_2\beta = \beta. \tag{5.56}$$

Between two bicolored Coxeter transformations (3.1) we select one such that

$$\mathbf{C} = w_2 w_1.$$

Consider, for example, the Dynkin diagram $E_6$. Here,

$$w_1 = \begin{pmatrix} 1 & & & & & \\ & 1 & & & & \\ & & 1 & & & \\ 1 & 1 & 0 & -1 & & \\ 1 & 0 & 1 & & -1 & \\ 1 & 0 & 0 & & & -1 \end{pmatrix} \begin{matrix} x_0 \\ x_1 \\ x_2 \\ y_1 \\ y_2 \\ y_3 \end{matrix}, \quad w_2 = \begin{pmatrix} -1 & & 1 & 1 & 1 \\ & -1 & 1 & 0 & 0 \\ & & -1 & 0 & 1 & 0 \\ & & 1 & & & \\ & & & 1 & & \\ & & & & 1 \end{pmatrix} \begin{matrix} x_0 \\ x_1 \\ x_2 \\ y_1 \\ y_2 \\ y_3 \end{matrix} \quad (5.57)$$

The vector $z \in \mathfrak{h}^\vee$ and the highest root $\beta$ are:

$$z = \begin{matrix} x_1 \; y_1 \; x_0 \; y_2 \; x_2 \\ y_3 \end{matrix}, \quad \beta = \begin{matrix} 1 \; 2 \; 3 \; 2 \; 1 \\ 2 \end{matrix}, \quad \text{or} \quad \beta = \begin{pmatrix} 3 \\ 1 \\ 1 \\ 2 \\ 2 \\ 2 \end{pmatrix} \begin{matrix} x_0 \\ x_1 \\ x_2 \\ y_1 \\ y_2 \\ y_3 \end{matrix} \quad (5.58)$$

Here,

$$w_1 \beta = \begin{matrix} 1 \; 2 \; 3 \; 2 \; 1 \\ 1 \end{matrix}, \quad \text{and} \quad w_2 \beta = \begin{matrix} 1 \; 2 \; 3 \; 2 \; 1 \\ 2 \end{matrix} = \beta \quad (5.59)$$

Further, following B.Kostant [Kos84, Th.1.5] consider the alternating products $\tau^{(n)}$:

$$\tau^{(1)} = w_1,$$
$$\tau^{(2)} = \mathbf{C} = w_2 w_1,$$
$$\tau^{(3)} = w_1 \mathbf{C} = w_1 w_2 w_1,$$
$$\dots,$$
$$\tau^{(n)} = \begin{cases} \mathbf{C}^k = w_2 w_1 \dots w_2 w_1 & \text{for } n = 2k, \\ w_1 \mathbf{C}^k = w_1 w_2 w_1 \dots w_2 w_1 & \text{for } n = 2k+1, \end{cases} \quad (5.60)$$

and the orbit of the highest root $\beta$ under the action of $\tau^{(n)}$:

$$\beta_n = \tau^{(n)} \beta, \quad \text{where } n = 1, \dots, h - 1$$

($h$ is the Coxeter number, see(4.1)).

**Theorem 5.14. (B. Kostant, [Kos84, Theorems 1.3, 1.4, 1.5, 1.8])**
   *1) There exist $z_j \in \tilde{\mathfrak{h}}^\vee$, where $j = 0, \dots, h$, and even integers $a$ and $b$ such that $2 \le a \le b \le h$ (see (5.32) and Table 5.1) and so the generating functions $P_G(t)$ (see (1.3), (1.10), or (5.26), (5.28) ) are as follows:*

$$[P_G(t)]_i = \begin{cases} \dfrac{1 + t^h}{(1 - t^a)(1 - t^b)} & \textit{for } i = 0, \\[2em] \dfrac{\displaystyle\sum_{j=0}^{h} z_j t^j}{(1 - t^a)(1 - t^b)} & \textit{for } i = 1, \dots, r. \end{cases} \qquad (5.61)$$

*For* $n = 1, \dots, h-1$, *one has* $z_n \in \mathfrak{h}^\vee$ *( not just* $\tilde{\mathfrak{h}}^\vee$*). The indices* $i = 1, \dots, r$
*enumerate the vertices of the Dynkin diagram and the coordinates of the vec-*
*tors* $z_n$*; The index* $i = 0$ *corresponds to the additional (affine) vertex, the one*
*that extends the Dynkin diagram to the extended Dynkin diagram. One has*
$z_0 = z_h = \alpha_0$, *where* $\alpha_0 \in \mathfrak{h}^\vee$ *is the added simple root corresponding to the*
*affine vertex.*

*2) The vectors* $z_n$ *(we call these vectors the assembling vectors) are ob-*
*tained as follows:*

$$z_n = \tau^{(n-1)}\beta - \tau^{(n)}\beta. \qquad (5.62)$$

*3) We have*

$$z_g = 2\alpha_*, \textit{ where } g = \frac{h}{2},$$

*and* $\alpha_*$ *is the simple root corresponding to the branch point for diagrams* $D_n$,
$E_n$ *and to the midpoint for the diagram* $A_{2m-1}$. *In all these cases* $h$ *is even,*
*and* $g$ *is an integer. The diagram* $A_{2m}$ *is excluded.*

*4) The series of assembling vectors* $z_n$ *is symmetric:*

$$z_{g+k} = z_{g-k} \textit{ for } k = 1, \dots, g. \qquad (5.63)$$

In the case of the Dynkin diagram $E_6$, the vectors $\tau^{(n)}\beta$ are given in
Table 5.3, and the assembling vectors $z_n$ are given in Table 5.4. The vector $z_6$
coincides with $2\alpha_{x_0}$, where $\alpha_{x_0}$ is the simple root corresponding to the vertex
$x_0$, see (5.58). From Table 5.4 we see that

$$z_1 = z_{11}, \qquad z_2 = z_{10}, \qquad z_3 = z_9, \qquad z_4 = z_8, \qquad z_5 = z_7.$$

Denote by $z(t)_i$ the polynomial $\sum\limits_{j=0}^{h} z_j t^j$ from (5.61). In the case of $E_6$ we
have:

$$\begin{aligned} z(t)_{x_0} &= t^2 + t^4 + 2t^6 + t^8 + t^{10}, \\ z(t)_{x_1} &= t^4 + t^8, \\ z(t)_{x_2} &= t^4 + t^8, \\ z(t)_{y_1} &= t^3 + t^5 + t^7 + t^9, \\ z(t)_{y_2} &= t^3 + t^5 + t^7 + t^9, \\ z(t)_{y_3} &= t + t^5 + t^7 + t^{11}. \end{aligned} \qquad (5.64)$$

The Kostant numbers $a, b$ (see Table 5.1) for $E_6$ are $a = 6$, $b = 8$. From (5.61)
and (5.64), we have

**Table 5.3.**   The orbit of the Coxeter transformation on the highest root

| $\beta$ | $\tau^{(1)}\beta = w_1\beta$ | $\tau^{(2)}\beta = \mathbf{C}\beta$ |
|---|---|---|
| 1 2 3 2 1 | 1 2 3 2 1 | 1 2 2 2 1 |
| 2 | 1 | 1 |
| $\tau^{(3)}\beta = w_1\mathbf{C}\beta$ | $\tau^{(4)}\beta = \mathbf{C}^2\beta$ | $\tau^{(5)}\beta = w_1\mathbf{C}^2\beta$ |
| 1 1 2 1 1 | 0 1 1 1 0 | 0 0 1 0 0 |
| 1 | 1 | 0 |
| $\tau^{(6)}\beta = \mathbf{C}^3\beta$ | $\tau^{(7)}\beta = w_1\mathbf{C}^3\beta$ | $\tau^{(8)}\beta = \mathbf{C}^4\beta$ |
| 0 0 −1 0 0 | 0 −1 −1 −1 0 | −1 −1 −2 −1 −1 |
| 0 | −1 | −1 |
| $\tau^{(9)}\beta = w_1\mathbf{C}^4\beta$ | $\tau^{(10)}\beta = \mathbf{C}^5\beta$ | $\tau^{(11)}\beta = w_1\mathbf{C}^5\beta$ |
| −1 −2 −2 −2 −1 | −1 −2 −3 −2 −1 | −1 −2 −3 −2 −1 |
| −1 | −1 | −2 |

**Table 5.4.**   The assembling vectors $z_n = \tau^{(n-1)}\beta - \tau^{(n)}\beta$

| $z_1 = \beta - w_1\beta$ | $z_2 = w_1\beta - \mathbf{C}\beta$ | $z_3 = \mathbf{C}\beta - w_1\mathbf{C}\beta$ |
|---|---|---|
| 0 0 0 0 0 | 0 0 1 0 0 | 0 1 0 1 0 |
| 1 | 0 | 0 |
| $z_4 = w_1\mathbf{C}\beta - \mathbf{C}^2\beta$ | $z_5 = \mathbf{C}^2\beta - w_1\mathbf{C}^2\beta$ | $z_6 = w_1\mathbf{C}^2\beta - \mathbf{C}^3\beta$ |
| 1 0 1 0 1 | 0 1 0 1 0 | 0 0 2 0 0 |
| 0 | 1 | 0 |
| $z_7 = \mathbf{C}^3\beta - w_1\mathbf{C}^3\beta$ | $z_8 = w_1\mathbf{C}^3\beta - \mathbf{C}^4\beta$ | $z_9 = \mathbf{C}^4\beta - w_1\mathbf{C}^4\beta$ |
| 0 1 0 1 0 | 1 0 1 0 1 | 0 1 0 1 0 |
| 1 | 0 | 0 |
| $z_{10} = w_1\mathbf{C}^4\beta - \mathbf{C}^5\beta$ | $z_{11} = \mathbf{C}^5\beta - w_1\mathbf{C}^5\beta$ | |
| 0 0 1 0 0 | 0 0 0 0 0 | |
| 0 | 1 | |

$$[P_G(t)]_{x_0} = \frac{t^2 + t^4 + 2t^6 + t^8 + t^{10}}{(1 - t^6)(1 - t^8)},$$

$$[P_G(t)]_{x_1} = [P_G(t)]_{x_2} = \frac{t^4 + t^8}{(1 - t^6)(1 - t^8)},$$

$$[P_G(t)]_{y_1} = [P_G(t)]_{y_2} = \frac{t^3 + t^5 + t^7 + t^9}{(1 - t^6)(1 - t^8)},$$

$$[P_G(t)]_{y_3} = \frac{t + t^5 + t^7 + t^{11}}{(1 - t^6)(1 - t^8)}.$$

Since

$$\frac{1}{1 - t^6} = \sum_{n=0}^{\infty} t^{6n}, \qquad \frac{1}{1 - t^8} = \sum_{n=0}^{\infty} t^{8n},$$

we have

$$[P_G(t)]_{x_1} = [P_G(t)]_{x_2} = \sum_{i,j=0}^{\infty} (t^{6i+8j+4} + t^{6i+8j+8}),$$

$$[P_G(t)]_{y_1} = [P_G(t)]_{y_2} = \sum_{i,j=0}^{\infty} (t^{6i+8j+3} + t^{6i+8j+5} + t^{6i+8j+7} + t^{6i+8j+9}),$$

$$[P_G(t)]_{y_3} = \sum_{i,j=0}^{\infty} (t^{6i+8j+1} + t^{6i+8j+5} + t^{6i+8j+7} + t^{6i+8j+11}),$$

$$(5.65)$$

Recall that $m_\alpha(n)$, where $\alpha = x_1, x_2, y_1, y_2, y_3$, are the multiplicities of the irreducible representations $\rho_\alpha$ of $G$ (considered in the context of the McKay correspondence, §A.4) in the decomposition of $\pi_n|G$ (5.20). These multiplicities are the coefficients of the Poincaré series (5.65), see (5.24), (5.26), (5.28). For example,

$$[P_G(t)]_{x_1} = [P_G(t)]_{x_2} = t^4 + t^8 + t^{10} + t^{12} + t^{14} + 2t^{16} + t^{18} + 2t^{20} + \dots$$

$$m_{x_1} = m_{x_2} = \begin{cases} 0 & \text{for } n = 1, 2, 3, 5, 6 \text{ and } n = 2k + 1, k \geq 3, \\ 1 & \text{for } n = 4, 8, 10, 12, 14, 18, \dots \\ 2 & \text{for } n = 16, 20, \dots \\ \dots & \end{cases}$$

In particular, the representations $\rho_{x_1}(n)$ and $\rho_{x_2}(n)$ do not enter in the decomposition of $\pi_n$ of $SU(2)$ (see §5.4.1) for any odd $n$.

In [Kos04], concerning the importance of the polynomials $z(t)_i$, B. Kostant points out: *"Unrelated to the Coxeter element, the polynomials $z(t)_i$ are also determined in Springer, [Sp87]. They also appear in another context in Lusztig, [Lus83] and [Lus99]. Recently, in a beautiful result, Rossmann, [Ros04], relates the character of $\gamma_i$ to the polynomial $z(t)_i$."*

### 5.5.2 One more observation of McKay

In this section we prove one more observation of McKay [McK99] relating the Molien-Poincaré series. In our context these series are the Kostant generating functions $P_G(t)$ corresponding to the irreducible representations of group $G$:

$$[P_G(t)]_i = \frac{z(t)_i}{(1 - t^a)(1 - t^b)}, \tag{5.66}$$

see (5.61), (5.64).

**Theorem 5.15.** (Observation of McKay [McK99, (*)]) *For diagrams* $\Gamma = D_n, E_n$ *and* $A_{2m-1}$, *the Kostant generating functions* $[P_G(t)]_i$ *are related as follows:*

$$(t + t^{-1})[P_G(t)]_i = \sum_{j \leftarrow i}[P_G(t)]_j, \tag{5.67}$$

*where $j$ runs over all successor[1] vertices to $i$, and $[P_G(t)]_0$ related to the affine vertex $\alpha_0$ occurs in the right side only: $i = 1, 2, \ldots, r$.*

By (5.66), observation of McKay (5.67) is equivalent to the following one:

$$(t + t^{-1})z(t)_i = \sum_{j \leftarrow i} z(t)_j, \text{ where } i = 1, \ldots, r. \tag{5.68}$$

So, we will prove (5.68).

*Remark 5.16.* J. McKay introduces in [McK99] so-called *semi-affine* graph which is defined in terms of the Dynkin diagram with an additional edge (two edges for $A$-type). This edge is directed toward to the affine node. Thus, the semi-affine graph is an oriented graph, one edge is unidirectional, all other edges are bidirectional. The sink of the given directed edge is called the *successor* node. □

The *adjacency matrix* $\mathcal{A}$ for types $ADE$ is the matrix containing non-diagonal entries $a_{ij}$ if and only if the vertices $i$ and $j$ are connected by an edge, and then $a_{ij} = 1$, and all diagonal entries $a_{ii}$ vanish:

$$\mathcal{A} = \begin{pmatrix} 0 & -2D \\ -2D^t & 0 \end{pmatrix}, \tag{5.69}$$

---

[1] See Remark 5.16.

see (3.3) and also §C.7.1. Let $\alpha_0$ be the affine vertex of the graph $\Gamma$, and $u_0$ be a vertex adjacent to $\alpha_0$. Extend the adjacency matrix $\mathcal{A}$ to the *semi-affine adjacency matrix* $\mathcal{A}^\gamma$ (in the style to the McKay definition of the semi-affine graph in [McK99], see Remark 5.16) by adding a row and a column corresponding to the affine vertex $\alpha_0$ as follows: 0 is set in the $(u_0, \alpha_0)$th slot and 1 is set in the$(\alpha_0, u_0)$th slot, all remaining places in the $\alpha_0$th row and the $\alpha_0$th column are 0, see Fig. 5.4. Note, that for the $A_n$ case, we set 1 in the two places: $(\alpha_0, u_0)$ and $(\alpha_0, u'_0)$ corresponding to vertices $u_0$ and $u'_0$ adjacent to $\alpha$.

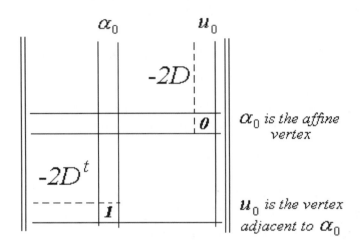

**Fig. 5.4.**     The semi-affine adjacent matrix $\mathcal{A}^\gamma$

Using the semi-affine adjacency matrix $\mathcal{A}^\gamma$ we express McKay's observation (5.68) in the matrix form:

$$(t + t^{-1})z(t)_i = (\mathcal{A}^\gamma z(t))_i, \quad \text{where } z(t) = \{z(t)_0, \ldots, z(t)_r\}$$
$$\text{and } i = 1, \ldots, r. \tag{5.70}$$

To prove (5.70), we consider the action of the adjacency matrix $\mathcal{A}$ and the semi-affine adjacency matrix $\mathcal{A}^\gamma$ related to the extended Dynkin diagram of types $ADE$ on assembling vectors $z_n$ (5.62).

**Proposition 5.17.** *1) For the vectors $z_i \in \mathfrak{h}^\vee$ from (5.62), we have*

$$(a) \quad \mathcal{A}z_i = z_{i-1} + z_{i+1} \quad for \quad 1 < i < h - 1,$$
$$(b) \quad \mathcal{A}z_1 = z_2 \quad and \quad \mathcal{A}z_{h-1} = z_{h-2}, \tag{5.71}$$

*2) Consider the same vectors $z_i$ as vectors from $\tilde{\mathfrak{h}}^\vee$, so we just add a zero coordinate to the affine vertex $\alpha_0$. We have*

$$(a) \quad \mathcal{A}^\gamma z_i = z_{i-1} + z_{i+1} \quad for \quad 1 < i < h - 1,$$
$$(b) \quad \mathcal{A}^\gamma z_1 = z_2 \quad and \quad \mathcal{A}^\gamma z_{h-1} = z_{h-2}, \qquad (5.72)$$
$$(c) \quad \mathcal{A}^\gamma z_0 = z_1 \quad and \quad \mathcal{A}^\gamma z_h = z_{h-1}.$$

*Proof.*

1a) Let us prove (5.71 (a)). According to (5.62) we have

$$z_{2n} = w_1 \mathbf{C}^{n-1}\beta - \mathbf{C}^n\beta = (1 - w_2)w_1 \mathbf{C}^{n-1}\beta,$$

and

$$z_{2n+1} = \mathbf{C}^n\beta - w_1 \mathbf{C}^n\beta = (1 - w_1)\mathbf{C}^n\beta.$$

Thus, for $i = 2n$, eq. (5.71 (a)) is equivalent to

$$\mathcal{A}(1 - w_2)w_1 \mathbf{C}^{n-1}\beta = (1 - w_1)\mathbf{C}^{n-1}\beta + (1 - w_1)\mathbf{C}^n\beta, \qquad (5.73)$$

and for $i = 2n + 1$, eq. (5.71 (a)) is equivalent to

$$\mathcal{A}(1 - w_1)\mathbf{C}^n\beta = (1 - w_2)w_1 \mathbf{C}^{n-1}\beta + (1 - w_2)w_1 \mathbf{C}^n\beta. \qquad (5.74)$$

To prove relations (5.73) and (5.74), it suffices to show that

$$\mathcal{A}(1 - w_2)w_1 = (1 - w_1) + (1 - w_1)\mathbf{C} = (1 - w_1)(1 + \mathbf{C}), \qquad (5.75)$$

and

$$\mathcal{A}(1 - w_1)\mathbf{C} = (1 - w_2)w_1 + (1 - w_2)w_1\mathbf{C} = (1 - w_2)w_1(1 + \mathbf{C}). \qquad (5.76)$$

In (5.56), (5.57), $w_1$ and $w_2$ are chosen as

$$w_1 = \begin{pmatrix} I & 0 \\ -2D^t & -I \end{pmatrix}, \qquad w_2 = \begin{pmatrix} -I & -2D \\ 0 & I \end{pmatrix},$$

So, by (5.69) we have

$$1 - w_1 = \begin{pmatrix} 0 & 0 \\ 2D^t & 2I \end{pmatrix}, \quad 1 - w_2 = \begin{pmatrix} 2I & 2D \\ 0 & 0 \end{pmatrix},$$

$$(1 - w_2)w_1 = \begin{pmatrix} 2I - 4DD^t & -2D \\ 0 & 0 \end{pmatrix}, \quad \mathbf{C} = \begin{pmatrix} 4DD^t - I & 2D \\ -2D^t & -I \end{pmatrix},$$

$$\mathcal{A}(1 - w_2)w_1 = \begin{pmatrix} 0 & 0 \\ -4D^t + 8D^t DD^t & 4D^t D \end{pmatrix},$$

$$1 + \mathbf{C} = \begin{pmatrix} 4DD^t & 2D \\ -2D^t & 0 \end{pmatrix}.$$

$$(5.77)$$

By (5.77) we have

$$(1 - w_1)(1 + \mathbf{C}) = \begin{pmatrix} 0 & 0 \\ -4D^t + 8D^t DD^t & 4D^t D \end{pmatrix},$$

and (5.75) is true. Further, we have

$$(1 - w_1)\mathbf{C} = \begin{pmatrix} 0 & 0 \\ 8D^t DD^t - 6D^t & 4D^t D - 2I \end{pmatrix},$$

$$\mathcal{A}(1 - w_1)\mathbf{C} = \begin{pmatrix} -16DD^t DD^t + 12DD^t & -8DD^t D + 4D \\ 0 & 0 \end{pmatrix}, \tag{5.78}$$

By (5.77) we obtain

$$(1 - w_2)w_1(1 + \mathbf{C}) = \begin{pmatrix} -16DD^t DD^t + 12DD^t & -8DD^t D + 4D \\ 0 & 0 \end{pmatrix},$$

and (5.74) is also true.

1b) Let us move on to (5.71 (b)). This is equivalent to

$$\mathcal{A}(\beta - w_1\beta) = w_1\beta - \mathbf{C}\beta, \quad \text{or}$$
$$\mathcal{A}(1 - w_1)\beta = (1 - w_2)w_1\beta. \tag{5.79}$$

By (5.69), (5.77) eq. (5.79) is equivalent to

$$\begin{pmatrix} -4DD^t & -4D \\ 0 & 0 \end{pmatrix} \begin{pmatrix} x \\ y \end{pmatrix} = \begin{pmatrix} 2I - 4DD^t & -2D \\ 0 & 0 \end{pmatrix} \begin{pmatrix} x \\ y \end{pmatrix}, \tag{5.80}$$

where

$$\beta = \begin{pmatrix} x \\ y \end{pmatrix} \tag{5.81}$$

is given in two-component form (3.8) corresponding to a bipartite graph. Eq. (5.80) is equivalent to

$$-2Dy = 2x. \tag{5.82}$$

Since the matrix $-2D$ contains a 1 at the $(i, j)$th slot if and only if the vertices $i$ and $j$ are connected, eq. (5.82) follows from the well-known fact formulated in Remark 5.18(b).

*Remark 5.18 (On the highest root and imaginary roots).* (a) The highest root $\beta$ for types $ADE$ coincides with the minimal positive *imaginary root* (of the corresponding extended Dynkin diagram) without affine coordinate $\alpha_0$, see (2.26) and §2.2.1. The coordinates of the imaginary vector $\delta$ are given in Fig. 2.6.

(b) For any vertex $x_i$ of the highest root $\beta$ (except the vertex $u_0$ adjacent to the affine vertex $\alpha_0$), the sum of coordinates of the adjacent vertices $y_j$ coincides with the doubled coordinate of $\alpha_{x_i}$:

$$\sum_{y_i \to x_j} \alpha_{y_j} = 2\alpha_{x_i}. \tag{5.83}$$

(c) For any vertex $x_i$ of the imaginary root $\delta$, the sum of coordinates of the adjacent vertices $y_j$ coincides with the doubled coordinate of $\alpha_{x_i}$ as above in (5.83).

Since we can choose partition (5.81) such that $\alpha_0$ belongs to subset $y$, we obtain (5.82). The second relation of (5.71 (b)) follows from the symmetry of assembling vectors, see (5.63) of the Kostant theorem (Theorem 5.14).    □

2) Relations (a), (b) of (5.72) follow from the corresponding relations in (5.71) since the addition of a "1" to the $(\alpha_0, u_0)$th slot of the matrix $\mathcal{A}^\gamma$ (5.4) is neutralized by the affine coordinate 0 of the vectors $z_i$, where $i = 1, \ldots, h - 1$.

Let us prove (c) of (5.72). First,

$$\mathcal{A}^\gamma z_0 = \alpha_{u_0}, \qquad \text{for } D_n, E_n,$$
$$\mathcal{A}^\gamma z_0 = \alpha_{u_0} + \alpha_{u'_0}, \qquad \text{for } A_{2m-1},$$

where $\alpha_{u_0}$ (resp. $\alpha_{u'_0}$) is the simple root with a "1" in the $u_0$th (resp. $u'_0$th) position. By (5.62) $z_1 = (1 - w_1)\beta$. Then, by (5.77), we have that eq. (5.72 (c)) is equivalent to

$$\alpha_{u_0} = (1 - w_1)\beta = \begin{pmatrix} 0 \\ 2D^t x + 2y \end{pmatrix}, \qquad \text{for } D_n, E_n,$$

$$\alpha_{u_0} + \alpha_{u'_0} = (1 - w_1)\beta = \begin{pmatrix} 0 \\ 2D^t x + 2y \end{pmatrix}, \qquad \text{for } A_{2m-1},$$

Again, by Remark 5.18 (b), we have $2D^t x + 2y = 0$ for all coordinates excepting coordinate $u_0, u'_0$. For coordinate $u_0$ (resp. $u'_0$), by Remark 5.18 (c), we have $(\alpha_{u_0} - 2D^t x)_{u_0} = 2y_{u_0}$ (resp. $(\alpha_{u'_0} - 2D^t x)_{u'_0} = 2y_{u'_0}$ for $A_{2m-1}$).    □

*Proof of (5.70)* . Since,

$$z(t) = \sum_{j=0}^{h} z_j t^j, \quad z(t)_i = (\sum_{j=0}^{h} z_j t^j)_i, \text{ where } i = 1, \ldots, r,$$

by (5.72) we have

$$\mathcal{A}^\gamma z(t) = \sum_{j=0}^{h} \mathcal{A}^\gamma z_j t^j =$$

$$z_1 + z_2 t + (z_1 + z_3)t^2 + \cdots +$$
$$(z_{h-3} + z_{h-1})t^{h-2} + z_{h-2}t^{h-1} + z_{h-1}t^h =$$
$$(z_1 + z_2 t + z_3 t^2 + \cdots + z_{h-1}t^{h-2}) +$$
$$(z_1 t^2 + \cdots + z_{h-2}t^{h-1} + z_{h-1}t^h) =$$
$$t^{-1}(z_1 t + z_2 t^2 + z_3 t^3 + \cdots + z_{h-1}t^{h-1}) +$$
$$t(z_1 t + \cdots + z_{h-2}t^{h-2} + z_{h-1}t^{h-1}) =$$
$$(t + t^{-1})(z_1 t + z_2 t^2 + z_3 t^3 + \cdots + z_{h-1}t^{h-1}) =$$
$$(t + t^{-1})(z(t) - z_0 - z_h t^h).$$

Since $z_0 = z_h$ we have

$$\mathcal{A}^\gamma z(t) = (t + t^{-1})z(t) - (t + t^{-1})(1 + t^h)z_0.$$

Coordinates $(z_0)_i = (z_h)_i$ are zeros for $i = 1, \dots, r$, and

$$(\mathcal{A}^\gamma z(t))_i = (t + t^{-1})z(t)_i, \quad i = 1, \dots, r. \tag{5.84}$$

For the coordinate $i = 0$, corresponding to affine vertex $\alpha_0$, by definition of $\mathcal{A}^\gamma$, see Fig. 5.4, and by (5.61), we have $(\mathcal{A}^\gamma z(t))_0 = 0$, and $z(t)_0 = (1 + t^h)z_0$.
□

Let us check the observation of McKay for the Kostant generating functions for the case of $E_6$. According to (5.64) we should get the following relations:

$$
\begin{aligned}
&\text{1) For } x_0: \quad (t + t^{-1})z(t)_{x_0} = z(t)_{y_1} + z(t)_{y_2} + z(t)_{y_3}, \\
&\text{2) For } y_1: \quad (t + t^{-1})z(t)_{y_1} = z(t)_{x_0} + z(t)_{x_1}, \\
&\text{3) For } y_2: \quad (t + t^{-1})z(t)_{y_2} = z(t)_{x_0} + z(t)_{x_2}, \\
&\text{4) For } x_1: \quad (t + t^{-1})z(t)_{x_1} = z(t)_{y_1}, \\
&\text{5) For } x_2: \quad (t + t^{-1})z(t)_{x_2} = z(t)_{y_2}, \\
&\text{6) For } y_3: \quad (t + t^{-1})z(t)_{y_3} = z(t)_{x_0} + z(t)_{\alpha_0}.
\end{aligned}
$$

1) For $x_0$, we have

$$
\begin{aligned}
&(t + t^{-1})(t^2 + t^4 + 2t^6 + t^8 + t^{10}) = \\
&2(t^3 + t^5 + t^7 + t^9) + (t + t^5 + t^7 + t^{11}),
\end{aligned}
$$

or

$$
\begin{aligned}
&(t^3 + t^5 + 2t^7 + t^9 + t^{11}) + (t + t^3 + 2t^5 + t^7 + t^9) = \\
&t + 2t^3 + 3t^5 + 3t^7 + 2t^9 + t^{11}.
\end{aligned}
$$

2) For $y_1$ and for $y_2$, we have

$$
\begin{aligned}
&(t + t^{-1})(t^3 + t^5 + t^7 + t^9) = \\
&(t^2 + t^4 + 2t^6 + t^8 + t^{10}) + (t^4 + t^8),
\end{aligned}
$$

or

$$
\begin{aligned}
&(t^4 + t^6 + t^8 + t^{10}) + (t^2 + t^4 + t^6 + t^8) = \\
&(t^2 + 2t^4 + 2t^6 + 2t^8 + t^{10}).
\end{aligned}
$$

3) For $x_1$ and for $x_2$, we have

$$(t + t^{-1})(t^4 + t^8) = t^3 + t^5 + t^7 + t^9.$$

4) For $y_3$, we have

$$(t + t^{-1})(t + t^5 + t^7 + t^{11}) =$$
$$(t^2 + t^4 + 2t^6 + t^8 + t^{10}) + (1 + t^{12}),$$

or

$$(t^2 + t^6 + t^8 + t^{12}) + (1 + t^4 + t^6 + t^{10}) =$$
$$(t^2 + t^4 + 2t^6 + t^8 + t^{10}) + (1 + t^{12}).$$

# 6

## The affine Coxeter transformation

## 6.1 The Weyl Group and the affine Weyl group

### 6.1.1 The semidirect product

Let $G$ be a group, $N$ a normal subgroup of $G$ (i.e., $N \triangleleft G$) and $H$ a subgroup of $G$. We say, that $G$ is the semidirect product of $N$ and $H$, if

$$G = NH \text{ and } N \cap H = e, \tag{6.1}$$

where $e$ is the unit element of $G$, [CR62]. The semidirect product is frequently denoted by

$$N \rtimes H.$$

**Proposition 6.1.** *The following statements are equivalent:*

*a) Every element of $G$ can be written in one and only one way as a product of an element of $N$ and an element of $H$.*

*b) Every element of $G$ can be written in one and only one way as a product of an element of $H$ and an element of $N$.*

*c) The natural embedding $H \to G$ composed with the natural projection $G \to G/N$, yields an isomorphism*

$$H \simeq G/N.$$

$\square$

The group $G$ is the semidirect product of $N$ and $H$ if and only if one (and therefore all) of statements of Proposition 6.1 hold.

A group $G$ is isomorphic to a semidirect product of the two groups $N$ and $H$ if and only if there exists a *short exact sequence*

$$0 \longrightarrow N \xrightarrow{\beta} G \xrightarrow{\alpha} H \longrightarrow 0 \tag{6.2}$$

and a group homomorphism $\gamma : H \longrightarrow G$ such that $\alpha \circ \gamma = \text{Id}_H$, the identity map on H, i.e., the exact sequence (6.2) *splits*.

Consider the product of two elements $n_1 h_1$ and $n_2 h_2$ in the semidirect product $G = N \rtimes H$. Here

$$n_1, n_2 \in N, \text{ and } h_1, h_2 \in H.$$

Then we have

$$(n_1 h_1) * (n_2 h_2) = n_1 (h_1 n_2 h_1^{-1})(h_1 h_2).$$

Since $N \triangleleft G$, then $h_1 n_2 h_1^{-1} \in N$, and the product of $n_1 h_1$ and $n_2 h_2$ is the element $n_0 h_0$, where

$$n_0 = n_1 h_1 n_2 h_1^{-1} \in N, \text{ and } h_0 = h_1 h_2.$$

According to Proposition 6.1, the element $n_0 h_0$ is well-defined.

*Example 6.2.* Let $C_n$ and $C_2$ be cyclic groups with $n$ and 2 elements, respectively. The dihedral group $D_n$ with $2n$ elements is isomorphic to a semidirect product of $C_n$ and $C_2$. Here,

$$D_n \simeq C_n \rtimes C_2,$$

and the non-identity element of $C_2$ acts on $C_n$ by inverting elements. The presentation for this group is:

$$\{a, b \mid a^2 = e, b^n = e, aba^{-1} = b^{-1}\}.$$

### 6.1.2 Two representations of the affine Weyl group

The *Weyl group* $W$ corresponds to the Dynkin diagram and it is generated by orthogonal reflections, i.e., reflections leaving the origin in $V$ fixed. The *affine Weyl group* $W_a$ corresponds to the extended Dynkin diagram and it is also generated by orthogonal reflections, but contains also *affine reflections* relative to hyperplanes which do not necessary pass through the origin, see [Hu90, Ch. 4], [Bo, Ch.6, §2], [Mac72]. Below we consider a relation between the Weyl group and the affine Weyl group.

Consider the Weyl group $W$ acting in the euclidean space $V$. In the Dynkin diagram case, the Weyl group $W$ is generated by orthogonal reflections $s_i$, i.e., reflections leaving the origin in $V$ fixed:

$$s_i(z) = z - 2 \frac{(z, \alpha_i)}{(\alpha_i, \alpha_i)} \alpha_i,$$

$\alpha_1, \alpha_2, \ldots, \alpha_n$ are the simple roots enumerated by vertices of the corresponding Dynkin diagram.

*Remark 6.3.* For every root system corresponding to some Dynkin diagram, there are not more than two different lengths of roots. In the case of two

different lengths of roots one says about *short roots* and *long roots*. If all roots
have the same length, then these roots are considered long. The number

$$m = \frac{\max{(\alpha, \alpha)}}{\min{(\alpha, \alpha)}}$$

is equal to the maximal multiplicity of an edge in the Dynkin diagram:

$$m = \begin{cases} 1, & \text{for the } ADE \text{ diagrams,} \\ 2, & \text{for the } BCF \text{ diagrams,} \\ 3, & \text{for the } G_2 \text{ diagrams,} \end{cases}$$

see [Kir04, §8.8]. For the Dynkin diagram with the root system $\Delta$, we denote
the set of the real short (resp. long) roots by $\Delta_s$ (resp. $\Delta_l$).     □

**Proposition 6.4. [Kac93, Prop.6.3]** *The set of the real roots $\Delta^{re}$ for different types of the extended Dynkin diagrams is as follows:*

$$\Delta^{re} = \begin{cases} \{\alpha + k\omega \mid \alpha \in \Delta\} \text{ for } r = 1, \\ \\ \{\alpha + k\omega \mid \alpha \in \Delta_s\} \coprod \{\alpha + rk\omega \mid \alpha \in \Delta_l\} \\ \qquad \text{for } r = 2 \text{ or } 3, \text{ and the diagram is not } A_{2l}^{(2)}, \\ \\ \{\alpha + k\omega \mid \alpha \in \Delta_s\} \coprod \{\alpha + 2k\omega \mid \alpha \in \Delta_l\} \coprod \\ \{\frac{1}{2}(\alpha + (2k-1)\omega \mid \alpha \in \Delta_l\} \\ \qquad \text{for the diagram } A_{2l}^{(2)}, \end{cases} \qquad (6.3)$$

*where $k \in \mathbb{Z}$, $\omega$ is the nil-root, and $r$ is the order of the diagram automorphism*[1].     □

*Remark 6.5.* 1) There are three types of affine Lie algebras: Aff1, Aff2, Aff3
identified by $r = 1, 2, 3$, see [Kac93, Ch. 4.8]. Diagrams Aff1 are obtained by
adding the vertex $\alpha_0$. Diagrams Aff2 and Aff3 correspond to the so-called
*twisted affine Lie algebras*, see Remark 4.4.
   2) For $r = 1$ in (6.3), we have the diagrams Aff1:

$$A_1^{(1)}, B_n^{(1)}, C_n^{(1)} D_n^{(1)}, E_n^{(1)}, F_4^{(1)}, G_2^{(1)}. \qquad (6.4)$$

   The cases $r = 2, 3$ (if the diagram is not $A_{2l}^{(2)}$) correspond to the diagrams
Aff2, Aff3 as follows:

$$\begin{aligned} r = 2 &: A_{2n-1}^{(2)}, D_{n+1}^{(2)}, E_6^{(2)}; \\ r = 3 &: D_4^{(3)}; \end{aligned} \qquad (6.5)$$

---

[1] The order $r$ in the notation of twisted affine Lie algebras has an invariant sense,
see Table 4.2, Remark 4.4 and Proposition 4.5.

The last case in (6.3) contains only the diagram $A_{2l}^{(2)}$ of the class Aff2, see Table 4.2.

3) For the case Aff1, from eq. (6.4) we have

$$\alpha \in \Delta \Longleftrightarrow \omega - \alpha \in \Delta$$

It is not correct for $\alpha \in \Delta_l$ in the cases Aff2 and Aff3. In particular, for the case Aff1, if $\beta$ is the highest root, the vector $\alpha_0 = \omega - \beta$ also is a root.     $\square$

The affine Weyl group $W_a$ can be represented in two different forms: by means of the coroot lattice or by means of the root lattice, see [Hu97, §12]. Consider these forms more closely.

(I) The affine Weyl group $W_a$ is the semidirect product of $W$ by the group of translations relative to the *coroot lattice*. This is Bourbaki's approach, see [Bo, Ch.6, §2.1]. Here, reflections $s_{\alpha,k}$ are given as follows:

$$s_{\alpha,k} := z - ((z,\alpha) - k)\alpha^{\vee}, \quad \alpha \in \Delta_+$$

where the *coroot* $\alpha^{\vee}$ is as follows:

$$\alpha^{\vee} = \frac{2\alpha}{(\alpha,\alpha)}.$$

The reflection $s_{\alpha,k}$ fixes the hyperplane

$$H_{\alpha,k} = \{z \mid (z,\alpha) = k\}.$$

According to Bourbaki's definition [Bo, Ch.6, §2.1], the affine Weyl group $W_a$ is generated by reflections $s_{\alpha,k}$, where $\alpha \in \Delta_+$, $k \in \mathbb{Z}$.

(II) The affine Weyl group $W_a$ is also the Coxeter group, and thus is also represented as the group of linear transformations generated by reflections in central hyperplanes

$$H_{\alpha} = \{z \mid (z,\alpha) = 0\},$$

where $\alpha$ runs over roots given by (6.3). The affine Weyl group $W_a$ is the semidirect product of $W$ with the group of translations relative to the *root lattice*. Here, the reflections $s_{\alpha+k\omega}$ are given as follows:

$$s_{\alpha+k\omega} := z - \frac{2(z, \alpha + k\omega)}{(\alpha + k\omega, \alpha + k\omega)}(\alpha + k\omega),$$

i.e.,

$$s_{\alpha+k\omega}(z) = z - \frac{2(z,\alpha)}{(\alpha,\alpha)}(\alpha + k\omega) = s_{\alpha}(z) - k\frac{2(z,\alpha)}{(\alpha,\alpha)}\omega.$$

or

$$s_{-\alpha+k\omega}(z) = s_{\alpha}(z) + k\frac{2(z,\alpha)}{(\alpha,\alpha)}\omega,$$

where $\alpha \in \Delta_+$, $k \in \mathbb{Z}$. For equivalence of these two forms, see [Wa98, Ch.3, §8], [Kir05].

## 6.1.3 The translation subgroup

In the extended Dynkin diagram case, the Weyl group $W_a$ is generated by orthogonal reflections $s_i$, where $i = 0, 1, 2, \ldots, n$. Let

$$\beta = \omega - \alpha_0, \tag{6.6}$$

where $\omega$ is the nil-root, i.e., the vector from the one-dimensional kernel of the Tits form $\mathcal{B}$, see (2.26), and $\alpha_0$ is the root corresponding to the additional (affine) vertex 0, the one that extends the Dynkin diagram to the extended Dynkin diagram.

*Remark 6.6.* According to Remark 6.5, heading 3), $\beta$ and $\alpha_0$ are roots for the case Aff1. In order to study the spectra of the affine Coxeter transformation, it suffices to consider only the Aff1 diagrams because extended Dynkin diagrams from the lists Aff2 and Aff3 are dual to some diagrams from the list Aff1:

$$D_{n+1}^{(2)} \text{ is dual to } C_n^{(1)},$$
$$D_4^{(3)} \text{ is dual to } G_2^{(1)},$$
$$E_6^{(2)} \text{ is dual to } F_4^{(1)},$$

see (6.4), (6.5), Table 4.2, and the spectra of the Coxeter transformations for dual diagrams coincide, see (3.7), (3.10).

From now on, we consider only the Aff1 diagrams, see (6.4).  □

Note, that the nil-root $\omega$ coincides with the fixed point $z^1$ of the Coxeter transformation, see §2.2.1.

For any vector $z \in V$, let $t_\lambda$ be the translation

$$t_\lambda(z) = z - 2 \frac{(\lambda, z)}{(\lambda, \lambda)} \omega. \tag{6.7}$$

**Proposition 6.7.** *The translations $t_\lambda$ generate a group of translations, and the following properties hold:*

*1) For roots $\alpha$ and $\gamma$ such that $\alpha + \gamma = \omega$, we have*

$$t_\alpha = t_{-\gamma}.$$

*2) The translation $t_\omega$ is the 0-translation:*

$$t_\omega(z) = z. \tag{6.8}$$

*Proof.* 1) Indeed,

$$t_\alpha(z) = z - 2 \frac{(\alpha, z)}{(\alpha, \alpha)} \omega = z - 2 \frac{(-\gamma, z)}{(\gamma, \gamma)} \omega = t_{-\gamma}(z).$$

2) Eq. (6.8) follows from eq. (6.7) since the nil-root $\omega$ is the kernel of the bilinear form $(\cdot, \cdot)$, see (2.26) and Proposition 3.2.  □

**Proposition 6.8.** *1) In the cases of the Dynkin diagram or the extended Dynkin diagram, every real root is conjugate under $W$ to a simple root.*

*2) The reflection $s_\beta$ given by*

$$s_\beta(z) = z - 2\frac{(\beta, z)}{(\beta, \beta)}\beta,$$

*belongs to $W_a$.*

*3) The element $t_\beta = t_{-\alpha_0}$ is as follows*

$$t_\beta = s_{\alpha_0}s_\beta, \ \text{and} \ s_{\alpha_0} = t_\beta s_\beta, \tag{6.9}$$

*and the element $t_{-\beta} = t_{\alpha_0}$ is as follows*

$$t_{-\beta} = s_\beta s_{\alpha_0}, \ \text{and} \ s_{\alpha_0} = s_\beta t_{-\beta}. \tag{6.10}$$

*Proof.* 1) It is proved by induction on *height* $ht(\alpha)$ (the sum of coordinates of the real root $\alpha$). We may assume $\alpha > 0$. If $ht(\alpha) = 1$, then $\alpha$ is a simple root and all is done. Since $\alpha$ is the real root, i.e., $(\alpha, \alpha) > 0$, it follows that $(\alpha, \alpha_i) > 0$ for some $i$. Then $s_i(z) = z - (z, \alpha_i)\alpha_i$ is another positive root of strictly smaller height.

2) Since, $\beta = w\alpha_i$ for some simple root $\alpha_i$ and some $w \in W_a$, we have

$$s_{w\alpha_i}(z) = z - 2\frac{(w\alpha_i, z)}{(w\alpha_i, w\alpha_i)}w\alpha_i = z - 2\frac{(\alpha_i, w^{-1}z)}{(\alpha_i, \alpha_i)}w\alpha_i =$$

$$w(w^{-1}z - 2\frac{(\alpha_i, w^{-1}z)}{(\alpha_i, \alpha_i)}\alpha_i) = ws_{\alpha_i}(w^{-1}z) = ws_{\alpha_i}w^{-1}(z).$$

So,

$$s_\beta = s_{w\alpha_i} = ws_{\alpha_i}w^{-1} \in W_a.$$

3) We have

$$s_{\alpha_0}(z) = z - 2\frac{(\alpha_0, z)}{(\alpha_0, \alpha_0)}\alpha_0, \ \text{and} \ s_\beta(z) = z - 2\frac{(\beta, z)}{(\beta, \beta)}\beta.$$

Further,

$$s_{\alpha_0}s_\beta(z) = z - 2\frac{(\beta, z)}{(\beta, \beta)}\beta - 2\frac{(\alpha_0, z)}{(\alpha_0, \alpha_0)}\alpha_0 + 4\frac{(\beta, z)(\beta, \alpha_0)}{(\alpha_0, \alpha_0)(\beta, \beta)}\alpha_0. \tag{6.11}$$

Since $\beta = w - \alpha_0$ and $(\beta, \beta) = (\alpha_0, \alpha_0)$, it follows from (6.11) that

$$s_{\alpha_0}s_\beta(z) = z + 2\frac{(\alpha_0, z)}{(\alpha_0, \alpha_0)}(w - \alpha_0) - 2\frac{(\alpha_0, z)}{(\alpha_0, \alpha_0)}\alpha_0 + 4\frac{(\alpha_0, z)}{(\alpha_0, \alpha_0)}\alpha_0 =$$

$$z + 2\frac{(\alpha_0, z)}{(\alpha_0, \alpha_0)}w = t_{-\alpha_0}(z) = t_\beta(z). \tag{6.12}$$

Similarly,

$$s_\beta s_{\alpha_0}(z) = z - 2\frac{(\alpha_0, z)}{(\alpha_0, \alpha_0)}\alpha_0 - 2\frac{(\beta, z)}{(\beta, \beta)}\beta + 4\frac{(\beta, \alpha_0)(\alpha_0, z)}{(\alpha_0, \alpha_0)(\alpha_0, \alpha_0)}\beta =$$

$$z - 2\frac{(\alpha_0, z)}{(\alpha_0, \alpha_0)}\alpha_0 + 2\frac{(\alpha_0, z)}{(\alpha_0, \alpha_0)}\beta - 4\frac{(\alpha_0, z)}{(\alpha_0, \alpha_0)}\beta =$$

$$z - 2\frac{(\alpha_0, z)}{(\alpha_0, \alpha_0)}w = t_{\alpha_0}(z) = t_{-\beta}(z). \quad \square$$

Let $T$ be the *translations subgroup* of $W_a$ generated by the translations $t_\alpha$, where $\alpha$ runs over all roots.

**Proposition 6.9.** *The subgroup $T \subset W_a$ is normal:*

$$T \triangleleft W_a$$

*Proof.* We have

$$wt_\alpha w^{-1}(z) = wt_\alpha(w^{-1}z) =$$

$$w(w^{-1}z - 2\frac{(\alpha, w^{-1}z)}{(\alpha, \alpha)}w) = z - 2\frac{(\alpha, w^{-1}z)}{(\alpha, \alpha)}w\omega =$$

$$z - 2\frac{(w\alpha, z)}{(w\alpha, w\alpha)}w = t_{w\alpha},$$

i.e., for every $w \in W \subset W_a$, we have

$$wt_\alpha w^{-1} = t_{w\alpha}. \quad \square \tag{6.13}$$

Recall that, according to Remark 6.6, we consider only the Aff1 diagrams, see (6.4).

**Proposition 6.10.** *The affine Weyl group $W_a$ is the semidirect product of the Weyl group $W$ and the translation group $T$.*

*Proof.* 1) First, consider the case where all roots have the same length, i.e., all roots are long. By (6.9) we have $t_\beta \in W_a$. By (6.13), $t_\alpha \in W_a$ for all roots of the root system corresponding to $W$, because

$$\alpha = w\beta \text{ and } t_\alpha = t_{w\beta} = wt_\beta w^{-1} \in W_a.$$

Let $R$ be the $\mathbb{Z}$-span of the root system $\Delta$, i.e.,

$$R = \mathbb{Z}\alpha_0 \oplus \mathbb{Z}\alpha_1 \oplus \cdots \oplus \mathbb{Z}\alpha_n.$$

Since, in the simply-laced case,

$$t_\alpha t_\beta = t_{\alpha+\beta},$$

we see that $t_\alpha \in W_a$ for every $\alpha \in R$. Thus, $T \subset W_a$. By Proposition 6.9 $T$ is a normal subgroup in $W_a$. Every generator $s_i \neq s_{\alpha_0}$ belongs to $W$. By

(6.9), the element $s_{\alpha_0}$ is expressed in terms of $t_\beta \in T$ and $s_\beta \in W$, so $W_a$ is generated by $T$ and $W$, and $T \cap W = \{1\}$. By (6.1) the proposition is proved for the case without short roots.

2) Let there be long and short roots. For every long root $\alpha$, we have $\alpha = w\beta$ for some $w \in W_a$. As above, $t_\alpha \in W_a$.

If $\alpha$ is a short root, then by Proposition 6.4 we see that $w - \alpha$ is also a root for this root system. (Note, that for the long root, this is not so.) Then, as in (6.11) and (6.12) we have

$$t_\alpha = s_\alpha s_{w-\alpha} \in W_a.$$

and we finish the proof as in 1). $\square$

### 6.1.4 The affine Coxeter transformation

The Coxeter transformation $\mathbf{C}_a$ associated with an extended Dynkin diagram is called the *affine Coxeter transformation*, see §2.2.6. We add the word "affine" because the affine Coxeter transformation is an element of the affine Weyl group, see §6.1.2. In the orientation where $\alpha_0$ is source-admissible (or sink-admissible), see §2.2.6, we have

$$\mathbf{C}_a = s_{\alpha_0} s_{\alpha_1} ... s_{\alpha_n} = s_{\alpha_0} \mathbf{C},$$

where $\mathbf{C}$ is the Coxeter transformation of the corresponding Dynkin diagram.

Let $S = S_1 \coprod S_2$ be a bicolored partition of the vertices of the Dynkin diagram $\Gamma$ and let $w_1$ (resp. $w_2$) be the product of reflections corresponding to vertices $S_1$ (resp. $S_2$), see §3.1.1. We choose the notation in such a way that the highest root $\beta$ is orthogonal to the roots of $S_2$, so

$$w_2\beta = \beta, \tag{6.14}$$

see (5.58), (5.59) for the case $E_6$, §5.5. Let $\Pi = \Pi_1 \coprod \Pi_2$ be corresponding partition of simple roots.

Consider the decomposition of the Coxeter transformation

$$\mathbf{C} = w_2 w_1,$$

corresponding to a bicolored partition of the Dynkin diagram. Then

$$\mathbf{C}_a = s_{\alpha_0} w_2 w_1 \tag{6.15}$$

is the affine Coxeter transformation. Consider also

$$\mathbf{C}' = s_\beta w_2 w_1 \tag{6.16}$$

called the *linear part* of the affine Coxeter transformation[1], see [Stb85, p. 595].

---

[1] $\alpha_0$ is the root corresponding to the affine vertex, $\beta$ is the highest root in the root system, see (6.6).

**Proposition 6.11. [Stb85, p. 595]** *1) The affine Coxeter transformation* $\mathbf{C}_a$ *and the linear part of the affine Coxeter transformation* $\mathbf{C}'$ *are connected by a translation[1]* $t_{\alpha_0}$ *as follows:*

$$\mathbf{C}' = t_{\alpha_0} \mathbf{C}_a. \tag{6.17}$$

*2) Let* $W_a$ *be the affine Weyl group that acts on the linear space* $V$, *and let* $V' \subset V$ *be the hyperplane of vectors orthogonal to* $z^1$. *The spectrum of the affine Coxeter transformation* $\mathbf{C}_a$ *with deleted eigenvalue 1 coincides with the spectrum of the linear part of the affine Coxeter transformation* $\mathbf{C}'$ *with restricted action on the* $V'$.

*Proof.* 1) Eq. (6.17) follows from eq. (6.15), (6.16) and (6.10).

2) Let $z_i$ be the eigenvector with eigenvalue $\lambda_i$ of the affine Coxeter transformation $\mathbf{C}_a$:

$$\mathbf{C}_a z_i = \lambda_i z_i.$$

By (6.17) and (6.7) we have

$$\mathbf{C}' z_i = t_{\alpha_0} \mathbf{C}_a(z_i) = t_{\alpha_0}(\lambda_i z_i) = \lambda_i z_i - 2\frac{(\lambda_i z_i, \alpha_0)}{(\alpha_0, \alpha_0)} z^1.$$

On the vector space $V'$ the coefficient of $z^1$ vanishes. $\square$

Thanks to Proposition 6.11, 2), instead of the spectrum of the affine Coxeter transformation $\mathbf{C}_a$, the spectrum of the linear part $\mathbf{C}'$ can be calculated. We will do this in §6.2.

For more details of affine root systems and affine Weyl groups, see [BS06], [BS06a], [Max98], [Stm04], [Cip00], [Br03].

## 6.2 R. Steinberg's theorem again

In this section we restore a remarkable proof of R. Steinberg's theorem (Th. 5.1), [Stb85]. The R. Steinberg trick is to get, instead of the affine Coxeter transformation, the product of reflections without the reflection corresponding to the branch point, see Proposition 6.27. For this purpose, instead of the affine Coxeter transformation, R. Steinberg considers the *linear part* of the affine Coxeter transformation having the same spectrum, Proposition 6.11. The passage from the highest root to the root corresponding to the branch point is executed by means of the alternating products $\tau^{(n)}$ (Propositions 6.17, 6.21). We use the alternating products $\tau^{(n)}$ also for a description of orbits of the Coxeter transformation, generating functions, and Poincaré series, see §5.4.1.

---

[1] See (6.7) and Proposition 6.7.

## 6.2.1 The element of the maximal length in the Weyl group

The *length* of an element $w \in W$, denoted by $l(w)$, is the minimal $k$ such that $w$ can be written as a product of $k$ generators.

**Proposition 6.12 ([Stb59]).** *1) In the Weyl group $W$, there is only one element $w_0$ of the maximal length.*
   *2) The element $w_0$ makes all positive roots negative:*

$$w_0 \alpha < 0 \quad \text{for all} \quad \alpha \in \Delta_+, \tag{6.18}$$

*see §2.2.1.*
   *3) The length $l(w_0)$ coincides with the number of positive roots $|\Delta_+|$.*
   *4) The element $w_0$ is an involution:*

$$w_0^2 = 1.$$

This classical result is due to R. Steinberg, the proof can be found in [Bo, Ch.6, §1.6 , Corol. 3] or [Stb59].   □

*Remark 6.13.* Note that the Coxeter number $h$ is even for every Dynkin diagram except for $A_{2n+1}$, see Table 4.2. Following R. Steinberg [Stb85, §2] we consider in what follows only the case of even Coxeter number: $h = 2g$. Then

$$w_0 = \mathbf{C}^g.$$

We use also alternating products $\tau^{(n)}$, see (5.60):

$$
\begin{aligned}
\tau^{(1)} &= w_1, \\
\tau^{(2)} &= w_2 w_1, \\
\tau^{(3)} &= w_1 w_2 w_1, \\
&\ldots \\
\tau^{(2n)} &= (w_2 w_1)^n, \\
\tau^{(2n+1)} &= w_1 (w_2 w_1)^n,
\end{aligned}
\tag{6.19}
$$

$$\ldots$$

The following relation between the length function $l(w)$ in the Weyl group and actions of Weyl group holds:

**Proposition 6.14.** *1) The lengths of elements $w$ and $w^{-1}$ coincide:*

$$l(w) = l(w^{-1}). \tag{6.20}$$

*2) Let $w \in W$ and let $\alpha$ be a simple root. Then*

$$l(ws_\alpha) > l(w) \quad \text{if and only if} \quad w(\alpha) > 0. \tag{6.21}$$

For the proof, see [Hu90, Ch. 5.2 (L1), Th. 5.4 and Prop. 5.7].   □
    From (6.20), (6.21) we have

$$l(s_\alpha w) = l(w^{-1}s_\alpha) > l(w^{-1}) = l(w) \quad \text{if and only if} \quad w^{-1}(\alpha) > 0. \quad (6.22)$$

*Remark 6.15.* We need also the following property of the highest root $\beta$. For any positive root $\alpha$, we have

$$(\beta, \alpha) > 0. \quad (6.23)$$

Indeed, if $\alpha_i$ is a simple positive root with $(\beta, \alpha_i) < 0$, then

$$s_\alpha(\beta) = \beta - 2\frac{(\beta, \alpha_i)}{(\alpha_i, \alpha_i)}\alpha > \beta.$$

This contradicts to the maximality of $\beta$. Thus, $(\beta, \alpha_i) > 0$ for every simple positive root $\alpha_i$. Since $\alpha = n_1\alpha_1 + n_2\alpha_2 + \cdots + n_l\alpha_l$ with $n_i \geq 0$, the relation (6.23) holds.

    The following proposition establishes a connection between the partial ordering in the Weyl group and the partial ordering of the root system.

**Proposition 6.16. [Stb85, §2]** *1) We have*

$$1 < l(\tau^{(1)}) < l(\tau^{(2)}) < \cdots < l(\tau^{(2g)}) = l(w_0). \quad (6.24)$$

*2) For the highest root $\beta$, we have*

$$\beta \geq \tau^{(1)}\beta \geq \tau^{(2)}\beta \geq \cdots \geq \tau^{(2g)}\beta. \quad (6.25)$$

*3) For the alternating products $\tau^{(i)}$, where $0 \leq i < g$, we have*

$$\tau^{(i)}\beta = -\tau^{(2g-i-1)}\beta. \quad (6.26)$$

*Proof.* 1) Let $|\Pi|$ be the number of simple roots (equal to the rank of the Cartan subalgebra and to the number of vertices $|\Gamma_0|$ in the Dynkin diagram). Consider the decomposition of $w_0$ as a product of $w_1$, $w_2$:

$$w_0 = \tau^{(2g)} = \underbrace{w_2w_1 \ldots w_2w_1}_{g \text{ pairs}} \quad (6.27)$$

Calculate the number of reflections entering the decomposition of $w_0$:

$$l(w_0) = l(\tau^{(2g)}) = gl(w_2w_1) = g|\Pi|.$$

By (1.1) we have

$$hl = h|\Pi| = |\Delta|,$$

and

$$l(w_0) = g|\Pi| = \frac{h}{2}|\Pi| = |\Delta_+|,$$

i.e., the length of $w_0$ coincides with the number of positive simple roots. Therefore, by Proposition 6.12, 3), the decomposition (6.27) is minimal, and the decomposition of every segment $\tau^{(i)}$, where $1 \leq i \leq 2g$, is also minimal, and (6.24) holds.

2) It suffices, by the induction, to prove that

$$l(s_\alpha w) > l(w) \Longrightarrow s_\alpha w\beta \leq w\beta, \qquad (6.28)$$

where $s_\alpha$ is the reflection corresponding to some simple root $\alpha$. We have

$$s_\alpha w\beta = w\beta - 2\frac{(w\beta, \alpha)}{(\alpha, \alpha)}\alpha = w\beta - 2\frac{(\beta, w^{-1}\alpha)}{(\alpha, \alpha)}\alpha.$$

By (6.22) we have $w^{-1}\alpha > 0$. Then by (6.23), we have $(\beta, w^{-1}\alpha) > 0$ and (6.28) holds.

3) Consider

$$\tau^{(i)}\beta = w_i \ldots w_2 w_1 \beta, \text{ where } w_i \in \{w_2, w_1\},$$

By (6.27) and (6.19) we have

$$\tau^{(i)}\beta = (w_{i+1} \ldots w_{2g})(w_{2g} \ldots w_{i+1})w_i \ldots w_2 w_1 \beta,$$
$$(w_{i+1} \ldots w_{2g})w_0 \beta = -w_{i+1} \ldots w_{2g}\beta.$$

Since $w_{2g} = w_2$, by (6.14) we have

$$\tau^{(i)}\beta = -\underbrace{w_{i+1} \ldots w_{2g-1}}_{2g - i - 1 \text{ factors}}\beta =$$

$$- w_{2g-i-1} \ldots w_1 \beta = -\tau^{(2g-i-1)}\beta. \qquad \Box$$

## 6.2.2 The highest root and the branch point

The vertex $b$ of the Dynkin diagram connected to another vertex $\tilde{b}$ by means of a weighted edge is called a *point with a multiple bond*. In other words, if $b$ is the vertex corresponding to short (resp. long) root, then the vertex $\tilde{b}$ corresponds to long (resp. short) root. The corresponding simple root is called the *root with a multiple bond*. We say also, that the vertex $b$ is connected to vertex $\tilde{b}$ by a *multiple bond*.

In this section we prove that $\tau^{(g-1)}\beta$ is the simple root corresponding to the branch point of the Dynkin diagram or a multiple bond point.

**Proposition 6.17. [Stb85, §2,(5)]** *The element* $b = \tau^{(g-1)}\beta$ *is a simple root,* $b \in \Pi_g$.

*Proof.* First, by (6.26), we have

$$\tau^{(g)}\beta = -\tau^{(g-1)}\beta,$$

and

$$\tau^{(g-1)}\beta = (\tau^{(g-1)}\beta - \tau^{(g)}\beta)/2.$$

By (6.25) $b = \tau^{(g-1)}\beta > 0$, so $b$ is a positive root. Since

$$\tau^{(g)}\beta = w_g \tau^{(g-1)}\beta = -\tau^{(g-1)}\beta,$$

we see that $\tau^{(g-1)}\beta$ is a simple root lying in $\Pi_g$. $\quad\square$

The hyperplanes

$$H_\alpha = \{x \in V \mid (x,\alpha) = 0\}, \quad \text{where } \alpha \in \Pi$$

divide the space $V$ into regions called *Weyl chambers*, and the chamber

$$\mathfrak{C} = \{x \in V \mid (x,\alpha) > 0 \text{ for all } \alpha \in \Pi\}$$

is called the *fundamental Weyl chamber*.

**Proposition 6.18.** *The Weyl group permutes the Weyl chambers, and this action is simply transitive*[1], *see*

For the proof, see [Hu78, §10.3] or [Bo, Ch.6, §1.5, Th.2] $\quad\square$

**Proposition 6.19.** *The highest root $\beta$ is long, i.e.,*

$$(\beta,\beta) \geq (\alpha,\alpha) \text{ for all } \alpha \in \Delta.$$

*Proof.* Since the length $(\alpha,\alpha)$ is fixed under the action of the Weyl group, according to Proposition 6.18 the root $\alpha$ can be chosen from the fundamental Weyl chamber $\mathfrak{C}$. The inequality

$$(\beta - \alpha, \beta) \geq 0 \tag{6.29}$$

holds because the highest root $\beta \in \mathfrak{C}$, see (6.23), and the inequality

$$(\beta - \alpha, \alpha) \geq 0 \tag{6.30}$$

holds because $\alpha \in \mathfrak{C}$.

By (6.29) and (6.30) we have

$$(\beta,\beta) \geq (\beta,\alpha) \geq (\alpha,\alpha). \quad\square$$

**Proposition 6.20.** [Stb85, §2,(5)] *Let $\alpha_0 = z^1 - \beta$ be the root corresponding to the affine vertex, let $\alpha_0$ be connected to just one simple root $\alpha$. Let $\alpha$ be of the same length as $\beta$. Then the following formula holds:*

$$(\tau^{(g)}\beta, \tau^{(g+1)}\beta) = -\frac{1}{2}|\beta|^2. \tag{6.31}$$

---

[1] The action of a given group $G$ on a given set $X$ is called *simply transitive* if for any two elements $x, y \in X$, there exists precisely one $g$ in $G$ such that $gx = y$.

*Proof.* By (6.19) we have

$$(\tau^{(g)}\beta, \tau^{(g+1)}\beta) = (w_g\tau^{(g)}\beta, w_g\tau^{(g+1)}\beta) = (\tau^{(g-1)}\beta, \tau^{(g+2)}\beta) =$$
$$(w_{g-1}\tau^{(g-1)}\beta, w_{g-1}\tau^{(g+2)}\beta) = (\tau^{(g-2)}\beta, \tau^{(g+3)}\beta) = \dots$$
$$(\beta, \tau^{(2g+1)}\beta).$$

So, by (6.27) and (6.18)

$$(\tau^{(g)}\beta, \tau^{(g+1)}\beta) = (\beta, \tau^{(2g+1)}\beta) = (\beta, w_1\tau^{(2g)}\beta) = -(\beta, w_1\beta). \qquad (6.32)$$

Since $\alpha_0 = z^1 - \beta$ corresponds to the affine vertex, we have

$$(\alpha_0, \alpha_i) = 0 \text{ for each simple root } \alpha_i \neq \alpha,$$
$$(\alpha_0, \alpha) \neq 0.$$

The same holds for $\beta$:

$$(\beta, \alpha_i) = 0 \text{ for each simple root } \alpha_i \neq \alpha,$$
$$(\beta, \alpha) \neq 0.$$

So, $\beta$ is not orthogonal only to $\alpha$, and

$$w_1\beta = s_\alpha\beta,$$

and (6.32) is equivalent to

$$(\tau^{(g)}\beta, \tau^{(g+1)}\beta) = -(\beta, s_\alpha\beta).$$

Further, $\alpha_0$ and $\alpha$ form an angle of $120°$. Since

$$(\alpha_0, \alpha) = (z^1 - \beta, \alpha) = -(\beta, \alpha),$$

the roots $\beta$ and $\alpha$ form an angle of $60°$, and

$$(\beta, \alpha) = \frac{1}{2}|\beta|^2 = \frac{1}{2}|\alpha|^2. \qquad (6.33)$$

By (6.33) we have

$$(\beta, s_\alpha\beta) = (\beta, \beta - 2\frac{(\alpha, \beta)}{(\alpha, \alpha)}\alpha) = (\beta, \beta) - 2\frac{(\alpha, \beta)^2}{(\alpha, \alpha)} =$$
$$|\beta|^2 - 2 \cdot \frac{1}{4}|\beta|^2 = \frac{1}{2}|\beta|^2,$$

and by (6.32)

$$(\tau^{(g)}\beta, \tau^{(g+1)}\beta) = -\frac{1}{2}|\beta|^2. \qquad \square$$

**Proposition 6.21.** [Stb85, §2,(5)] *The element $b = \tau^{(g-1)}\beta$ is the long simple root at which there is a branch point or a point with a multiple bond.*

*Proof.* By Proposition 6.17 the element $b$ is the simple root. By Proposition 6.19 the highest root $\beta$ is long, so is $b$. Further, since $A_n$ is excluded, the root $\beta$ is connected to just one simple root $\alpha$. Consider the two cases:

(1) The root $\alpha$ is shorter that $\beta$. It means that, in the extended Dynkin diagram, the affine root $\alpha_0 = z^1 - \beta$ is connected to $\alpha$ by a multiple bond because

$$(\alpha_0, \alpha_0) = (\beta, \beta) > (\alpha, \alpha).$$

It happens only in the case $\widetilde{C}_n$, see [Bo, Table III]. In this case there is only one simple long root $\alpha_l$, so the simple long root $b$ coincides with $\alpha_l$.

(2) The root $\alpha$ has the same length as $\beta$. Then, by (6.31) we have:

$$|\tau^{(g)}\beta - \tau^{(g+1)}\beta| =$$
$$|\tau^{(g)}\beta|^2 + |\tau^{(g+1)}\beta)|^2 - 2(\tau^{(g)}\beta, \tau^{(g+1)}\beta) = 3|\beta|^2.$$

On the other hand, since $b = \tau^{(g-1)}\beta$, we have

$$\tau^{(g)}\beta - \tau^{(g+1)}\beta =$$
$$(1 - w_{g+1})w_g\beta = (1 - w_{g+1})w_g\tau^{(g-1)}\beta =$$
$$(1 - w_{g+1})w_g(b) = (1 - w_{g+1})(-b) =$$
$$w_{g+1}b - b = \sum_{\gamma \in \Delta_{g+1}} -\frac{2(b,\gamma)}{(\gamma,\gamma)}\gamma,$$

where the sum runs over the neighbors of $b$.

(2a) If $b$ has the same length as all its neighbors, then we have

$$-\frac{2(b,\gamma)}{(\gamma,\gamma)} = 1,$$

and

$$w_{g+1}b - b = \sum_{\gamma \in \Delta_{g+1}} \gamma,$$

and by case (2) $b$ is the branch point.

(2b) Otherwise, $b$ is connected to the short root with a multiple bond.  □

### 6.2.3 The orbit of the highest root. Examples

We consider three examples of orbits of the highest root: $E_6$, $F_4$, $C_4$. These cases correspond to cases (2a), (2b), (1) of Proposition 6.21, respectively.

*Example 6.22.* Case $E_6$. All roots have the same length. The vectors $\tau^{(i)}\beta$ are given in Table 5.3, §5.5.1. Then $h = 12$, $g = \frac{h}{2} = 6$, $g - 1 = 5$.

$$\beta = \begin{matrix} 1\ 2\ 3\ 2\ 1 \\ 2 \end{matrix}, \qquad \tau^{(5)}\beta = \begin{matrix} 0\ 0\ 1\ 0\ 0 \\ 0 \end{matrix}.$$

This case corresponds to Proposition 6.21, (2a).

*Example 6.23.* Case $F_4$. The roots $a_1$ and $b_1$ are of length 2, the roots $a_2$ and $b_2$ are of length 1, see Fig. 6.1, Table 6.1. This case corresponds to Proposition 6.21, (2b). Here, the long simple root is $b = b_1$ and it is connected to the short root $a_2$. Then $h = 12$, and $g - 1 = 5$.

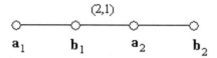

**Fig. 6.1.**    The highest root for the Dynkin diagram $F_4$

$$w_1 = \begin{pmatrix} -1 & 1 & 0 \\ & -1 & 2 & 1 \\ & & 1 \\ & & & 1 \end{pmatrix} \begin{matrix} x_1 \\ x_2 \\ y_1 \\ y_2 \end{matrix}, \qquad w_2 = \begin{pmatrix} 1 \\ & 1 \\ & 1 & 1 & -1 \\ & 0 & 1 & & -1 \end{pmatrix} \begin{matrix} x_1 \\ x_2 \\ y_1 \\ y_2 \end{matrix},$$

$$\beta = \begin{pmatrix} 2 \\ 4 \\ 3 \\ 2 \end{pmatrix} \begin{matrix} x_1 \\ x_2 \\ y_1 \\ y_2 \end{matrix}, \qquad w_1\beta = \begin{pmatrix} 1 \\ 4 \\ 3 \\ 2 \end{pmatrix} \begin{matrix} x_1 \\ x_2 \\ y_1 \\ y_2 \end{matrix}, \qquad w_2\beta = \begin{pmatrix} 2 \\ 4 \\ 3 \\ 2 \end{pmatrix} \begin{matrix} x_1 \\ x_2 \\ y_1 \\ y_2 \end{matrix} = \beta.$$

*Example 6.24.* Case $C_4$. The roots $a_1, b_1$ and $a_2$ are of length 1, the root $b_2$ is of length 2, see Fig. 6.2, Table 6.2. This case corresponds to Proposition 6.21, (1). Here, the long simple root $b = b_2$ and it is connected to the short root $a_2$. Then $h = 2 \cdot 4 = 8$, $g - 1 = 3$.

**Table 6.1.**    The orbit of the highest root for $F_4$

| $\beta$ | $\tau^{(1)}\beta = w_1\beta$ | $\tau^{(2)}\beta = \mathbf{C}\beta$ |
|---|---|---|
| 2 4 3 2 | 1 4 3 2 | 1 4 2 2 |
| $\tau^{(3)}\beta = w_1\mathbf{C}\beta$ | $\tau^{(4)}\beta = \mathbf{C}^2\beta$ | $\tau^{(5)}\beta = w_1\mathbf{C}^2\beta$ |
| 1 2 2 2 | 1 2 1 0 | 0 0 1 0 |
| $\tau^{(6)}\beta = \mathbf{C}^3\beta$ | $\tau^{(7)}\beta = w_1\mathbf{C}^3\beta$ | $\tau^{(8)}\beta = \mathbf{C}^4\beta$ |
| 0 0 −1 0 | −1 −2 −1 0 | −1 −2 −2 −2 |
| $\tau^{(9)}\beta = w_1\mathbf{C}^4\beta$ | $\tau^{(10)}\beta = \mathbf{C}^5\beta$ | $\tau^{(11)}\beta = w_1\mathbf{C}^5\beta$ |
| −1 −4 −2 −2 | −1 −4 −3 −2 | −2 −4 −3 −2 |

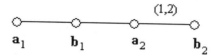

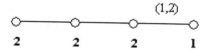

**Fig. 6.2.**    The highest root for the Dynkin diagram $C_4$

$$w_1 = \begin{pmatrix} -1 & 1 & 0 & \\ -1 & 1 & 2 & \\ & & 1 & \\ & & & 1 \end{pmatrix} \begin{matrix} x_1 \\ x_2 \\ y_1 \\ y_2 \end{matrix}, \qquad w_2 = \begin{pmatrix} 1 & & & \\ & 1 & & \\ 1 & 1 & -1 & \\ 0 & 1 & & -1 \end{pmatrix} \begin{matrix} x_1 \\ x_2 \\ y_1 \\ y_2 \end{matrix},$$

$$\beta = \begin{pmatrix} 2 \\ 2 \\ 2 \\ 1 \end{pmatrix} \begin{matrix} x_1 \\ x_2 \\ y_1 \\ y_2 \end{matrix}, \qquad w_1\beta = \begin{pmatrix} 0 \\ 2 \\ 2 \\ 1 \end{pmatrix} \begin{matrix} x_1 \\ x_2 \\ y_1 \\ y_2 \end{matrix}, \qquad w_2\beta = \begin{pmatrix} 2 \\ 2 \\ 2 \\ 1 \end{pmatrix} \begin{matrix} x_1 \\ x_2 \\ y_1 \\ y_2 \end{matrix} = \beta.$$

## 6.2.4 The linear part of the affine Coxeter transformation

**Proposition 6.25.** *Two roots $\alpha$ and $\gamma$ of the same length are connected by means of the some element $w \in W$ if and only if reflections $s_\alpha$ and $s_\gamma$ are conjugate in the Weyl group by means of $w$:*

**Table 6.2.**    The orbit of the highest root for $C_4$

| $\beta$ | $\tau^{(1)}\beta = w_1\beta$ | $\tau^{(2)}\beta = \mathbf{C}\beta$ | $\tau^{(3)}\beta = w_1\mathbf{C}\beta$ |
|---|---|---|---|
| 2 2 2 1 | 0 2 2 1 | 0 2 0 1 | 0 0 0 1 |
| $\tau^{(4)}\beta = \mathbf{C}\beta$ | $\tau^{(5)}\beta = w_1\mathbf{C}^2\beta$ | $\tau^{(6)}\beta = \mathbf{C}^2\beta$ | $\tau^{(7)}\beta = w_1\mathbf{C}^2\beta$ |
| 0 0 0 −1 | 0 −2 0 1 | 0 −2 −2 −1 | −2 −2 −2 −1 |

$$\gamma = w\alpha \iff w s_\alpha w^{-1} = s_\gamma.$$

*Proof.* 1) Let $\gamma = w\alpha$. Then

$$w s_\alpha w^{-1}(x) = w s_\alpha (w^{-1}x) = w(w^{-1}x - 2\frac{(\alpha, w^{-1}x)}{(\alpha, \alpha)}\alpha) =$$

$$x - 2\frac{(\alpha, w^{-1}x)}{(\alpha, \alpha)}w\alpha = x - 2\frac{(w\alpha, x)}{(w\alpha, w\alpha)}w\alpha(x) = s_{w\alpha}.$$

2) Conversely, let $w s_\alpha w^{-1} = s_\gamma$. Then

$$w(w^{-1}x - 2\frac{(\alpha, w^{-1}x)}{(\alpha, \alpha)}\alpha) = x - 2\frac{(\alpha, w^{-1}x)}{(\alpha, \alpha)}w\alpha = x - 2\frac{(\gamma, x)}{(\gamma, \gamma)}\gamma,$$

i.e., $w\alpha$ and $\gamma$ are proportional: $w\alpha = k\gamma$. Since $s_\gamma = s_{-\gamma}$, we can consider $k > 0$. Since $(w\alpha, w\alpha) = (\alpha, \alpha) = (\gamma, \gamma)$, we have $k = 1$.   □

**Corollary 6.26.** *Let $\beta$ be the highest root and let $b$ be the root corresponding to the branch point (or the unique long root $b$ connected to the short root with a multiple bond). Then corresponding reflections $s_\beta$ and $s_b$ are conjugate as follows:*

$$s_b = w s_\beta w^{-1}, \quad \text{where } w = \tau^{(g-1)}. \tag{6.34}$$

The corollary follows from Proposition 6.25 and Proposition 6.21.   □

**Proposition 6.27. [Stb85, p. 595]** *1) The following relation holds*

$$\tau^{(g-1)}w_2w_1\tau^{(g-1)^{-1}} = \begin{cases} w_1w_2 \text{ for } g = 2k \\ w_2w_1 \text{ for } g = 2k - 1. \end{cases} \tag{6.35}$$

*2) The linear part $\mathbf{C}'$ is conjugate to $w_2w_1$ (and also $w_1w_2$) with canceled reflection $s_b$ corresponding to the branch point $b$.*

*Proof.* 1) For the case $g = 2k$:

$$\tau^{(g-1)}w_2w_1\tau^{(g-1)^{-1}} =$$

$$w_1 \underbrace{w_2w_1 \dots w_2w_1}_{k-1 \text{ pairs}}(w_2w_1)w_1 \underbrace{w_2w_1 \dots w_2w_1}_{k-1 \text{ pairs}} = w_1w_2,$$

For the case $g = 2k - 1$:

$$\tau^{(g-1)} w_2 w_1 \tau^{(g-1)^{-1}} =$$

$$\underbrace{w_2 w_1 \ldots w_2 w_1}_{k-1 \text{ pairs}} (w_2 w_1) \underbrace{w_1 w_2 \ldots w_1 w_2}_{k-1 \text{ pairs}} = w_2 w_1.$$

2) By (6.34)

$$\tau^{(g-1)} \mathbf{C}' \tau^{(g-1)^{-1}} = \tau^{(g-1)} s_\beta w_2 w_1 \tau^{(g-1)^{-1}} =$$

$$\tau^{(g-1)} s_\beta \tau^{(g-1)^{-1}} \tau^{(g-1)} w_2 w_1 \tau^{(g-1)^{-1}} =$$

$$s_b \big( \tau^{(g-1)} w_2 w_1 \tau^{(g-1)^{-1}} \big).$$

and by (6.35)

$$\tau^{(g-1)} \mathbf{C}' \tau^{(g-1)^{-1}} = \begin{cases} s_b w_1 w_2 \text{ for } g = 2k, \\ s_b w_2 w_1 \text{ for } g = 2k - 1. \end{cases}$$

If the branch point $b$ belongs to the subset of vertices $S_1$, the reflection $s_b$ cancels with $s_b$ in $w_1$. If the branch point $b$ belongs to the subset of vertices $S_2$ instead of $s_b w_1 w_2$, we consider the conjugate element $w_2 s_b w_1$, and the reflection $s_b$ cancels with $s_b$ in $w_2$.  □

From Proposition 6.27, 2) we see that the spectrum of $\mathbf{C}_a$ can be calculated as the spectrum of the product of three Coxeter transformations of type $A_n$, where $n = p - 1, q - 1, r - 1$ are the lengths of branches of the corresponding Dynkin diagram, see Theorem 5.1.

### 6.2.5 Two generalizations of the branch point

Back again to the point $b$ from Corollary 6.26 and Proposition 6.21. Let the point $b$ of the Dynkin diagram be either a branch point or the node corresponding to the unique long root connected to the short root with multiple bond. We call this point a *Steinberg's generalized branch point*, or simply *generalized branch point*.

*Remark 6.28.* Before, in §5.3.1 we used another concept of generalization of the branch point — non-homogeneous point connected to the folded Dynkin diagrams, see Remark 5.4. This generalization does not coincide with generalization of the branch point by Steinberg. The generalized branch point by Steinberg corresponds to the long root of the weighted edge, and the non-homogeneous branch point from §5.3.1 corresponds to the short root of the weighted edge. This difference does not, however, affect the end result — the generalized Steinberg theorem (Theorem 5.5).

For the cases of $\widetilde{E}_n$, $\widetilde{D}_n$, the corresponding Dynkin diagram satisfy Steinberg's theorem (Theorem 5.1). In these cases the point $b$ is a usual branch point. Consider the remaining cases[1] from the diagrams Aff1:

$$F_4^{(1)}, G_2^{(1)}, B_n^{(1)}, C_n^{(1)}.$$

1) The diagram $F_4^{(1)}$, see Example 6.23. We discard the generalized branch point $b = b_1$ on the Dynkin diagram $F_4$, see Fig. 6.1. Then the characteristic polynomial of the affine Coxeter transformation for $F_4^{(1)}$ (without eigenvalue 1) is[2]

$$\mathcal{X}_2 \mathcal{X}_1,$$

see (5.16).

2) The diagram $G_2^{(1)}$, see Table 4.2. Discard the generalized branch point corresponding to the long root on the Dynkin diagram $G_2$, see Fig. 2.3. The characteristic polynomial of the affine Coxeter transformation for $G_2^{(1)}$ (without eigenvalue 1) is

$$\mathcal{X}_1,$$

see (5.18).

3) The diagram $B_n^{(1)}$, another notation: $\widetilde{CD}_n$, see Table 4.2. Discard the generalized branch point corresponding to the long root on the Dynkin diagram $B_n$, see Fig. 2.3. The characteristic polynomial of the affine Coxeter transformation for $B_n^{(1)}$ (without eigenvalue 1) is

$$\mathcal{X}_{n-2}\mathcal{X}_1,$$

see (5.17).

4) The diagram $C_n^{(1)}$, another notation: $\widetilde{C}_n$, see Table 4.2. Discard the generalized branch point corresponding to the long root on the Dynkin diagram $C_n$, see Fig. 2.3. The characteristic polynomial of the affine Coxeter transformation for $C_n^{(1)}$ (without eigenvalue 1) is

$$\mathcal{X}_{n-1},$$

see (5.19).

## 6.3 The defect

### 6.3.1 The affine Coxeter transformation and defect

There is the important characteristic associated with the affine Coxeter transformation. It is a linear form called *defect*. For the extended Dynkin diagram

---

[1] By Remark 6.6, in order to study the spectra of affine Coxeter transformations to consider only the cases of Aff1, see (6.4).

[2] Recall, that $\mathcal{X}_n$ is the characteristic polynomial corresponding to the Dynkin diagram $A_n$, see Remark 4.13.

it defines the hyperplane of regular representations. For the Dyknin diagrams, this characteristic does not exist, and for indefinite Tits form, the defect defines a cone lying between two planes.

The defect was introduced by Dlab and Ringel [DR76] for the classification of *tame type quivers* in the representation theory of quivers. For the case of the extended Dynkin diagram $\widetilde{D}^4$, the defect $\delta$ (for bicolored orientation) was applied by Gelfand and Ponomarev in [GP72] in the study of quadruples of subspaces.

Dlab and Ringel introduced in [DR76] the *defect* $\delta_\Omega$ for an arbitrary orientation $\Omega$ as a vector from $\mathcal{E}_\Gamma^*$ obtained as a solution of the equation

$$\mathbf{C}_\Omega^* \delta_\Omega = \delta_\Omega. \tag{6.36}$$

Here, $\mathcal{E}_\Gamma^*$ is the dual vector space of all linear forms on $\mathcal{E}_\Gamma$, and $\mathbf{C}_\Omega^*$ means the dual operator:

$$\langle \mathbf{C}_\Omega^* z, x \rangle = \langle z, \mathbf{C}_\Omega x \rangle, \tag{6.37}$$

and in the matrix form $\mathbf{C}_\Omega^*$ is given as the transposed matrix of $\mathbf{C}_\Omega$.

For an arbitrary orientation $\Omega$, another $\Omega$-defect denoted $\rho_\Omega$, was introduced in [SuSt75], [SuSt78]. We will prove below in Proposition 6.35 that the defect $\delta_\Omega$ of Dlab-Ringel and the $\Omega$-defect $\rho_\Omega$ coincide (up to a factor).

Let $\mathbf{C}_\Omega$ be the Coxeter operator corresponding to a given orientation $\Omega$ and let $\mathbf{C}_\Lambda$ be the Coxeter operator corresponding to a bicolored orientation $\Lambda$. Then $\mathbf{C}_\Omega$ and $\mathbf{C}_\Lambda$ are conjugate if the quiver is a tree. Let $T$ be an element in the Weyl group interrelating $\mathbf{C}_\Omega$ and $\mathbf{C}_\Lambda$:

$$\mathbf{C}_\Omega = T^{-1} \mathbf{C}_\Lambda T. \tag{6.38}$$

**Definition 6.29.** The linear form $\rho_\Omega(z)$ defined as follows

$$\rho_\Omega(z) = \begin{cases} \langle Tz, \tilde{z}^1 \rangle & \text{in the simply-laced case,} \\ \langle Tz, \tilde{z}^{1\vee} \rangle & \text{in the multiply-laced case} \end{cases} \tag{6.39}$$

is said to be the $\Omega$-*defect* of the vector $z$. Here, $T$ is any element from (6.38); $\tilde{z}^1$ is the adjoint vector corresponding to the eigenvalue $\lambda = 1$ of the Coxeter transformation, see §3.3.1; $z^{1\vee}$ is the eigenvector corresponding to eigenvalue 1 of the Coxeter transformation for the dual diagram $\Gamma^\vee$. Here $\tilde{v}$ denotes the *conjugate vector*[1] to $v$.

The following remarkable formula is due to V. Dlab and C. M. Ringel, see [DR76]:

$$\mathbf{C}_\Omega^h z = z + h \delta_\Omega(z) z^1, \tag{6.40}$$

where $h$ is the Coxeter number. This formula is proved below in Proposition 6.34.

---

[1] See Remark 3.1 and Definition 6.31.

### 6.3.2 The necessary regularity conditions

The regular representations are the most complicated in the category of all representations of a given quiver. For every Dynkin diagram, the category of regular representations is empty, there is only a finite number of non-regular representations (P. Gabriel's theorem, Th. 2.14). For this reason, in the representation theory of quivers, the Dynkin diagrams are called *finite type quivers*. The regular representations are completely described only for the extended Dynkin diagrams, which, for this reason, were dubbed *tame quivers* in the representation theory of quivers, ([Naz73], [DR76]), see §2.2.3.

**Definition 6.30.** 1) The vector $z \in \mathcal{E}_\Gamma$ is said to be $\Omega$-*regular* or *regular in the orientation* $\Omega$ if

$$\mathbf{C}_\Omega^k z > 0 \text{ for all } k \in \mathbb{Z}. \tag{6.41}$$

2) The representation $V$ is said to be a *regular representation in the orientation* $\Omega$ of the quiver $\Gamma$ if its dimension $\dim V$ is a $\Omega$-regular vector, i.e.,

$$\mathbf{C}_\Omega^k(\dim V) > 0 \text{ for all } k \in \mathbb{Z}. \tag{6.42}$$

For motivation and details of regularity conditions, see Ch.B, §B.1.5.

Let $(\alpha_1, \tilde{\alpha}_1, \alpha_1^{\varphi_2}, \alpha_2^{\varphi_2}, ...)$ be coordinates of the vector $Tz$ in the Jordan basis of eigenvectors and adjoint vectors (3.22) – (3.24):

$$Tz = \alpha_1 z^1 + \tilde{\alpha}_1 \tilde{z}^1 + \alpha_1^{\varphi_2} z_1^{\varphi_2} + \alpha_2^{\varphi_2} z_2^{\varphi_2} + \ldots \tag{6.43}$$

According to (3.25), (3.26), (3.27), we have

$$\mathbf{C}_\Lambda^k Tz = \alpha_1 z^1 + \tilde{\alpha}_1(kz^1 + \tilde{z}^1) + \alpha_1^{\varphi_2}(\lambda_1^{\varphi_2})^k z_1^{\varphi_2} + \alpha_2^{\varphi_2}(\lambda_2^{\varphi_2})^k z_2^{\varphi_2} + \ldots$$

and

$$\begin{aligned}
\mathbf{C}_\Omega^k z = & T^{-1}\mathbf{C}_\Lambda^k Tz = \alpha_1 z^1 + \tilde{\alpha}_1(kz^1 + T^{-1}\tilde{z}^1) + \\
& \alpha_1^{\varphi_2}(\lambda_1^{\varphi_2})^k T^{-1} z_1^{\varphi_2} + \alpha_2^{\varphi_2}(\lambda_2^{\varphi_2})^k T^{-1} z_2^{\varphi_2} + \ldots
\end{aligned} \tag{6.44}$$

Since $z^1 > 0$, $k \in \mathbb{Z}$, and $|\lambda_{1,2}^{\varphi_i}| = 1$, we have the following necessary condition of $\Omega$-regularity:

$$\tilde{\alpha}_1 = 0, \tag{6.45}$$

where $\tilde{\alpha}_1$ is the coordinate of the vector $Tz$ corresponding to the adjoint basis vector $\tilde{z}^1$.

a) Simply-laced case. Recall, that

$$z^1 = \begin{pmatrix} \mathbb{X} \\ -D^t\mathbb{X} \end{pmatrix}, \quad \tilde{z}^1 = \frac{1}{4}\begin{pmatrix} \mathbb{X} \\ D^t\mathbb{X} \end{pmatrix}. \tag{6.46}$$

The adjoint vector $\tilde{z}^1$ is orthogonal to vectors $z_i^{\varphi_j}$ ($i = 1, 2$, and $\varphi_j$ are eigenvalues of $DD^t$) since the corresponding components $\mathbb{X}$ and $\mathbb{Y}$ are orthogonal, see Proposition 3.9. Further, $\tilde{z}^1$ is also orthogonal to $z^1$. Indeed,

$$4\langle z^1, \tilde{z}^1 \rangle = \langle \mathbb{X}^1, \mathbb{X}^1 \rangle - \langle D^t \mathbb{X}^1, D^t \mathbb{X}^1 \rangle =$$
$$\langle \mathbb{X}^1, \mathbb{X}^1 \rangle - \langle \mathbb{X}^1, DD^t \mathbb{X}^1 \rangle = \langle \mathbb{X}^1, \mathbb{X}^1 \rangle - \langle \mathbb{X}^1, \mathbb{X}^1 \rangle = 0.$$

Thus, from (6.43) we have

$$\langle Tz, \tilde{z}^1 \rangle = \tilde{\alpha}_1 \langle \tilde{z}^1, \tilde{z}^1 \rangle \tag{6.47}$$

and (6.45) is equivalent to the following relation:

$$\langle Tz, \tilde{z}^1 \rangle = 0. \tag{6.48}$$

b) Multiply-laced case. The normal basis of $DF$ (resp. $FD$) is not orthogonal; however, the dual graph will help us.

**Definition 6.31.** Let $u$ be an eigenvector of the Coxeter transformation given by (3.22), (3.23). The vector $\tilde{u}$ is said to be *conjugate* to $u$, if it is obtained from $u$ by changing the sign of the $\mathbb{Y}$-component and replacing the eigenvalue $\lambda$ by $\frac{1}{\lambda}$.

It is easy to see that the vectors $z^1$ and $4\tilde{z}^1$ are conjugate.

**Proposition 6.32.** *1) Let $\varphi$ be an eigenvalue of $DF$ and $\varphi^\vee \neq \varphi$ an eigenvalue of $D^\vee F^\vee$. The eigenvectors of $DF$ and $D^\vee F^\vee$ corresponding to these eigenvalues are orthogonal.*

*2) Let $z_\varphi$ be an eigenvector with eigenvalue $\varphi \neq 0$ for the extended Dynkin diagram $\Gamma$ and $z_\varphi^\vee$ an eigenvector with eigenvalue $\varphi$ for the dual diagram $\Gamma^\vee$. Let $\tilde{z}_\varphi^\vee$ be conjugate to $z_\varphi^\vee$. Then $z_\varphi$ and $\tilde{z}_\varphi^\vee$ are orthogonal.*

*Proof.* 1) Let $DFx = \varphi x$ and $D^\vee F^\vee x = \varphi^\vee x^\vee$. Since $\varphi \neq \varphi^\vee$, one of these eigenvalues is $\neq 0$. Let, for example, $\varphi \neq 0$. Then

$$\langle x, x^\vee \rangle = \frac{1}{\varphi} \langle DFx, x^\vee \rangle = \frac{1}{\varphi} \langle x, (DF)^t x^\vee \rangle. \tag{6.49}$$

From (6.49) and (3.7) we have

$$\langle x, x^\vee \rangle = \frac{1}{\varphi} \langle x, D^\vee F^\vee x^\vee \rangle = \frac{\varphi^\vee}{\varphi} \langle x, x^\vee \rangle.$$

Since $\varphi \neq \varphi^\vee$, we have $\langle x, x^\vee \rangle = 0$.

2) Let us express vectors $z_\varphi$ and $\tilde{z}_\varphi^\vee$ as follows:

$$z_\varphi = \begin{pmatrix} x \\ -\dfrac{2}{\lambda+1} Fx \end{pmatrix}, \qquad \tilde{z}_\varphi^\vee = \begin{pmatrix} x^\vee \\ \dfrac{2\lambda}{\lambda+1} F^\vee x^\vee \end{pmatrix}. \tag{6.50}$$

According to (3.6) and (3.7) we have

$$\langle z_\varphi, \tilde{z}_\varphi^\vee \rangle = \langle x, x^\vee \rangle - \frac{4\lambda}{(\lambda+1)^2} \langle Fx, F^\vee x^\vee \rangle =$$

$$\langle x, x^\vee \rangle - \frac{1}{\varphi} \langle DFx, x^\vee \rangle = \langle x, x^\vee \rangle - \langle x, x^\vee \rangle = 0. \quad \square$$

Thus, by Proposition 6.32 heading 1), the vector $\tilde{z}^{1\vee}$ is orthogonal to the $z_i^\varphi$ for $i = 1, 2$ and by Proposition 6.32 heading 2), the vector $\tilde{z}^{1\vee}$ is orthogonal to $z^1$. Therefore from (6.43) we deduce that

$$\langle Tz, \tilde{z}^{1\vee} \rangle = \tilde{\alpha}_1 \langle \tilde{z}^{1\vee}, \tilde{z}^{1\vee} \rangle \tag{6.51}$$

and (6.45) is equivalent to the following relation:

$$\langle Tz, \tilde{z}^{1\vee} \rangle = 0. \tag{6.52}$$

From (6.48) and (6.52) we derive the following theorem:

**Theorem 6.33 ([SuSt75],[SuSt78]).** *If $z$ is a regular vector for the extended Dynkin diagram $\Gamma$ in the orientation $\Omega$, then*

$$\rho_\Omega(z) = 0. \tag{6.53}$$

In Proposition B.9 we will show that the condition (6.53) is also sufficient, if $z$ is a root in the root system related to the given extended Dynkin diagram.

### 6.3.3 The Dlab-Ringel formula

**Proposition 6.34 ([DR76]).** *The following formula holds*

$$\mathbf{C}_\Omega^h z = z + h\tilde{\alpha}_1(z)z^1, \tag{6.54}$$

*where $h$ is the Coxeter number and $\tilde{\alpha}_1 = \tilde{\alpha}_1(z)$ is the linear form proportional to $\rho_\Omega(z)$. By (6.47) in the simply-laced case and by (6.51) in the multiply-laced case the form $\tilde{\alpha}_1(z)$ can be calculated as follows:*

$$\tilde{\alpha}_1(z) = \begin{cases} \frac{\langle Tz, \tilde{z}^1 \rangle}{\langle \tilde{z}^1, \tilde{z}^1 \rangle} & \text{in the simply-laced case,} \\[2ex] \frac{\langle Tz, \tilde{z}^{1\vee} \rangle}{\langle \tilde{z}^{1\vee}, \tilde{z}^{1\vee} \rangle} & \text{in the multiply-laced case.} \end{cases}$$

*Proof.* The vector $z^1$ from the kernel of the Tits form is the fixed point for the Weyl group, so $T^{-1}z^1 = z^1$, and from (6.43) we deduce

$$z = \alpha_1 z^1 + \tilde{\alpha}_1 T^{-1} \tilde{z}^1 + \alpha_1^{\varphi_2} T^{-1} z_1^{\varphi_2} + \alpha_2^{\varphi_2} T^{-1} z_2^{\varphi_2} + \dots \tag{6.55}$$

and from (6.44) for $k = h$ (the Coxeter number, i.e., $(\lambda_{1,2}^{\varphi_2})^k = 1$), we obtain

$$\mathbf{C}_\Omega^h z = \alpha_1 z^1 + \tilde{\alpha}_1(hz^1 + T^{-1}\tilde{z}^1) + \alpha_1^{\varphi_2} T^{-1} z_1^{\varphi_2} + \alpha_2^{\varphi_2} T^{-1} z_2^{\varphi_2} + \dots \tag{6.56}$$

From (6.55) and (6.56) we get (6.54).    $\square$

### 6.3.4 The Dlab-Ringel defect and the $\Omega$-defect coincide

**Proposition 6.35 ([St85]).** *The Dlab-Ringel defect $\delta_\Omega$ given by (6.36) coincides (up to a factor) with the $\Omega$-defect $\rho_\Omega$ given by Definition 6.29.*

*Proof.* 1) Let us show that $\delta_\Omega$ is obtained from $\delta_\Lambda$ in the same way as $\rho_\Omega$ is obtained from $\rho_\Lambda$, so it suffices to prove the proposition only for bicolored orientations $\Lambda$. Indeed,

$$\mathbf{C}_\Omega = T^{-1}\mathbf{C}_\Lambda T \qquad \text{implies that} \qquad \mathbf{C}_\Omega^* = T^*\mathbf{C}_\Lambda^* T^{-1*}.$$

Since $\mathbf{C}_\Omega^* \delta_\Omega = \delta_\Omega$, we obtain

$$\mathbf{C}_\Lambda^* T^{-1*}\delta_\Omega = T^{-1*}\delta_\Omega,$$

and

$$T^{-1*}\delta_\Omega = \delta_\Lambda, \qquad \text{i.e.,} \qquad \delta_\Omega = T^*\delta_\Lambda.$$

Further,

$$\langle \delta_\Lambda, Tz \rangle = \langle T^*\delta_\Lambda, z \rangle = \langle \delta_\Omega, z \rangle.$$

Thus, $\delta_\Omega$ is obtained from $\delta_\Lambda$ in the same way as $\rho_\Omega$ is obtained from $\rho_\Lambda$.

2) Now, let us prove the proposition for the bicolored orientation $\Lambda$, i.e., let us prove that $\rho_\Lambda$ is proportional to $\delta_\Lambda$. According to (3.1), (3.2) and (3.4) the relation

$$\mathbf{C}_\Lambda^* z = z \qquad \text{is equivalent to} \qquad w_1^* z = w_2^* z,$$

which, in turn, is equivalent to the following:

$$\begin{pmatrix} -I & 0 \\ -2D^t & I \end{pmatrix} \begin{pmatrix} x \\ y \end{pmatrix} = \begin{pmatrix} I & -2F^t \\ 0 & -I \end{pmatrix} \begin{pmatrix} x \\ y \end{pmatrix}.$$

By (3.6) we have

$$\begin{cases} x = F^t y = D^\vee y, \\ y = D^t x = F^\vee x. \end{cases}$$

Thus,

$$z = \begin{pmatrix} x \\ F^\vee x \end{pmatrix} = \begin{pmatrix} x \\ D^t x \end{pmatrix}, \quad \text{where } x = F^t D^t x = D^\vee F^\vee x,$$

i.e., $x$ corresponds to $\lambda = 1$, an eigenvalue of $\mathbf{C}$, and to $\varphi = 1$, an eigenvalue of $(DF)^\vee$. By (6.50) we have $z = \tilde{z}^{1\vee}$, i.e., $\delta_\Lambda$ is proportional to $\rho_\Lambda$, see (6.39). For the simply-laced case, compare with (3.23).    $\square$

# A

## The McKay correspondence and the Slodowy correspondence

> We have seen during the past few years a major assault on the problem of determining all the finite simple groups. ... If I am right, I foresee new proofs of classification which will owe little or nothing to the current proofs. They will be much shorter and will help us to understand the simple groups in a context much wider than finite group theory.

J. McKay, [McK80, p.183], 1980

## A.1 Finite subgroups of $SU(2)$ and $SO(3,\mathbb{R})$

Let us consider the special unitary group $SU(2)$, the subgroup of unitary transformations in $GL(2,\mathbb{C})$ with determinant $1$ and its quotient group $PSU(2) = SU(2)/\{\pm 1\}$ acting on the complex projective line $\mathbb{CP}^1$, see §C.5. The projective line $\mathbb{CP}^1$ can be identified with the sphere $S^2 \subset \mathbb{R}^3$, see §C.5 and (C.20).

The transformations of the sphere $S^2$ induced by elements of $SO(3,\mathbb{R})$ (orientation preserving rotations of $\mathbb{R}^3$) correspond under these identifications to transformations of $\mathbb{CP}^1$ in the group $PSU(2)$:

$$PSU(2) \cong SO(3,\mathbb{R}), \tag{A.1}$$

see [Sp77, Prop.4.4.3], [PV94, §0.13].

It is well known that the finite subgroups of $SO(3,\mathbb{R})$ are precisely the rotation groups of the following polyhedra: the regular $n$-angled pyramid, the $n$-angled dihedron (a regular plane $n$-gon with two faces), the tetrahedron, the cube (its rotation group coincides with the rotation group of the octahedron), the icosahedron (its rotation group coincides with the rotation group of the dodecahedron). By (A.1) we have a classification of the finite subgroups of $PSU(2)$, see Table A.1.

To get a classification of all finite subgroups of $SU(2)$ we consider the double covering
$$\pi : SU(2) \longrightarrow SO(3, \mathbb{R}).$$

If $G$ is a finite subgroup of $SO(3, \mathbb{R})$, we see that the preimage $\pi^{-1}(G)$ is a finite subgroup of $SU(2)$ and $|\pi^{-1}(G)| = 2|G|$. The finite subgroups of $SO(3, \mathbb{R})$ are called *polyhedral groups*, see Table A.1. The finite subgroups of $SU(2)$ are naturally called *binary polyhedral groups*, see Table A.2.

*Remark A.1.* The cyclic group is an exceptional case. The preimage of the cyclic group $G = \mathbb{Z}/n\mathbb{Z}$ is the even cyclic group $\mathbb{Z}/2n\mathbb{Z}$, which is a binary cyclic group. The cyclic group $\mathbb{Z}/(2n-1)\mathbb{Z}$ is not a preimage with respect to $\pi$, so $\mathbb{Z}/(2n-1)\mathbb{Z}$ is not a binary polyhedral group. Anyway, the cyclic groups $\mathbb{Z}/n\mathbb{Z}$ complete the list of finite subgroups of $SU(2)$.

For every polyhedral group $G$, the axis of rotations under an element $\gamma \in G$ passes through either the mid-point of a face, or the mid-point of an edge, or a vertex. We denote the orders of symmetry of these axes by $p$, $q$, and $r$, respectively, see [Sl83]. These numbers are listed in Table A.1. The triples $p, q, r$ listed in Table A.1 are exactly the solutions of the diophantine inequality
$$\frac{1}{p} + \frac{1}{q} + \frac{1}{r} > 1, \tag{A.2}$$

see [Sp77].

## A.2 The generators and relations in polyhedral groups

The *quaternion group* introduced by W. R. Hamilton is defined as follows: It is generated by three generators $i$, $j$, and $k$ subject to the relations

$$i^2 = j^2 = k^2 = ijk = -1; \tag{A.3}$$

for references, see [Cox40], [CoxM84]. A natural generalization of the quaternion group is the group generated by three generators $R$, $S$, and $T$ subject to the relations
$$R^p = S^q = T^r = RST = -1. \tag{A.4}$$

Denote by $\langle p, q, r \rangle$ the group defined by (A.4).

W. Threlfall, (see [CoxM84]) has observed that

$$\langle 2, 2, n \rangle, \langle 2, 3, 3 \rangle, \langle 2, 3, 4 \rangle, \langle 2, 3, 5 \rangle$$

are the *binary polyhedral groups* of order

$$\frac{4}{\dfrac{1}{p} + \dfrac{1}{q} + \dfrac{1}{r} - 1} \quad .$$

**Table A.1.**   The polyhedral groups in $\mathbb{R}^3$

| Polyhedron | Orders of symmetries | Rotation group | Group order |
|:---:|:---:|:---:|:---:|
| Pyramid | $-$ | cyclic | $n$ |
| Dihedron | $n$ 2 2 | dihedral | $2n$ |
| Tetrahedron | 3 2 3 | $\mathcal{A}_4$ | 12 |
| Cube | 4 2 3 | $\mathcal{S}_4$ | 24 |
| Octahedron | 3 2 4 | $\mathcal{S}_4$ | 24 |
| Dodecahedron | 5 2 3 | $\mathcal{A}_5$ | 60 |
| Icosahedron | 3 2 5 | $\mathcal{A}_5$ | 60 |

Here, $\mathcal{S}_m$ (resp. $\mathcal{A}_m$) denotes the symmetric, (resp. alternating) group of all (resp. of all even) permutations of $m$ letters.

**Table A.2.**   The finite subgroups of $SU(2)$

| $\langle l, m, n \rangle$ | Order | Denotation | Well-known name |
|:---:|:---:|:---:|:---:|
| $-$ | $n$ | $\mathbb{Z}/n\mathbb{Z}$ | cyclic group |
| $\langle 2, 2, n \rangle$ | $4n$ | $\mathcal{D}_n$ | binary dihedral group |
| $\langle 2, 3, 3 \rangle$ | 24 | $\mathcal{T}$ | binary tetrahedral group |
| $\langle 2, 3, 4 \rangle$ | 48 | $\mathcal{O}$ | binary octahedral group |
| $\langle 2, 3, 5 \rangle$ | 120 | $\mathcal{J}$ | binary icosahedral group |

H. S. M. Coxeter proved in [Cox40, p.370] that, having added one more generator $Z$, one can replace (A.4) by the relations:

$$R^p = S^q = T^r = RST = Z. \tag{A.5}$$

For the groups $\langle 2, 2, n \rangle$ and $\langle 2, 3, n \rangle$, where $n = 3, 4, 5$, the relation (A.5) implies

$$Z^2 = 1.$$

The polyhedral groups from Table A.1 are described by generations and relations as follows:

$$R^p = S^q = T^r = RST = 1, \qquad (A.6)$$

see [Cox40], [CCS72], [CoxM84].


## A.3 The Kleinian singularities and the Du Val resolution

Consider the quotient variety $\mathbb{C}^2/G$, where $G$ is a binary polyhedral group $G \subset SU(2)$ from Table A.2. According to (§C.6.4 and (C.44)) $X = \mathbb{C}^2/G$ is the orbit space given by the prime spectrum on the algebra of invariants $R^G$:

$$X := \mathrm{Spec}(R^G), \qquad (A.7)$$

where $R = \mathbb{C}[z_1, z_2]$ (see eq.(2.28)) which coincides with the symmetric algebra $\mathrm{Sym}((\mathbb{C}^2)^*)$ (see §2.3.1; for more details, see [Sp77], [PV94], [Sl80].)

F. Klein [Kl1884] observed that the algebra of invariants $\mathbb{C}[z_1, z_2]^G$ for every binary polyhedral group $G \subset SU(2)$ from Table A.2 can be considered uniformly:

**Theorem A.2 (F. Klein, [Kl1884]).** *The algebra of invariants $\mathbb{C}[z_1, z_2]^G$ is generated by 3 variables $x, y, z$, subject to one essential relation*

$$R(x, y, z) = 0, \qquad (A.8)$$

*where $R(x, y, z)$ is defined in Table A.3, col. 2. In other words, the algebra of invariants $\mathbb{C}[z_1, z_2]^G$ coincides with the coordinate algebra (see §C.6.3) of the curve defined by the eq. (A.8), i.e.,*

$$\mathbb{C}[z_1, z_2]^G \simeq \mathbb{C}[x, y, z]/(R(x, y, z)). \qquad (A.9)$$

The quotient $X$ from (A.7) has no singularity except at the origin $O \in \mathbb{C}^3$. The quotient variety $X$ is called a *Kleinian singularity* also known as a *Du Val singularity*, a *simple surface singularity* or a *rational double point*. The quotient variety $X$ can be embedded as a surface $X \subset \mathbb{C}^3$ with an isolated singularity at the origin, see [Sl83, §5].

*Remark A.3.* According to Theorem 2.19 (Shephard-Todd-Chevalley-Serre) the algebra of invariants $k[V]^G$ is isomorphic to a polynomial algebra in some number of variables if the image of $G$ in $GL(V)$ is generated by reflections. Every binary polyhedral group $G$ is generated by reflections, see, e.g., [CoxM84], therefore the algebra of invariants $k[V]^G$ for the binary polyhedral group is a polynomial algebra.

**Table A.3.**   The relations $R(x, y, z)$ describing the algebra of invariants $\mathbb{C}[z_1, z_2]^G$

| Finite subgroup of $SU(2)$ | Relation $R(x, y, z)$ | Dynkin diagram |
|---|---|---|
| $\mathbb{Z}/n\mathbb{Z}$ | $x^n + yz$ | $A_{n-1}$ |
| $\mathcal{D}_n$ | $x^{n+1} + xy^2 + z^2$ | $D_{n+2}$ |
| $\mathcal{T}$ | $x^4 + y^3 + z^2$ | $E_6$ |
| $\mathcal{O}$ | $x^3 y + y^3 + z^2$ | $E_7$ |
| $\mathcal{J}$ | $x^5 + y^3 + z^2$ | $E_8$ |

*Example A.4.* Consider the cyclic group $G = \mathbb{Z}/r\mathbb{Z}$ of order $r$. The group $G$ acts on $\mathbb{C}[z_1, z_2]$ as follows:

$$(z_1, z_2) \mapsto (\varepsilon z_1, \varepsilon^{r-1} z_2), \qquad (A.10)$$

where $\varepsilon = e^{2\pi i/r}$, and the polynomials

$$x = z_1 z_2, \qquad y = -z_1^r, \qquad z = z_2^r \qquad (A.11)$$

are invariant polynomials in $\mathbb{C}[x, y, z]$ which satisfy the following relation

$$x^r + yz = 0, \qquad (A.12)$$

see Table A.3. We have

$$k[V]^G = \mathbb{C}[z_1 z_2, z_1^r, z_2^r] \simeq \mathbb{C}[x, y, z]/(x^r + yz).$$

see, e.g., [Sp77, pp.95-97], [PV94, p.143].

Du Val obtained the following description of the minimal resolution

$$\pi : \tilde{X} \longrightarrow X$$

of a Kleinian singularity $X = \mathbb{C}^2/G$, see [DuVal34], [Sl80, §6.1, §6.2] [Sl83, §5][1]. The *exceptional divisor* (the preimage of the singular point $O$) is a finite union of complex projective lines:

$$\pi^{-1}(O) = L_1 \cup \cdots \cup L_n, \qquad L_i \simeq \mathbb{CP}^1 \text{ for } i = 1, \dots, n.$$

The intersection $L_i \cap L_j$ is empty or consists of exactly one point for $i \neq j$.

---

[1] For more details, see also [Gb02], [Rie02], [Hob02], [Cr01].

To each complex projective line $L_i$ (which can be identified with the sphere $S^2 \subset \mathbb{R}^3$, see §C.5) we assign a vertex $i$, and two vertices are connected by an edge if the corresponding projective lines intersect. The corresponding diagrams are Dynkin diagrams (this phenomenon was observed by Du Val in [DuVal34] ), see Table A.3.

In the case of the binary dihedral group $\mathcal{D}_2$ the real resolution of the real variety
$$\mathbb{C}^3/R(x, y, z) \cap \mathbb{R}^3$$
gives a quite faithful picture of the complex situation, the minimal resolution $\pi^{-1} : \tilde{X} \longrightarrow X$ for $X = \mathcal{D}_2$ depicted in Fig. A.1. Here $\pi^{-1}(O)$ consists of four circles, the corresponding diagram is the Dynkin diagram $D_4$.

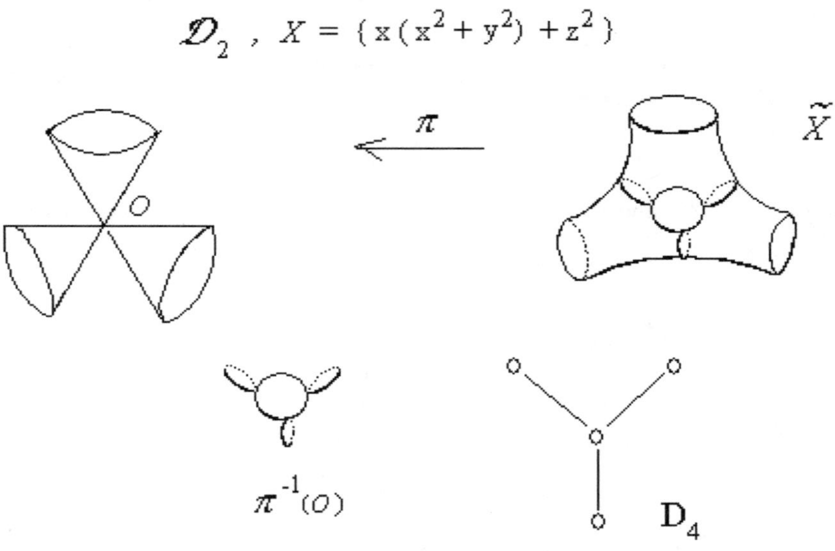

$$\mathcal{D}_2 \ , \ X = \{\, x\,(\,x^2 + y^2\,) + z^2 \,\}$$

Fig. A.1.    The minimal resolution $\pi^{-1} : \tilde{X} \longrightarrow X$ for $X = \mathcal{D}_2$

## A.4 The McKay correspondence

Let $G$ be a finite subgroup of $SU(2)$. Let $\{\rho_0, \rho_1, \ldots, \rho_n\}$ be the set of all distinct irreducible finite dimensional complex representations of $G$, of which $\rho_0$ is the trivial one. Let $\rho : G \longrightarrow SU(2)$ be a faithful representation, then, for each group $G$, we define a matrix $A(G) = (a_{ij})$, by decomposing the tensor products:

$$\rho \otimes \rho_j = \bigoplus_{k=0}^{r} a_{jk} \rho_k, \qquad j = 0, 1, ..., r, \qquad (A.13)$$

where $a_{jk}$ is the multiplicity of $\rho_k$ in $\rho \otimes \rho_j$. McKay [McK80] observed that

*The matrix $2I - A(G)$ is the Cartan matrix of the extended Dynkin diagram $\tilde{\Gamma}(G)$ associated to $G$. There is a one-to-one correspondence between finite subgroups of $SU(2)$ and simply-laced extended Dynkin diagrams.*

This remarkable observation, called the *McKay correspondence*, was based first on an explicit verification [McK80].

For the multiply-laced case, the McKay correspondence was extended by D. Happel, U. Preiser, and C. M. Ringel in [HPR80], and by P. Slodowy in [Sl80, App.III]. We consider P. Slodowy's approach in §A.5.

The systematic proof of the McKay correspondence based on the study of affine Coxeter transformations was given by R. Steinberg in [Stb85].

Other proofs of the McKay correspondence were given by G. Gonzalez-Sprinberg and J.-L. Verdier in [GV83], by H. Knörrer in [Kn85]. A nice review is given by J. van Hoboken in [Hob02].

B. Kostant used the *McKay matrix* $A(G)$ (or *McKay operator*) in [Kos84] and showed that the multiplicities $m_i(n)$ in the decomposition

$$\pi_n | G = \sum_{i=0}^{r} m_i(n) \rho_i,$$

(see §5.4.1) come in an amazing way from the orbit structure of the Coxeter transformation on the highest root of the corresponding Lie algebra $\mathfrak{g}$, see §5.5. To calculate these multiplicities, Kostant employed generating functions and Poincaré series. We applied Kostant's technique in §5.4 in order to show a relation between Poincaré series and the ratio of characteristic polynomials of the Coxeter transformations.

## A.5 The Slodowy generalization of the McKay correspondence

We consider here the Slodowy generalization [Sl80] of the McKay correspondence to the multiply-laced case and illustrate Slodowy's approach with the diagrams $\widetilde{F}_{41}$ and $\widetilde{F}_{42}$, see Fig. 2.6.

Slodowy's approach is based on the consideration of *restricted representations* and *induced representations* instead of an original representation. Let $\rho : G \longrightarrow GL(V)$ be a representation of a group $G$. We denote the *restricted representation* of $\rho$ to a subgroup $H \subset G$ by $\rho \downarrow_H^G$, or, briefly, $\rho^\downarrow$ for fixed $G$ and $H$. Let $\tau : H \longrightarrow GL(V)$ be a representation of a subgroup $H$. We denote by $\tau \uparrow_H^G$ the representation *induced* by $\tau$ to a representation of the group $G$ containing $H$; we briefly write $\tau^\uparrow$ for fixed $G$ and $H$. For a detailed definition on restricted and induced representations, see, for example, [Kar92] or [Bak04].

Let us consider pairs of groups $H \triangleleft G$, where $H$ and $G$ are binary polyhedral groups from Tables A.2 and A.3. See, e.g., [Sl80, p.163], [Sp77, p.89], [Hob02, p.25].

**Table A.4.**    The pairs $H \triangleleft G$ of binary polyhedral groups

| Subgroup $H$ | Dynkin diagram $\Gamma(H)$ | Group $G$ | Dynkin diagram $\Gamma(G)$ | Index $[G : H]$ |
|---|---|---|---|---|
| $\mathcal{D}_2$ | $D_4$ | $\mathcal{T}$ | $E_6$ | 3 |
| $\mathcal{T}$ | $E_6$ | $\mathcal{O}$ | $E_7$ | 2 |
| $\mathcal{D}_{n-1}$ | $D_{n+1}$ | $\mathcal{D}_{2(n-1)}$ | $D_{2n}$ | 2 |
| $\mathbb{Z}/2n\mathbb{Z}$ | $A_{2n-1}$ | $\mathcal{D}_n$ | $D_{n+2}$ | 2 |

### A.5.1 The Slodowy correspondence

Let us fix a pair $H \triangleleft G$ from Table A.4. We formulate now the essence of the Slodowy correspondence [Sl80, App.III].

1) Let $\rho_i$, where $i = 1, \ldots, n$, be irreducible representations of $G$; let $\rho_i^{\downarrow}$ be the corresponding restricted representations of the subgroup $H$. Let $\rho$ be a faithful representation of $H$, which may be considered as the restriction of the fixed faithful representation $\rho_f$ of $G$. Then the following decomposition formula makes sense

$$\rho \otimes \rho_i^{\downarrow} = \bigoplus_j a_{ji} \rho_j^{\downarrow} \tag{A.14}$$

and uniquely determines an $n \times n$ matrix $\widetilde{A} = (a_{ij})$ such that

$$K = 2I - \widetilde{A} \tag{A.15}$$

(see [Sl80, p.163]), where $K$ is the Cartan matrix of the corresponding folded extended Dynkin diagram given in Table A.5

2) Let $\tau_i$, where $i = 1, \ldots, n$, be irreducible representations of the subgroup $H$, let $\tau_i^{\uparrow}$ be the induced representations of the group $G$. Then the following decomposition formula makes sense

$$\rho \otimes \tau_i^{\uparrow} = \bigoplus a_{ij} \tau_j^{\uparrow}, \tag{A.16}$$

i.e., the decomposition of the induced representation is described by the matrix $A^{\vee} = A^t$ which satisfies the relation

$$K^{\vee} = 2I - \widetilde{A}^{\vee} \tag{A.17}$$

(see [Sl80, p.164]), where $K^{\vee}$ is the Cartan matrix of the dual folded extended Dynkin diagram given in Table A.5.

We call matrices $\widetilde{A}$ and $\widetilde{A}^{\vee}$ the *Slodowy matrices*, they are analogs of the McKay matrix. The *Slodowy correspondence* is an analogue to the McKay correspondence for the multiply-laced case, so one can speak about the *McKay-Slodowy correspondence*.

**Table A.5.**   The pairs $H \lhd G$ and folded extended Dynkin diagrams

| Groups $H \lhd G$ | Dynkin diagram $\tilde{\Gamma}(H)$ and $\tilde{\Gamma}(G)$ | Folded extended Dynkin diagram |
|---|---|---|
| $\mathcal{D}_2 \lhd \mathcal{T}$ | $D_4$ and $E_6$ | $\widetilde{G}_{21}$ and $\widetilde{G}_{22}$ |
| $\mathcal{T} \lhd \mathcal{O}$ | $E_6$ and $E_7$ | $\widetilde{F}_{41}$ and $\widetilde{F}_{42}$ |
| $\mathcal{D}_{r-1} \lhd \mathcal{D}_{2(r-1)}$ | $D_{n+1}$ and $D_{2n}$ | $\widetilde{DD}_n$ and $\widetilde{CD}_n$ |
| $\mathbb{Z}/2r\mathbb{Z} \lhd \mathcal{D}_r$ | $A_{n-1}$ and $D_{r+2}$ | $\widetilde{B}_n$ and $\widetilde{C}_n$ |

## A.5.2 The binary tetrahedral group and the binary octahedral group

Now we will illustrate the Slodowy correspondence for the binary tetrahedral and octahedral groups, i.e., $H$ is the binary tetrahedral group $\mathcal{T}$ and $G$ is the binary octahedral group $\mathcal{O}$, $\mathcal{T} \triangleleft \mathcal{O}$. These groups have orders $|\mathcal{T}| = 24$ and $|\mathcal{O}| = 48$, see Table A.2.

We will use the Springer formula for elements of the group $\mathcal{O}$ from [Sp77, §4.4.11]. Let

$$a = \begin{pmatrix} \varepsilon & 0 \\ 0 & \varepsilon^{-1} \end{pmatrix}, \quad b = \begin{pmatrix} 0 & i \\ i & 0 \end{pmatrix}, \quad c = \frac{1}{\sqrt{2}} \begin{pmatrix} \varepsilon^{-1} & \varepsilon^{-1} \\ -\varepsilon & \varepsilon \end{pmatrix}, \tag{A.18}$$

where $\varepsilon = e^{\pi i/4}$. Then each of the 48 different elements $x \in \mathcal{O}$ may be expressed as follows:

$$x = a^h b^j c^l, \quad 0 \le h < 8, \quad 0 \le j < 2, \quad 0 \le l < 3. \tag{A.19}$$

The elements $x \in \mathcal{O}$ and their traces are collected in Table A.6. Observe that every element $u \in SL(2, \mathbb{C})$ from Table A.6 is of the form

$$u = \begin{pmatrix} \alpha & \beta \\ -\bar{\beta} & \bar{\alpha} \end{pmatrix}, \tag{A.20}$$

see (5.36). We can now distinguish the elements of $\mathcal{O}$ by their traces and by means of the 1-dimensional representation $\rho_1$ such that

$$\rho_1(a) = -1, \quad \rho_1(b) = 1, \quad \rho_1(c) = 1. \tag{A.21}$$

**Proposition A.5.** *There are 8 conjugacy classes in the binary octahedral group $\mathcal{O}$. Rows of Table A.7 constitute these conjugacy classes.*

For the proof, see [St05, Prop. A.5].     $\square$

*Remark A.6.* We denote by $Cl(g)$ the conjugacy class containing the element $g \in \mathcal{O}$. The union $Cl(b) \cup \{1, -1\}$ constitutes the 8-element subgroup

$$\{1, a^2, a^4, a^6, b, a^2b, a^4b, a^6b\} = \{1, a^2, -1, -a^2, b, a^2b, -b, -a^2b\}. \tag{A.22}$$

Setting $i = b$, $j = a^2$, $k = a^2b$ we see that

$$\begin{aligned} i^2 = j^2 = k^2 = -1, \\ ji = -ij = k, \quad ik = -ki = j, \quad kj = -jk = i, \end{aligned} \tag{A.23}$$

i.e., group (A.22) is the quaternion group $Q_8$, see (A.3).

It is easy to check that $Q_8$ is a normal subgroup in $\mathcal{O}$:

$$Q_8 \triangleleft \mathcal{O}. \tag{A.24}$$

**Table A.6.**    The elements of the binary octahedral group

| Elements $x = a^p b^j c^l$ $0 \le p < 8$ | Matrix form $u$ | Trace $tr(x)$ |
|---|---|---|
| $a^p$ | $\begin{pmatrix} \varepsilon^p & 0 \\ 0 & \varepsilon^{-p} \end{pmatrix}$ | $2\cos\dfrac{\pi p}{4}$ |
| $a^p b$ | $\begin{pmatrix} 0 & i\varepsilon^p \\ i\varepsilon^{-p} & 0 \end{pmatrix}$ | $0$ |
| $a^p c$ | $\dfrac{1}{\sqrt{2}}\begin{pmatrix} \varepsilon^{p-1} & \varepsilon^{p-1} \\ -\varepsilon^{-(p-1)} & \varepsilon^{-(p-1)} \end{pmatrix} = \dfrac{1}{\sqrt{2}}\begin{pmatrix} \varepsilon^{p-1} & \varepsilon^{p-1} \\ \varepsilon^{-p+5} & \varepsilon^{-p+1} \end{pmatrix}$ | $\sqrt{2}\cos\dfrac{\pi(p-1)}{4}$ |
| $a^p bc$ | $\dfrac{1}{\sqrt{2}}\begin{pmatrix} -i\varepsilon^{p+1} & i\varepsilon^{p+1} \\ i\varepsilon^{-(p+1)} & i\varepsilon^{-(p+1)} \end{pmatrix} = \dfrac{1}{\sqrt{2}}\begin{pmatrix} \varepsilon^{p-1} & \varepsilon^{p+3} \\ \varepsilon^{-p+1} & \varepsilon^{-p+1} \end{pmatrix}$ | $\sqrt{2}\cos\dfrac{\pi(p-1)}{4}$ |
| $a^p c^2$ | $\dfrac{1}{\sqrt{2}}\begin{pmatrix} -\varepsilon^{p+1} & \varepsilon^{p-1} \\ -\varepsilon^{-(p-1)} & -\varepsilon^{-(p+1)} \end{pmatrix} = \dfrac{1}{\sqrt{2}}\begin{pmatrix} \varepsilon^{p+5} & \varepsilon^{p-1} \\ \varepsilon^{-p+5} & \varepsilon^{-p+3} \end{pmatrix}$ | $-\sqrt{2}\cos\dfrac{\pi(p+1)}{4}$ |
| $a^p bc^2$ | $\dfrac{1}{\sqrt{2}}\begin{pmatrix} -i\varepsilon^{p+1} & -i\varepsilon^{p-1} \\ -i\varepsilon^{-(p-1)} & i\varepsilon^{-(p+1)} \end{pmatrix} = \dfrac{1}{\sqrt{2}}\begin{pmatrix} \varepsilon^{p-1} & \varepsilon^{p-3} \\ \varepsilon^{-p-1} & \varepsilon^{-p+1} \end{pmatrix}$ | $\sqrt{2}\cos\dfrac{\pi(p-1)}{4}$ |

Now consider conjugacy classes in the binary tetrahedral group $\mathcal{T}$. According to the Springer formula [Sp77, §4.4.10], the elements of the group $\mathcal{T}$ are given by (A.19) with even numbers $h$. In other words, each of the 24 different

**Table A.7.**    The conjugacy classes in the binary octahedral group

| Trace $tr(x)$ | Representation $\rho_1$ | Conjugacy class $Cl(g)$ | Class order | Representative $g$ |
|---|---|---|---|---|
| 2 | 1 | $\begin{pmatrix} 1 & 0 \\ 0 & 1 \end{pmatrix}$ | 1 | 1 |
| $-2$ | 1 | $b^2 = c^3 = a^4 = \begin{pmatrix} -1 & 0 \\ 0 & -1 \end{pmatrix}$ | 1 | $-1$ |
| 0 | $-1$ | $ab, a^3b, a^5b, a^7b,$ $a^3bc, a^7bc, ac^2, a^5c^2,$ $a^3c, a^7c, a^3bc^2, a^7bc^2$ | 12 | $ab$ |
| 0 | 1 | $a^2, a^6, b, a^2b, a^4b, a^6b$ | 6 | $b$ |
| $-1$ | 1 | $a^4c, a^6c, a^4bc, a^6bc,$ $c^2, a^6c^2, a^4bc^2, a^6bc^2$ | 8 | $c^2$ |
| 1 | 1 | $c, a^2c, bc, a^2bc,$ $a^4c^2, a^2c^2, bc^2, a^2bc^2$ | 8 | $c$ |
| $\sqrt{2}$ | $-1$ | $a, a^7, ac, abc, a^3c^2, abc^2$ | 6 | $a$ |
| $-\sqrt{2}$ | $-1$ | $a^3, a^5, a^5c, a^5bc, a^7c^2, a^5bc^2$ | 6 | $a^3$ |

elements $x \in \mathcal{T}$ may be given as follows:

$$x = a^{2h}b^jc^l, \quad 0 \le h < 4, \quad 0 \le j < 2, \quad 0 \le l < 3. \tag{A.25}$$

There are 24 elements of type (A.25). In the octahedral group $\mathcal{O}$, the elements (A.25) constitute 5 conjugacy classes:

$$\{1\}, \ \{-1\}, \ Cl(b), \ Cl(c) \text{ and } Cl(c^2),$$

see Table A.7. We will see now, that in the tetrahedral group $\mathcal{T}$, the elements (A.25) constitute 7 conjugacy classes:

$$\{1\}, \ \{-1\}, \ Cl(b), \ Cl(c), \ Cl(a^4c) \text{ and } Cl(a^4c^2).$$

The elements $x \in \mathcal{T}$ and their traces are collected in Table A.8. We can now distinguish the elements of $\mathcal{T}$ by their traces and by means of two 1-dimensional representations $\tau_1$ and $\tau_2$ such that

$$\begin{aligned}
\tau_1(a) &= 1, \quad \tau_1(b) = 1, \quad \tau_1(c) = \omega_3, \\
\tau_2(a) &= 1, \quad \tau_2(b) = 1, \quad \tau_2(c) = \omega_3^2,
\end{aligned} \tag{A.26}$$

where $\omega_3 = e^{2\pi i/3}$.

**Proposition A.7.** *There are 7 conjugacy classes in the binary tetrahedral group $\mathcal{T}$. The rows of Table A.8 constitute these conjugacy classes.*

For the proof, see [St05, Prop. A.7]. □

### A.5.3 Representations of the binary octahedral and tetrahedral groups

**Proposition A.8.** *The group $\mathcal{O}$ has the following 8 irreducible representations.*

*1) Two 1-dimensional representations:*

$$\begin{aligned}
\rho_0(a) &= \rho_0(b) = \rho_0(c) = 1; \\
\rho_1(a) &= -1, \quad \rho_1(b) = \rho_1(c) = 1,
\end{aligned} \tag{A.27}$$

*2) Two faithful 2-dimensional representations:*

$$\begin{aligned}
\rho_3(a) &= a, \quad \rho_3(b) = b, \quad \rho_3(c) = c; \\
\rho_4(a) &= -a, \quad \rho_4(b) = b, \quad \rho_4(c) = c.
\end{aligned} \tag{A.28}$$

*3) The 2-dimensional representation $\rho_2$ constructed by means of an epimorphism to the symmetric group $S_3$*

$$\mathcal{O} \longrightarrow \mathcal{O}/Q_8 \simeq S_3, \tag{A.29}$$

*where $Q_8$ is the quaternion group.*

*4) Two 3-dimensional representations $\rho_5$ and $\rho_6$ constructed by means of an epimorphism to the symmetric group $S_4$*

**Table A.8.**    The conjugacy classes in the binary tetrahedral group

| Trace $tr(x)$ | Repr. $\tau_1$ | Repr. $\tau_2$ | Conjugacy class $Cl(g)$ | Class order | Representative $g$ |
|---|---|---|---|---|---|
| 2 | 1 | 1 | $\begin{pmatrix} 1 & 0 \\ 0 & 1 \end{pmatrix}$ | 1 | 1 |
| $-2$ | 1 | 1 | $\begin{pmatrix} -1 & 0 \\ 0 & -1 \end{pmatrix}$ | 1 | $-1$ |
| 0 | 1 | 1 | $a^2, a^6, b, a^2b, a^4b, a^6b$ | 6 | $b$ |
| $-1$ | $\omega_3$ | $\omega_3^2$ | $a^4c, a^6c, a^4bc, a^6bc,$ | 4 | $a^4c = -c$ |
| $-1$ | $\omega_3^2$ | $\omega_3$ | $c^2, a^6c^2, a^4bc^2, a^6bc^2$ | 4 | $c^2$ |
| 1 | $\omega_3$ | $\omega_3^2$ | $c, a^2c, bc, a^2bc$ | 4 | $c$ |
| 1 | $\omega_3^2$ | $\omega_3$ | $a^4c^2, a^2c^2, bc^2, a^2bc^2$ | 4 | $a^4c^2 = -c^2$ |

$$\mathcal{O} \longrightarrow \mathcal{O}/\{1, -1\} \simeq S_4. \tag{A.30}$$

The representations $\rho_5$ and $\rho_6$ are related as follows

$$\rho_6(a) = -\rho_5(a), \quad \rho_6(b) = \rho_5(a), \quad \rho_6(c) = \rho_5(c). \tag{A.31}$$

5) The 4-dimensional representation $\rho_7$ constructed as the tensor product $\rho_2 \otimes \rho_3$ (it coincides with $\rho_2 \otimes \rho_4$).

**Table A.9.**   The characters of the binary octahedral group

| Character | Conjugacy class $Cl(g)$ and its order $|Cl(g)|$ under it | | | | | | | | Note on |
|-----------|---------|----------|----------|---------|-----------|---------|---------|-----------|---------|
| $\psi_i$ | $Cl(1)$ | $Cl(-1)$ | $Cl(ab)$ | $Cl(b)$ | $Cl(c^2)$ | $Cl(c)$ | $Cl(a)$ | $Cl(a^3)$ | represent. |
|  | 1 | 1 | 12 | 6 | 8 | 8 | 6 | 6 | $\rho_i$ |
| $\psi_0$ | 1 | 1 | 1 | 1 | 1 | 1 | 1 | 1 | trivial |
| $\psi_1$ | 1 | 1 | $-1$ | 1 | 1 | 1 | $-1$ | $-1$ | $\rho_1(a) = -a$ |
| $\psi_2$ | 2 | 2 | 0 | 2 | $-1$ | $-1$ | 0 | 0 | $\gamma_2 \pi_2$ |
| $\psi_3$ | 2 | $-2$ | 0 | 0 | $-1$ | 1 | $\sqrt{2}$ | $-\sqrt{2}$ | faithful |
| $\psi_4$ | 2 | $-2$ | 0 | 0 | $-1$ | 1 | $-\sqrt{2}$ | $\sqrt{2}$ | faithful |
| $\psi_5$ | 3 | 3 | $-1$ | $-1$ | 0 | 0 | 1 | 1 | $\gamma_5 \pi_{56}$ |
| $\psi_6$ | 3 | 3 | 1 | $-1$ | 0 | 0 | $-1$ | $-1$ | $\gamma_6 \pi_{56}$ |
| $\psi_7$ | 4 | $-4$ | 0 | 0 | 1 | $-1$ | 0 | 0 | $\rho_2 \otimes \rho_3$ |

*The characters of representations $\rho_i$ for $i = 0, \ldots, 7$ are collected in Table A.9.*

*Proof.* 1) and 2) are clear from constructions (A.27) and (A.28).

3) We construct the third 2-dimensional representation $\rho_2$ by using the homomorphism

$$\pi_2 : \mathcal{O} \longrightarrow \mathcal{O}/Q_8, \qquad (A.32)$$

see (A.24). The quotient group $\mathcal{O}/Q_8$ is isomorphic to the symmetric group $S_3$ consisting of 6 elements. The cosets of $\mathcal{O}/Q_8$ are

$$\{Q_8, cQ_8, c^2Q_8, abQ_8, acQ_8, ac^2Q_8\} = \\ \{\{1\}, \{c\}, \{c^2\}, \{ab\}, \{ac\}, \{ac^2\}\}. \qquad (A.33)$$

For more details about $\mathcal{O}/Q_8$, see [St05, Prop. A.8]. Further, we have

$$\mathcal{O}/Q_8 \simeq S_3, \\ \{ab\} \simeq (12), \quad \{ac\} \simeq (13), \quad \{ac^2\} \simeq (23), \qquad (A.34) \\ \{c\} \simeq (123), \quad \{c^2\} \simeq (132).$$

The symmetric group $S_3$ has a 2-dimensional representation $\gamma_2$ such that

$$\text{tr } \gamma_2(12) = \text{tr } \gamma_2(13) = \text{tr } \gamma_2(23) = 0, \quad \text{i.e.,} \\ \text{tr } \gamma_2\{ab\} = \text{tr } \gamma_2\{ac\} = \text{tr } \gamma_2\{ac^2\} = 0, \\ \text{tr } \gamma_2(123) = \text{tr } \gamma_2(132) = -1, \text{ i.e., } \text{tr } \gamma_2\{c\} = \text{tr } \gamma_2\{c^2\} = -1, \\ \text{tr } \gamma_2(1) = 2,$$

see, e.g., [CR62, §32]. We consider now the representation $\rho_2$ as the composition of epimorphism $\pi_2$ and the representation $\gamma_2$, i.e.,

$$\rho_2 = \gamma_2 \pi_2. \tag{A.35}$$

So, for all $u \in Cl(ab) \cup Cl(a) \cup Cl(a^3)$, we see that

$$\pi_2(u) \in \{ab\} \cup \{ac\} \cup \{ac^2\} \text{ and tr } \rho_2(u) = 0.$$

For all $u \in Cl(c) \cup Cl(c^2)$, we see that

$$\pi_2(u) \in \{c\} \cup \{c^2\} \text{ and tr } \rho_2(u) = -1.$$

Finally, for all $u \in Cl(1) \cup Cl(-1) \cup Cl(b)$ we see that

$$\pi_2(u) \in \{1\} \text{ and tr } \rho_2(u) = 2.$$

Thus we obtain the row of characters $\psi_2$.

**Table A.10.**    The 3-dimensional characters of $S_4$

| Character | Conjugacy class $C_i \subset S_4$ and its order $|C_i|$ under it | | | | |
|:---:|:---:|:---:|:---:|:---:|:---:|
| | $C_1$ | $C_2$ | $C_3$ | $C_4$ | $C_5$ |
| | 1 | 6 | 8 | 6 | 3 |
| $\gamma_5$ | 3 | $-1$ | 0 | 1 | $-1$ |
| $\gamma_6$ | 3 | 1 | 0 | $-1$ | $-1$ |

4) We construct representations $\rho_5$ and $\rho_6$ by means of the epimorphism

$$\pi_{56} : \mathcal{O} \longrightarrow \mathcal{O}/\{1, -1\}. \tag{A.36}$$

The epimorphism $\pi_{56}$ is well-defined because the subgroup $\{1, -1\}$ is normal:

$$\{1, -1\} \triangleleft \mathcal{O}. \tag{A.37}$$

The quotient group $\mathcal{O}/\{1, -1\}$ is the 24-element octahedral group coinciding with the symmetric group $S_4$. By [CR62, §32] $S_4$ has two 3-dimensional representations, $\gamma_5$ and $\gamma_6$, with characters as in Table A.10.

In Table A.10 we give the conjugacy classes $C_i$ of the group $S_4$ together with the number of elements of these classes:

$$C_1 = \{1\},$$
$$C_2 = \{ab, a^3b, a^3c, a^3bc, ac^2, a^3bc^2\},$$
$$C_3 = \{c, c^2, bc, a^2c, a^2bc, bc^2, a^2c^2, a^2bc^2\}, \qquad \text{(A.38)}$$
$$C_4 = \{a, a^3, ac, abc, a^3c^2, abc^2\},$$
$$C_5 = \{a^2, b, a^2b\}.$$

We have

$$\rho_5 = \gamma_5\pi_{56}, \quad \rho_6 = \gamma_6\pi_{56}. \qquad \text{(A.39)}$$

For any $u \in Cl(1) \cup Cl(-1)$, we see that

$$\pi_{56}(u) = 1 \text{ and tr } \rho_5(u) = 3.$$

For any $u \in Cl(ab)$, we see that

$$\pi_{56}(u) \in C_2 \text{ and tr } \rho_5(u) = -1.$$

For any $u \in Cl(b)$, we see that

$$\pi_{56}(u) \in C_5 \text{ and tr } \rho_5(u) = -1.$$

For any $u \in Cl(c) \cup Cl(c^2)$ , we see that

$$\pi_{56}(u) \in C_3 \text{ and tr } \rho_5(u) = 0.$$

Finally, for any $u \in Cl(a) \cup Cl(a^2)$ we see that

$$\pi_{56}(u) \in C_4 \text{ and tr } \rho_5(u) = 1.$$

Thus we obtain the row of characters $\psi_5$.

Note, that $\rho_6$ can be obtained from $\rho_5$ by the following relations:

$$\rho_6(a) = -\rho_5(a), \quad \rho_6(b) = \rho_5(b), \quad \rho_6(c) = \rho_5(c). \qquad \text{(A.40)}$$

5) Finally, the 4-dimensional representation $\rho_7$ is constructed as either of the tensor products $\rho_2 \otimes \rho_3$ or $\rho_2 \otimes \rho_4$. Observe that $\psi_2\psi_3$ and $\psi_2\psi_4$ have the same characters, see Table A.11.

**Table A.11.**    The character of the 4-dimensional representation $\rho_7$

| | $Cl(1)$ | $Cl(-1)$ | $Cl(ab)$ | $Cl(b)$ | $Cl(c^2)$ | $Cl(c)$ | $Cl(a)$ | $Cl(a^3)$ |
|---|---|---|---|---|---|---|---|---|
| $\psi_7 = \psi_2\psi_3 = \psi_2\psi_4$ | 4 | -4 | 0 | 0 | 1 | -1 | 0 | 0 |

The irreducibility of $\rho_2 \otimes \rho_3$ follows from the fact that

$$\langle \psi_2\psi_3, \psi_2\psi_3 \rangle = \frac{16 + 16 + 8 + 8}{48} = 1. \quad \square$$

*Example A.9 (The McKay correspondence for the binary octahedral group).*
Select $\rho_3$ as a faithful representation of $\mathcal{O}$ from the McKay correspondence.
All irreducible representations $\rho_i$ of $\mathcal{O}$ (see Proposition A.8 and Table A.9)
can be placed in vertices of the extended Dynkin diagram $\widetilde{E}_7$, see (A.41):

$$\rho_0 \relbar\joinrel\relbar \rho_3 \relbar\joinrel\relbar \rho_5 \relbar\joinrel\relbar \rho_7 \relbar\joinrel\relbar \rho_6 \relbar\joinrel\relbar \rho_4 \relbar\joinrel\relbar \rho_1$$

$$\begin{matrix} & & & | & & & \\ & & & \rho_2 & & & \end{matrix} \qquad\text{(A.41)}$$

Then, according to the McKay correspondence we have the following decompositions of the tensor products $\rho_3 \otimes \rho_i$:

$$
\begin{aligned}
\rho_3 \otimes \rho_0 &= \rho_3, \\
\rho_3 \otimes \rho_1 &= \rho_4, \\
\rho_3 \otimes \rho_2 &= \rho_7, \\
\rho_3 \otimes \rho_3 &= \rho_0 + \rho_5, \\
\rho_3 \otimes \rho_4 &= \rho_1 + \rho_6, \\
\rho_3 \otimes \rho_5 &= \rho_3 + \rho_7, \\
\rho_3 \otimes \rho_6 &= \rho_4 + \rho_7, \\
\rho_3 \otimes \rho_7 &= \rho_2 + \rho_5 + \rho_6. \quad \square
\end{aligned}
\qquad\text{(A.42)}
$$

**Table A.12.**    The characters of the binary tetrahedral group. Here, $\omega_3 = e^{2\pi i/3}$.

| Character | Conjugacy class $Cl(g)$ and its order $|Cl(g)|$ under it | | | | | | Note on |
|-----------|----------|-----------|----------|----------|-------------|-----------|-------------|
| $\chi_i$ | $Cl(1)$ | $Cl(-1)$ | $Cl(b)$ | $Cl(c)$ | $Cl(c^2)$ | $Cl(-c)$ | $Cl(-c^2)$ | $\tau_i$ |
|  | 1 | 1 | 6 | 4 | 4 | 4 | 4 |  |
| $\chi_0$ | 1 | 1 | 1 | 1 | 1 | 1 | 1 | trivial |
| $\chi_1$ | 1 | 1 | 1 | $\omega_3$ | $\omega_3^2$ | $\omega_3$ | $\omega_3^2$ | $\tau_1(c) = \omega_3$ |
| $\chi_2$ | 1 | 1 | 1 | $\omega_3^2$ | $\omega_3$ | $\omega_3^2$ | $\omega_3$ | $\tau_2(c) = \omega_3^2$ |
| $\chi_3$ | 2 | $-2$ | 0 | 1 | $-1$ | $-1$ | 1 | faithful |
| $\chi_4$ | 2 | $-2$ | 0 | $\omega_3$ | $-\omega_3^2$ | $-\omega_3$ | $\omega_3^2$ | $\tau_3 \otimes \tau_1$ |
| $\chi_5$ | 2 | $-2$ | 0 | $\omega_3^2$ | $-\omega_3$ | $-\omega_3^2$ | $\omega_3$ | $\tau_3 \otimes \tau_2$ |
| $\chi_6$ | 3 | 3 | $-1$ | 0 | 0 | 0 | 0 | $\gamma_6 \pi_6$ |

**Proposition A.10.** *The group $\mathcal{T}$ has the following 7 irreducible representations:*

*1) Three 1-dimensional representations*

$$\tau_0(a) = \tau_0(b) = \tau_0(c) = 1,$$
$$\tau_1(a) = \tau_1(b) = 1, \quad \tau_1(c) = \omega_3, \qquad (A.43)$$
$$\tau_2(a) = \tau_2(b) = 1, \quad \tau_2(c) = \omega_3^2,$$

*Representations $\tau_1$ and $\tau_2$ can be constructed by using an epimorphism onto the alternating group $A_4$:*

$$\mathcal{T} \longrightarrow \mathcal{T}/\{1, -1\} = A_4. \qquad (A.44)$$

*2) The faithful 2-dimensional representation*

$$\tau_3(a) = a, \quad \tau_3(b) = b, \quad \tau_3(c) = c, \qquad (A.45)$$

*3) Two 2-dimensional representation $\tau_4$ and $\tau_5$ constructed as tensor products*

$$\tau_4 = \tau_3 \otimes \tau_1, \quad \tau_5 = \tau_3 \otimes \tau_2. \qquad (A.46)$$

*4) The 3-dimensional representation $\tau_6$ constructed by using an epimorphism (A.44) onto the alternating group $A_4$.*
*The characters of representations $\tau_i$ for $i = 0, \ldots, 6$ are collected in Table A.12.*

*Proof.* 1), 2) and 3) are easily checked.
4) Consider the 3-dimensional representation of the alternating group $A_4$ with the character given in Table A.13, see, e.g., [CR62, §32]:

**Table A.13.**    The 3-dimensional character of $A_4$

| Character | Conjugacy class $C_i \subset A_4$ and its order $|C_i|$ under it | | | |
|:---:|:---:|:---:|:---:|:---:|
| | $C_1$ | $C_2$ | $C_3$ | $C_4$ |
| | 1 | 3 | 4 | 4 |
| $\gamma_6$ | 3 | $-1$ | 0 | 0 |

In Table A.13 we have $C_1 = \{1\}$, $C_2$ contains only elements of order 2, and $C_3$, $C_4$ contain only elements of order 3, see (A.47).

$$C_1 = \{1\},$$
$$C_2 = \{a^2, b, a^2b\},$$
$$C_3 = \{c, a^2c, bc, a^2bc\}, \qquad (A.47)$$
$$C_4 = \{c^2, a^2c^2, bc^2, a^2bc^2\}.$$

If $\pi_6$ is an epimorphism (A.44), then

$$\tau_6 = \gamma_6 \pi_6, \tag{A.48}$$

and

$$
\begin{aligned}
&\tau_6 : Cl(1) \cup Cl(-1) \longrightarrow C_1, \quad \text{tr } \tau_6(u) = 3, \\
&\tau_6 : Cl(b) \longrightarrow C_2, \qquad\qquad\ \text{tr } \tau_6(u) = -1, \\
&\tau_6 : Cl(c) \cup Cl(-c) \longrightarrow C_3, \quad \text{tr } \tau_6(u) = 0, \\
&\tau_6 : Cl(c^2) \cup Cl(-c^2) \longrightarrow C_4, \text{tr } \tau_6(u) = 0.
\end{aligned}
\tag{A.49}
$$

Thus we get the last row in Table A.12.   □

*Example A.11 (The McKay correspondence for the binary tetrahedral group).*
Select $\tau_3$ as a faithful representation of $\mathcal{T}$ from the McKay correspondence.
All irreducible representations $\tau_i$ of $\mathcal{T}$ (see Proposition A.10 and Table A.12)
can be placed in vertices of the extended Dynkin diagram $\widetilde{E}_6$, see (A.50):

$$
\begin{array}{ccccccccc}
\tau_1 & \!\!\!—\!\!\! & \tau_4 & \!\!\!—\!\!\! & \tau_6 & \!\!\!—\!\!\! & \tau_5 & \!\!\!—\!\!\! & \tau_2 \\
 & & & & | & & & & \\
 & & & & \tau_3 & & & & \\
 & & & & | & & & & \\
 & & & & \tau_0 & & & &
\end{array}
\tag{A.50}
$$

Then, according to the McKay correspondence, we have the following decompositions of the tensor products $\tau_3 \otimes \tau_i$:

$$
\begin{aligned}
\tau_3 \otimes \tau_0 &= \tau_3, \\
\tau_3 \otimes \tau_1 &= \tau_4, \\
\tau_3 \otimes \tau_2 &= \tau_5, \\
\tau_3 \otimes \tau_3 &= \tau_0 + \tau_6, \\
\tau_3 \otimes \tau_4 &= \tau_1 + \tau_6, \\
\tau_3 \otimes \tau_5 &= \tau_2 + \tau_6, \\
\tau_3 \otimes \tau_6 &= \tau_3 + \tau_4 + \tau_5. \quad □
\end{aligned}
\tag{A.51}
$$

### A.5.4 The induced and restricted representations

Let us denote the characters of induced and restricted representations of $\mathcal{T}$
and $\mathcal{O}$ as follows:

$$
\begin{aligned}
&\text{the irreducible representations } \tau \text{ of } \mathcal{T} & \Longleftrightarrow & \quad \text{char}(\tau) := \chi, \\
&\text{the irreducible representations } \rho \text{ of } \mathcal{O} & \Longleftrightarrow & \quad \text{char}(\rho) := \psi, \\
&\text{the induced representations } \tau \uparrow^{\mathcal{O}}_{\mathcal{T}} \text{ of } \mathcal{O} & \Longleftrightarrow & \quad \text{char}(\tau \uparrow^{\mathcal{O}}_{\mathcal{T}}) := \chi^{\uparrow}, \\
&\text{the restricted representations } \rho \downarrow^{\mathcal{O}}_{\mathcal{T}} \text{ of } \mathcal{T} & \Longleftrightarrow & \quad \text{char}(\rho \downarrow^{\mathcal{O}}_{\mathcal{T}}) := \psi^{\downarrow}.
\end{aligned}
\tag{A.52}
$$

**Table A.14.**   The restricted characters of the binary octahedral group

| Character $\psi_i^{\downarrow}$ | Conjugacy class $Cl(g)$ and order of class $|Cl(g)|$ | | | | | | |
|---|---|---|---|---|---|---|---|
| | $Cl_{\mathcal{O}}(1)$ | $Cl_{\mathcal{O}}(-1)$ | $Cl_{\mathcal{O}}(b)$ | $Cl_{\mathcal{O}}(c^2)$ | | $Cl_{\mathcal{O}}(c)$ | |
| | 1 | 1 | 6 | 8 | | 8 | |
| | $Cl_{\mathcal{T}}(1)$ | $Cl_{\mathcal{T}}(-1)$ | $Cl_{\mathcal{T}}(b)$ | $Cl_{\mathcal{T}}(-c)$ | $Cl_{\mathcal{T}}(c^2)$ | $Cl_{\mathcal{T}}(c)$ | $Cl_{\mathcal{T}}(-c^2)$ |
| | 1 | 1 | 6 | 4 | 4 | 4 | 4 |
| $\psi_1^{\downarrow} = \psi_0^{\downarrow}$ | 1 | 1 | 1 | 1 | 1 | 1 | 1 |
| $\psi_2^{\downarrow}$ | 2 | 2 | 2 | $-1$ | $-1$ | $-1$ | $-1$ |
| $\psi_4^{\downarrow} = \psi_3^{\downarrow}$ | 2 | $-2$ | 0 | $-1$ | $-1$ | 1 | 1 |
| $\psi_6^{\downarrow} = \psi_5^{\downarrow}$ | 3 | 3 | $-1$ | 0 | 0 | 0 | 0 |
| $\psi_7^{\downarrow}$ | 4 | $-4$ | 0 | 1 | 1 | $-1$ | $-1$ |

Consider the restriction of the binary octahedral group $\mathcal{O}$ onto the binary tetrahedral subgroup $\mathcal{T}$. Then the conjugacy classes $Cl(a)$, $Cl(a^3)$ and $Cl(ab)$ disappear, and the remaining 5 classes split into 7 conjugacy classes, see Table A.14. We denote the conjugacy classes of $\mathcal{O}$ by $Cl_{\mathcal{O}}$ and conjugacy classes of $\mathcal{T}$ by $Cl_{\mathcal{T}}$.

Now consider the restricted representations $\psi_i^{\downarrow}$ from $\mathcal{O}$ onto $\mathcal{T}$. By Table A.9 $\psi_0$ (resp. $\psi_3$ or $\psi_5$) differs from $\psi_1$ (resp. $\psi_4$ or $\psi_6$) only on $Cl(a)$, $Cl(a^3)$, and $Cl(ab)$, so we have the following coinciding pairs of restricted representations:

$$\psi_0^{\downarrow} = \psi_1^{\downarrow}, \quad \psi_3^{\downarrow} = \psi_4^{\downarrow}, \quad \psi_5^{\downarrow} = \psi_6^{\downarrow}. \tag{A.53}$$

The values of the characters $\psi_i^{\downarrow}$ for $i = 0, 2, 3, 5, 7$ are easily obtained from the corresponding characters $\psi_i$.

Observe that $\rho_3^{\downarrow} = \tau_3$ is a faithful representation of $\mathcal{T}$ with character $\psi_3^{\downarrow} = \chi_3$. All irreducible representations $\rho_i^{\downarrow}$ for $i = 0, 2, 3, 5, 7$ can be placed in vertices of the extended Dynkin diagram $\widetilde{F}_{42}$, see Fig. A.2. From Table A.14 we have

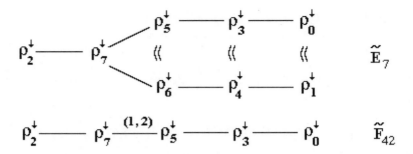

Fig. A.2.   The induced and restricted representations of $\mathcal{T} \lhd \mathcal{O}$

$$\tau_3 \otimes \rho_0^{\downarrow} = \rho_3^{\downarrow} \otimes \rho_0^{\downarrow} = \rho_3^{\downarrow},$$
$$\tau_3 \otimes \rho_2^{\downarrow} = \rho_3^{\downarrow} \otimes \rho_2^{\downarrow} = \rho_7^{\downarrow},$$
$$\tau_3 \otimes \rho_3^{\downarrow} = \rho_3^{\downarrow} \otimes \rho_3^{\downarrow} = \rho_0^{\downarrow} + \rho_5^{\downarrow}, \qquad (A.54)$$
$$\tau_3 \otimes \rho_5^{\downarrow} = \rho_3^{\downarrow} \otimes \rho_5^{\downarrow} = \rho_3^{\downarrow} + \rho_7^{\downarrow},$$
$$\tau_3 \otimes \rho_7^{\downarrow} = \rho_3^{\downarrow} \otimes \rho_7^{\downarrow} = \rho_2^{\downarrow} + 2\rho_5^{\downarrow}.$$

Thus, the decompositions (A.54) constitute the following matrix

$$\widetilde{A} = \begin{pmatrix} 0 & 1 & 0 & 0 & 0 \\ 1 & 0 & 2 & 0 & 0 \\ 0 & 1 & 0 & 1 & 0 \\ 0 & 0 & 1 & 0 & 1 \\ 0 & 0 & 0 & 1 & 0 \end{pmatrix} \begin{matrix} \rho_2^{\downarrow} \\ \rho_7^{\downarrow} \\ \rho_5^{\downarrow} \\ \rho_3^{\downarrow} \\ \rho_0^{\downarrow} \end{matrix} \qquad (A.55)$$

where the row associated with $\rho_i^{\downarrow}$ for $i = 2, 7, 5, 3, 0$ consists of the decomposition coefficients (A.54) of $\tau_3 \otimes \rho_i^{\downarrow}$. The matrix $\widetilde{A}$ satisfies the relation

$$\widetilde{A} = 2I - K, \tag{A.56}$$

where $K$ is the Cartan matrix for the extended Dynkin diagram $\widetilde{F}_{42}$, see (2.22). We call the matrix $\widetilde{A}$ the *Slodowy matrix*; it is an analog of the McKay matrix for the multiply-laced case.

Now we move on to the dual case and consider induced representations. To obtain induced representations $\chi \uparrow_T^{\mathcal{O}}$, we use the *Frobenius reciprocity formula* connecting restricted and induced representations, see, e.g., [JL2001, 21.16]

$$\langle \psi, \chi \uparrow_H^G \rangle_G = \langle \psi \downarrow_H^G, \chi \rangle_H. \tag{A.57}$$

By (A.57) we have the following expression for the characters of induced representations

$$\chi \uparrow_H^G = \sum_{\psi_i \in \mathrm{Irr}(G)} \langle \psi_i, \chi \uparrow_H^G \rangle_G \psi_i = \sum_{\psi_i \in \mathrm{Irr}(G)} \langle \psi \downarrow_H^G, \chi \rangle_H \psi_i. \tag{A.58}$$

Thus, to calculate the characters $\chi^{\uparrow} = \chi \uparrow_T^{\mathcal{O}}$, we only need to calculate the inner products

$$\langle \psi^{\downarrow}, \chi \rangle_T = \langle \psi \downarrow_T^{\mathcal{O}}, \chi \rangle_T. \tag{A.59}$$

**Table A.15.**   The inner products $\langle \psi^{\downarrow}, \chi \rangle$

|            | $\psi_1^{\downarrow} = \psi_0^{\downarrow}$ | $\psi_2^{\downarrow}$ | $\psi_3^{\downarrow} = \psi_4^{\downarrow}$ | $\psi_5^{\downarrow} = \psi_6^{\downarrow}$ | $\psi_7^{\downarrow}$ |
|------------|------|------|------|------|------|
| $\chi_0$   | 1    | 0    | 0    | 0    | 0    |
| $\chi_1$   | 0    | 1    | 0    | 0    | 0    |
| $\chi_2$   | 0    | 1    | 0    | 0    | 0    |
| $\chi_3$   | 0    | 0    | 1    | 0    | 0    |
| $\chi_4$   | 0    | 0    | 0    | 0    | 1    |
| $\chi_5$   | 0    | 0    | 0    | 0    | 1    |
| $\chi_6$   | 0    | 0    | 0    | 1    | 0    |

One can obtain the inner products (A.59) from Tables A.12 and A.14. The results are given in Table A.15. Further, from Table A.15 and (A.58) we deduce

$$\chi_0^\uparrow = \psi_0 + \psi_1, \qquad\qquad\qquad \tau_0^\uparrow = \rho_0 + \rho_1,$$
$$\chi_1^\uparrow = \chi_2^\uparrow = \psi_2, \qquad\qquad \tau_1^\uparrow = \tau_2^\uparrow = \rho_2,$$
$$\chi_3^\uparrow = \psi_3 + \psi_4, \qquad\qquad \tau_3^\uparrow = \rho_3 + \rho_4, \qquad\text{(A.60)}$$
$$\chi_4^\uparrow = \chi_5^\uparrow = \psi_7, \qquad\qquad \tau_4^\uparrow = \tau_5^\uparrow = \rho_7,$$
$$\chi_6^\uparrow = \psi_5 + \psi_6, \qquad\qquad \tau_6^\uparrow = \rho_5 + \rho_6.$$

Let us find the tensor products $\rho_f \otimes \tau_i^\uparrow = \rho_3 \otimes \tau_i^\uparrow$, where $\rho_f$ means the faithful representation of $\mathcal{O}$. By (A.42) and (A.60) we have

$$\rho_3 \otimes \tau_0^\uparrow = \rho_3 \otimes (\rho_0 + \rho_1) = \rho_3 \otimes \rho_0 + \rho_3 \otimes \rho_1 = \rho_3 + \rho_4 = \tau_3^\uparrow,$$
$$\rho_3 \otimes \tau_1^\uparrow = \rho_3 \otimes \tau_2^\uparrow = \rho_3 \otimes \rho_2 = \rho_7 = \tau_4^\uparrow = \tau_5^\uparrow,$$
$$\rho_3 \otimes \tau_3^\uparrow = \rho_3 \otimes (\rho_3 + \rho_4) =$$
$$\rho_3 \otimes \rho_3 + \rho_3 \otimes \rho_4 = (\rho_0 + \rho_5) + (\rho_1 + \rho_6) = \tau_0^\uparrow + \tau_6^\uparrow, \qquad\text{(A.61)}$$
$$\rho_3 \otimes \tau_4^\uparrow = \rho_3 \otimes \tau_5^\uparrow = \rho_3 \otimes \rho_7 = \rho_2 + \rho_5 + \rho_6 = \tau_1^\uparrow + \tau_6^\uparrow = \tau_2^\uparrow + \tau_6^\uparrow,$$
$$\rho_3 \otimes \tau_6^\uparrow = \rho_3 \otimes (\rho_5 + \rho_6) = \rho_3 \otimes \rho_5 + \rho_3 \otimes \rho_6 =$$
$$(\rho_3 + \rho_7) + (\rho_4 + \rho_7) = \tau_3^\uparrow + 2\tau_4^\uparrow = \tau_3^\uparrow + 2\tau_5^\uparrow.$$

Here $\rho_3$ is the faithful representation of $\mathcal{O}$ with character $\psi_3$. All irreducible representations $\tau_i^\uparrow$ for $i = 0, 2, 3, 5, 6$ can be placed in vertices of the extended Dynkin diagram $\widetilde{F}_{41}$, see Fig. A.2. For other details, see [Sl80, App.III, p. 164].

The decompositions (A.61) constitute the following matrix

$$\widetilde{A}^\vee = \begin{pmatrix} 0 & 1 & 0 & 0 & 0 \\ 1 & 0 & 1 & 0 & 0 \\ 0 & 2 & 0 & 1 & 0 \\ 0 & 0 & 1 & 0 & 1 \\ 0 & 0 & 0 & 1 & 0 \end{pmatrix} \begin{matrix} \tau_2^\uparrow \\ \tau_5^\uparrow \\ \tau_6^\uparrow \\ \tau_3^\uparrow \\ \tau_0^\uparrow \end{matrix} \qquad\text{(A.62)}$$

where the row associated with $\tau_i^\uparrow$ for $i = 2, 5, 6, 3, 0$ consists of the decomposition coefficients (A.61) of $\rho_3 \otimes \tau_i^\uparrow$. As in (A.56), the matrix $\widetilde{A}^\vee$ satisfies the relation

$$\widetilde{A}^\vee = 2I - K^\vee, \qquad\text{(A.63)}$$

where $K^\vee$ is the Cartan matrix for the extended Dynkin diagram $\widetilde{F}_{41}$, see (2.20). We see that the matrices $\widetilde{A}$ and $\widetilde{A}^\vee$ are mutually transposed:

$$\widetilde{A}^t = \widetilde{A}^\vee. \qquad\text{(A.64)}$$

As in (A.56), we call the matrix $\widetilde{A}^\vee$ the *Slodowy matrix*.

## A.6 The characters of the binary polyhedral groups

Representations of the binary octahedral and tetrahedral groups are considered in §A.5.3. For characters of the binary tetrahedral group, see Table A.12; for characters of the binary octahedral group, see Table A.9. Below, we give character tables of remaining three groups: cyclic groups, binary dihedral groups and binary dihedral group, see [Blu06], [Mon07], [Brn01], [IN99], [Hu75].

### A.6.1 The cyclic groups

The irreducible representations are

$$\tau_j : \mathbb{Z}/n\mathbb{Z} \longrightarrow \mathbb{C}^*, \quad a^i \longrightarrow (\xi^j)^i = \xi^{ji}, \text{ for } i, j = 0, 1, \ldots, n-1,$$
$$\text{where } \xi \text{ is a primitive } n\text{-th root of unity.} \tag{A.65}$$

**Table A.16.**  The characters of the cyclic group

| Character | Conjugacy class $Cl(g)$ and its order under it | | | | | | Note on |
|---|---|---|---|---|---|---|---|
| $\chi_i$ | $Cl(1)$ | $Cl(a)$ | $Cl(a^2)$ | $\ldots$ | $Cl(a^{n-2})$ | $Cl(a^{n-1})$ | represent. |
| | 1 | 1 | 1 | $\ldots$ | 1 | 1 | $\tau_i$ |
| $\chi_0$ | 1 | 1 | 1 | $\ldots$ | 1 | 1 | $\tau_0$ |
| $\chi_1$ | 1 | $\xi$ | $\xi^2$ | $\ldots$ | $\xi^{n-2}$ | $\xi^{n-1}$ | $\tau_1$ |
| $\chi_2$ | 1 | $\xi^2$ | $\xi^4$ | $\ldots$ | $\xi^{2(n-2)}$ | $\xi^{2(n-1)}$ | $\tau_2$ |
| $\ldots$ | $\ldots$ | $\ldots$ | $\ldots$ | $\ldots$ | $\ldots$ | $\ldots$ | $\ldots$ |
| $\chi_{n-2}$ | 1 | $\xi^{(n-2)}$ | $\xi^{2\times(n-2)}$ | $\ldots$ | $\xi^{(n-2)\times(n-2)}$ | $\xi^{(n-2)\times(n-1)}$ | $\tau_{n-2}$ |
| $\chi_{n-1}$ | 1 | $\xi^{(n-1)}$ | $\xi^{2\times(n-1)}$ | $\ldots$ | $\xi^{(n-2)\times(n-1)}$ | $\xi^{(n-1)\times(n-1)}$ | $\tau_{n-1}$ |

### A.6.2 The binary dihedral groups

The binary dihedral group $\mathcal{D}$ is also known as *dicyclic group*. The group $\mathcal{D}$ has the following presentation:

$$\mathcal{D} = \{a, b \mid a^n = b^2 = (ba)^2 = -1\}. \tag{A.66}$$

By setting

$$R := b, S := a, T := ba,$$

we deduce that presentation (A.66) is equivalent to

**Table A.17.**    The characters of the binary dihedral group, $n$ even

| Character $\chi_i$ | $Cl(1)$ | $Cl(-1)$ | $Cl(a^k)$ for $k = 1, \ldots, n-1$ | $Cl(b)$ | $Cl(ba)$ |
|---|---|---|---|---|---|
| | 1 | 1 | 2 | $n$ | $n$ |
| $\chi_1$ | 1 | 1 | 1 | 1 | 1 |
| $\chi_2$ | 1 | 1 | 1 | -1 | -1 |
| $\chi_3$ | 1 | -1 | $(-1)^k$ | $i$ | $-i$ |
| $\chi_4$ | 1 | -1 | $(-1)^k$ | $-i$ | $i$ |
| $\chi_1'$ | 2 | -2 | $\xi^k + \xi^{-k}$ | 0 | 0 |
| $\chi_2'$ | 2 | 2 | $\xi^{2k} + \xi^{-2k}$ | 0 | 0 |
| $\ldots$ | $\ldots$ | $\ldots$ | $\ldots$ | $\ldots$ | $\ldots$ |
| $\chi_{n-1}'$ | 2 | $(-2)^{n-1}$ | $\xi^{k(n-1)} + \xi^{-k(n-1)}$ | 0 | 0 |

**Table A.18.**    The characters of the binary dihedral group, $n$ odd

| Character $\chi_i$ | $Cl(1)$ | $Cl(-1)$ | $Cl(a^k)$ for $k = 1, \ldots, n-1$ | $Cl(b)$ | $Cl(ba)$ |
|---|---|---|---|---|---|
| | 1 | 1 | 2 | $n$ | $n$ |
| $\chi_1$ | 1 | 1 | 1 | 1 | 1 |
| $\chi_2$ | 1 | 1 | 1 | $-1$ | $-1$ |
| $\chi_3$ | 1 | -1 | $(-1)^k$ | 1 | $-1$ |
| $\chi_4$ | 1 | -1 | $(-1)^k$ | $-1$ | 1 |
| $\chi_1'$ | 2 | -2 | $\xi^k + \xi^{-k}$ | 0 | 0 |
| $\chi_2'$ | 2 | 2 | $\xi^{2k} + \xi^{-2k}$ | 0 | 0 |
| $\ldots$ | $\ldots$ | $\ldots$ | $\ldots$ | $\ldots$ | $\ldots$ |
| $\chi_{n-1}'$ | 2 | $(-2)^{n-1}$ | $\xi^{k(n-1)} + \xi^{-k(n-1)}$ | 0 | 0 |

$$R^2 = S^n = T^2 = RST = -1, \qquad (A.67)$$

i.e., eq. (A.5) holds for $\langle 2, 2, n \rangle$.

The characters of the binary dihedral group $\mathcal{D}$ are given in Table A.17 ($n$ is even) and Table A.18 ($n$ is odd). In these tables $\xi$ is a primitive $2n$-th root of unity.

### A.6.3 The binary icosahedral group

The binary icosahedral group $\mathcal{J}$ has the presentation:

$$\mathcal{J} = \{a, b \mid a^5 = b^3 = (ba)^2 = -1\}. \tag{A.68}$$

**Table A.19.** The characters of the binary icosahedral group

| Character | The conjugacy class $Cl(g)$ and its order (under it) | | | | | | | | |
|---|---|---|---|---|---|---|---|---|---|
| $\chi_i$ | $Cl(1)$ | $Cl(-1)$ | $Cl(a)$ | $Cl(a^2)$ | $Cl(a^3)$ | $Cl(a^4)$ | $Cl(b)$ | $Cl(b^2)$ | $Cl(ab)$ |
| | 1 | 1 | 12 | 12 | 12 | 12 | 20 | 20 | 30 |
| $\chi_1$ | 1 | 1 | 1 | 1 | 1 | 1 | 1 | 1 | 1 |
| $\chi_2$ | 2 | $-2$ | $\mu^+$ | $-\mu^-$ | $\mu^-$ | $-\mu^+$ | 1 | $-1$ | 0 |
| $\chi_3$ | 2 | $-2$ | $\mu^-$ | $-\mu^+$ | $\mu^+$ | $-\mu^-$ | 1 | $-1$ | 0 |
| $\chi_4$ | 3 | 3 | $\mu^+$ | $\mu^-$ | $\mu^-$ | $\mu^+$ | 0 | 0 | $-1$ |
| $\chi_5$ | 3 | 3 | $\mu^-$ | $\mu^+$ | $\mu^+$ | $\mu^-$ | 0 | 0 | $-1$ |
| $\chi_6$ | 4 | $-4$ | 1 | $-1$ | 1 | $-1$ | $-1$ | 1 | 0 |
| $\chi_7$ | 4 | 4 | $-1$ | $-1$ | $-1$ | $-1$ | 1 | 1 | 0 |
| $\chi_8$ | 5 | 5 | 0 | 0 | 0 | 0 | $-1$ | $-1$ | 1 |
| $\chi_9$ | 6 | $-6$ | $-1$ | 1 | $-1$ | 1 | 0 | 0 | 0 |

By setting

$$R := b, S := a, T := ba,$$

we deduce that presentation (A.68) is equivalent to the following presentation:

$$R^3 = S^5 = T^2 = RST = -1, \tag{A.69}$$

i.e., eq. (A.5) holds for $\langle 2, 3, 5 \rangle$.

Let $\mu^+$, $\mu^-$ be as follows:

$$\mu^+ = \frac{1}{2}(1 + \sqrt{5}), \text{ and } \mu^- = \frac{1}{2}(1 - \sqrt{5}).$$

The characters of the binary icosahedral group $\mathcal{J}$ are given in Table A.19.

# B

# Regularity conditions for representations of quivers

## B.1 The Coxeter functors and regularity conditions

Following Bernstein, Gelfand and Ponomarev [BGP73], given a quiver $Q$ and a field $K$, we define reflection functors and Coxeter functors. For details, see [ASS06], [Pie82] or [Ser05].

A vertex $a \in Q_0$ is said to be *sink-admissible* (resp. *source-admissible*) if all arrows containing $a$ have $a$ as a target (resp. as a source). By $\sigma_a Q$ we denote the quiver obtained from $Q$ by inverting all arrows containing $a$.

For each sink-admissible vertex $a$, we define the *reflection functor*

$$F_a^+ : \operatorname{rep}_K(Q) \longrightarrow \operatorname{rep}_K(\sigma_a Q)$$

between the categories of finite dimensional $K$-linear representations of the quivers $Q$ and $\sigma_a Q$, see §B.1.1, and for each source-admissible vertex $a$, we define the *reflection functor*

$$F_a^- : \operatorname{rep}_K(\sigma_a Q) \longrightarrow \operatorname{rep}_K(Q)$$

between the categories of finite dimensional $K$-linear representations of the quivers $\sigma_a Q$ and $Q$, see §B.1.2. After that, in §B.1.3 we give the definition of the Coxeter functors $\Phi^+$ and $\Phi^-$.

The connection between the Coxeter functors and the Coxeter transformations is as follows: the action of the Coxeter functor on the objects of the $\operatorname{rep}_K(Q)$ induces the action of the corresponding Coxeter transformation on dimensions of these objects. An essential difference is that the Coxeter functor is not always invertible, whereas the corresponding Coxeter is invertible. There are simple objects $V$, $W$ which are turned into $0$ under the action the Coxeter functors:

$$\Phi^+(V) = 0, \quad \Phi^-(V) \neq 0.$$
$$\Phi^-(W) = 0, \quad \Phi^+(W) \neq 0.$$

Only on the regular objects the behavior of the Coxeter functor is "more agreeable", and the regular representation never vanishes under the action of the Coxeter functor, see §B.1.4.

### B.1.1 The reflection functor $F_a^+$

Let

$$V = (V_i, \delta_\alpha)_{i \in Q_0, \alpha \in Q_1} \tag{B.1}$$

be an object of $\mathrm{rep}_K(Q)$. The object

$$F_a^+ V = (V_i', \delta_\alpha')_{i \in (\sigma_a Q)_0, \alpha \in (\sigma_a Q)_1}$$

in $\mathrm{rep}_K(\sigma_a Q)$ is defined as follows.

We put $V_i' = V_i$ for $i \neq a$. For $i = a$, we put $V_a' = \ker \nabla$, where

$$\nabla = \sum_{\alpha : s(\alpha) \to a} \delta_\alpha : \bigoplus_\alpha V_{s(\alpha)} \to V_\alpha,$$

and $s(\alpha)$ is the source of the arrow $\alpha$. The $K$-linear map $\nabla$ acts such that

$$\nabla(v_1, \ldots, v_n) = \sum_{\alpha : s(\alpha) \to a} \delta_\alpha(v_{s(\alpha)}),$$

for all collections $\{\eta_1, \ldots, \eta_n \mid \eta_i \in V_i\}$, where the indices $1, \ldots, n$ enumerate all arrows ending with $a$.

We put $\delta_\alpha' = \delta_\alpha$ for all arrows $\alpha$ whose sink $t(\alpha)$ differs from $a$. For $t(\alpha) = a$, we define the map

$$\delta_\alpha' : V_a' \to V_i' = V_i$$

to be the composition of the inclusion $V_a'$ into $\displaystyle\bigoplus_{\alpha : s(\alpha) \to a} V_{s(\alpha)}$ with the projection on the direct summand $V_i$.

Following [ASS06, §VII.5.5], we define the action of the reflection functor $F_a^+$ on the morphisms between representations in the category $\mathrm{rep}_K(Q)$. Let

$$f = (f_i)_{i \in Q_0} : V \to W$$

be a morphism in $\mathrm{rep}_K(Q)$, where $V = (V_i, \delta_\alpha)$ and $W = (W_i, \mu_\alpha)$. We define the morphism

$$F_a^+ f = f' = (f_i')_{i \in \sigma Q_0} : F_a^+ V \to F_a^+ W$$

as follows. For each $i \neq a$, we put $f_i' = f_i$. For $i = a$, we give the $K$-linear map $f_a'$, such that the following diagram is commutative

$$
\begin{array}{ccccccc}
0 & \longrightarrow & (F_a^+ V)_a & \longrightarrow & \displaystyle\bigoplus_{\alpha : s(\alpha) \to a} V_{s(\alpha)} & \xrightarrow{(\delta_\alpha)_\alpha} & V_\alpha \\
& & \Big\downarrow{f_a'} & & \Big\downarrow{\bigoplus_\alpha f_{s(\alpha)}} & & \Big\downarrow{f_\alpha} \\
0 & \longrightarrow & (F_a^+ W)_a & \longrightarrow & \displaystyle\bigoplus_{\alpha : s(\alpha) \to a} W_{s(\alpha)} & \xrightarrow{(\mu_\alpha)_\alpha} & W_\alpha
\end{array}
$$

## B.1.2 The reflection functor $F_a^-$

Let

$$V' = (V_i', \delta_\alpha')_{i \in \sigma_a Q_0, \alpha \in \sigma_a Q_1} \qquad (B.2)$$

be an object of $\text{rep}_K(\sigma_a Q)$. The object

$$F_a^- V' = (V_i, \delta_\alpha)_{i \in (\sigma_a Q)_0, \alpha \in (\sigma_a Q)_1}$$

in $\text{rep}_K(Q)$ is defined as follows.

We put $V_i = V_i'$ for $i \neq a$. For $i = a$, we put $V_a = \bigoplus_{\alpha:a \to t(\alpha)} V_{t(\alpha)}'/\text{Im}\delta_\alpha'$,

where

$$\delta_\alpha' : V_a' \to \bigoplus_{\alpha:a \to t(\alpha)} V_{t(\alpha)}'$$

and $t(\alpha)$ is the sink of the arrow $\alpha$.

We put $\delta_\alpha = \delta_\alpha'$ for all arrows $\alpha$ such that the source $s(\alpha) \neq a$. For $s(\alpha) = a$, we define the map

$$\delta_\alpha : V_i' = V_i \to V_a$$

to be the composition of the inclusion $V_i'$ into $\bigoplus_{\alpha:a \to t(\alpha)} V_{t(\alpha)}'$ with the projection

onto $V_a = \bigoplus_{\alpha:a \to t(\alpha)} V_{t(\alpha)}'/\text{Im}\delta_\alpha'$.

We define the action of the reflection functor $F_a^-$ on the morphisms between representations in the category $\text{rep}_K(\sigma_a Q)$. Let

$$f' = (f_i')_{i \in \sigma_a Q_0} : V' \to W'$$

be a morphism in $\text{rep}_K(\sigma_a Q)$, where $V' = (V_i', \delta_\alpha')$ and $W = (W_i', \mu_\alpha')$. We define the morphism

$$F_a^- f' = f = (f_i)_{i \in \sigma Q_0} : F_a^- V' \to F_a^- W'$$

as follows. For each $i \neq a$, we put $f_i = f_i'$. For $i = a$, we give the $K$-linear map $f_a$, such that the following diagram is commutative

$$
\begin{array}{ccccccc}
V_a' & \longrightarrow & \bigoplus_{\alpha:a \to t(\alpha)} V_{t(\alpha)}' & \xrightarrow{(\delta_\alpha)_\alpha} & (F_a^- V')_\alpha & \longrightarrow & 0 \\
\downarrow{f_a'} & & \downarrow{\bigoplus_\alpha f_{t(\alpha)}'} & & \downarrow{f_a} & & \\
W_a' & \longrightarrow & \bigoplus_{\alpha:a \to t(\alpha)} W_{t(\alpha)}' & \xrightarrow{(\mu_\alpha)_\alpha} & (F_a^- W')_\alpha & \longrightarrow & 0
\end{array}
$$

### B.1.3 The Coxeter functors $\Phi^+$, $\Phi^-$

The main property of the *reflection functors is the fact that the reflection functors preserve the indecomposability of representations*, namely, the following proposition takes place.

**Proposition B.1 ([BGP73]).** *Let $Q$ be a quiver without cycles and $(V, \delta)$ an indecomposable representation. Let $i$ be a sink-admissible (resp. source-admissible) representation. Then*

(i) *if $(V, \delta) \cong (P_i, 0)$, where $(P_i, 0)$ is the simple indecomposable representation, then*

$$F_i^+(V, \delta) = (0, 0) \quad (resp. \quad F_i^-(V, \delta) = (0, 0)).$$

(ii) *if $(V, \delta) \not\cong (P_i, 0)$, then the representation $F_i^+(V, \delta) = (V', \delta')$ (resp. $F_i^-(V, \delta) = (V', \delta')$) is indecomposable,*

$$F_i^- F_i^+(V, \delta) \cong (V, \delta) \quad (resp. \quad F_i^+ F_i^-(V, \delta) \cong (V, \delta))$$

*and*

$$\dim_K V_i' = \dim_K V_i \text{ for } i \neq a,$$
$$\dim_K V_a' = -\dim_K V_a + \sum_{\alpha: s(\alpha) \to a} \dim_K M_{s(\alpha)}.$$

For a proof, see, e.g., [Pie82, §8.8.7]. □

Every tree has a fully sink-admissible sequence $\mathcal{S}$, see §2.2.6, and for every sink-admissible sequence $\mathcal{S}$, we define the Coxeter functors $\Phi^+$ and $\Phi^-$ as follows:

$$\begin{aligned}
\Phi^+ &= F_{i_n}^+ F_{i_{n-1}}^+ ... F_{i_2}^+ F_{i_1}^+, \\
\Phi^- &= F_{i_1}^- F_{i_2}^- ... F_{i_{n-1}}^- F_{i_n}^-.
\end{aligned} \tag{B.3}$$

For every tree-shaped quiver $Q$, every fully sink-admissible sequence gives rise to the same Coxeter functor $\Phi^+$, and every fully source-admissible sequence gives rise to $\Phi^-$, thus the definition of the Coxeter functors do not depend on the order of vertices in $\mathcal{S}$.

The Coxeter functors $\Phi^+, \Phi^-$ are endofunctors, i.e.,

$$\Phi^+ : \operatorname{rep}_K L \longrightarrow \operatorname{rep}_K L, \qquad \Phi^- : \operatorname{rep}_K L \longrightarrow \operatorname{rep}_K L,$$

because every edge of the tree is twice reversed.

## B.1.4 The preprojective and preinjective representations

Let $V$ be an object (representation) of the category $\mathrm{rep}_K(Q)$, see (B.1). A given representation $V$ for which $\Phi^+V = 0$ (resp. $\Phi^-V = 0$) is said to be *projective* (resp. *injective*). For every indecomposable representation $V$, a new indecomposable representation $\Phi^+V$ (resp. $\Phi^-V$) can be constructed, except for the case where $V$ is *projective* (resp. *injective*).

By [GP79, Prop. 8,9] *the projective indecomposable representations of any quiver are naturally enumerated by the vertices of the graph and can be recovered from the orientation of the graph.*

If $V$ is indecomposable and not *projective*, i.e., $\Phi^+V \neq 0$, then $V = \Phi^-\Phi^+V$.

If $V$ is indecomposable and not *injective*, i.e., $\Phi^-V \neq 0$, then $V = \Phi^+\Phi^-V$.

If $V$ is indecomposable and $(\Phi^+)^kV \neq 0$, then $V = (\Phi^-)^k(\Phi^+)^kV$.

If $V$ is indecomposable and $(\Phi^-)^kV \neq 0$, then $V = (\Phi^+)^k(\Phi^-)^kV$.

A representation $V$ is called *preprojective* if, for some *projective* representation $\tilde{V}$,

$$(\Phi^+)^kV = \tilde{V}, \quad (\Phi^+)^{k+1}V = \Phi^+\tilde{V} = 0.$$

A representation $V$ is called *preinjective* if, for some *injective* representation $\tilde{V}$,

$$(\Phi^-)^kV = \tilde{V}, \quad (\Phi^-)^{k+1}V = \Phi^-\tilde{V} = 0.$$

A representation $V$ is called *regular* if

$$(\Phi^+)^kV \neq 0 \text{ and } (\Phi^-)^kV \neq 0 \text{ for every } k \in \mathbb{Z}. \tag{B.4}$$

## B.1.5 The regularity condition

The regularity condition given by the relation (B.4) can be reformulated in terms of dimensions of representations as follows:

**Lemma B.2 ([St75]).** *An indecomposable object* $V \in \mathrm{rep}_K(Q)$ *is regular if and only if, for each* $k \in \mathbb{Z}$, *we have*

$$\mathbf{C}^k\dim V > 0. \tag{B.5}$$

*Proof.* Note that the quiver $Q$ is considered with some orientation $\Omega$, and in the relation (B.5), we assume that $\mathbf{C} = \mathbf{C}_\Omega$.

1) Let $V$ be indecomposable and regular. Suppose there exists $k \in \mathbb{Z}$ such that

$$\mathbf{C}^k\dim V \not> 0. \tag{B.6}$$

Let $n$ be the number of vertices of the quiver $Q$ and $\{i_n, \ldots, i_1\}$ be a fully sink-admissible sequence corresponding to the orientation $\Omega$ as in §B.1. Consider the sequence

$$\{\beta_{nk}, \ldots, i_1\} = \{i_k, \ldots, i_1, \ldots, i_k, \ldots, i_1\}.$$

Then $\mathbf{C}^k = s_{\beta_{nk}} \dots s_{\beta_1}$.

Since $\dim V > 0$, there exists $i < kn$ such that

$$s_{\beta_i} \dots s_{\beta_1}(\dim V) > 0 \text{ and } s_{\beta_{i+1}} \dots s_{\beta_1}(\dim V) \not> 0.$$

According to [BGP73], Corollary 3.1, we have

$$F_{\beta_i}^+ \dots F_{\beta_1}^+(V) = L_{\beta_{i+1}},$$

where $L_{\beta_{i+1}}$ is the simplest object corresponding to the vertex $\beta_{i+1}$:

$$L_{\beta_{i+1}} = \begin{cases} \mathbb{R}^1 & \text{for the vertex } v_{\beta_{i+1}}, \\ 0 & \text{for the other vertices.} \end{cases}$$

Therefore, $F_{\beta_{i+1}}^+ F_{\beta_i}^+ \dots F_{\beta_1}^+(V) = 0$, and $\Phi^+ V = 0$.

2) Conversely, let, for each $k \in \mathbb{Z}$, we have

$$\mathbf{C}^k \dim V > 0. \tag{B.7}$$

Suppose there exists $k \in \mathbb{Z}$ such that $\Phi^+ V = 0$ or $\Phi^- V = 0$. We consider the case $\Phi^+ V = 0$; the case $\Phi^- V = 0$ is similarly considered. The relation $\Phi^+ V = 0$ means that for some $i < kn$, we have

$$F_{\beta_i}^+ \dots F_{\beta_1}^+(V) = L_{\beta_{i+1}},$$

and

$$\dim V = s_{\beta_1} \dots s_{\beta_i}(\beta_{i+1}),$$

where $\beta_{i+1}$ is the simple root corresponding to the vertex $v_{\beta_{i+1}}$.

For some $r \in \mathbb{Z}$, we have

$$\begin{aligned} \beta_{i+1} &= s_{\beta_i} \dots s_{\beta_1}(\dim V) = s_{\alpha_t} \dots s_{\alpha_1} \mathbf{C}^r(\dim V), \quad 1 \le t \le n, \\ z &= s_{\beta_{i+1}}(\beta_{i+1}) = s_{\alpha_{t+1}} s_{\alpha_t} \dots s_{\alpha_1} \mathbf{C}^r(\dim V) < 0, \\ \mathbf{C}^{r+1}(\dim V) &= s_{\alpha_n} \dots s_{\alpha_{t+2}} s_{\alpha_{t+1}} s_{\alpha_t} \dots s_{\alpha_1} \mathbf{C}^r(\dim V) > 0. \end{aligned} \tag{B.8}$$

The vector $z$ has only one non-zero coordinate $(z)_{\beta_{i+1}} < 0$, where $\beta_{i+1} = \alpha_t$. Since the reflections $s_{\alpha_n}, \dots, s_{\alpha_{t+1}}$ do not change this coordinate, we get a contradiction with the last relation in (B.8).   $\square$

## B.2 The necessary regularity conditions for diagrams with indefinite Tits form

Now, consider the case where $\mathcal{B}$ is indefinite, i.e., $\Gamma$ is any tree which is neither a Dynkin diagram nor an extended Dynkin diagram.

Let $(\alpha_1^m, \alpha_2^m, \alpha_1^{\varphi_2}, \alpha_2^{\varphi_2}, \dots)$ be coordinates of the vector $Tz$ in the Jordan basis of eigenvectors and adjoint vectors (3.22) – (3.24), where $\alpha_1^m$ and $\alpha_2^m$

are coordinates corresponding to the eigenvectors $z_1^m$ and $z_2^m$. The vectors $z_1^m$ and $z_2^m$ correspond to the maximal eigenvalue $\varphi^m = \varphi^{max}$ of $DD^t$ and $D^t D$, respectively. Let us decompose the vector $Tz$ as follows:

$$Tz = \alpha_1^m z_1^m + \alpha_2^m z_2^m + \alpha_1^{\varphi_2} z_1^{\varphi_2} + \alpha_2^{\varphi_2} z_2^{\varphi_2} + \dots \qquad (B.9)$$

According to (3.25), (3.26), (3.27) we have

$$\mathbf{C}_\Lambda^k Tz =$$
$$(\lambda_1^{\varphi_m})^k \alpha_1^m z_1^m + (\lambda_2^{\varphi_m})^k \alpha_2^m z_2^m + (\lambda_1^{\varphi_2})^k \alpha_1^{\varphi_2} z_1^{\varphi_2} + (\lambda_1^{\varphi_2})^k \alpha_2^{\varphi_2} z_2^{\varphi_2} + \dots \qquad (B.10)$$

and

$$\mathbf{C}_\Omega^k z = T^{-1}\mathbf{C}_\Lambda^k Tz = (\lambda_1^{\varphi_m})^k \alpha_1^m T^{-1} z_1^m + (\lambda_2^{\varphi_m})^k \alpha_2^m T^{-1} z_2^m +$$
$$(\lambda_1^{\varphi_2})^k \alpha_1^{\varphi_2} T^{-1} z_1^{\varphi_2} + (\lambda_1^{\varphi_2})^k \alpha_2^{\varphi_2} T^{-1} z_2^{\varphi_2} + \dots \qquad (B.11)$$

It will be shown in Theorem B.7 that the transforming element $T$ can be modified so that its decomposition does not contain any given reflection $\sigma_\alpha$. Since the coordinates of the eigenvectors $z_{1,2}^m$ are all positive, see Corollary 3.8 and (3.25), we see that each vector $T^{-1} z_{1,2}^m$ has at least one positive coordinate. Besides,

$$|\lambda_1^{\varphi_m}| > |\lambda_{1,2}^{\varphi_j}| > |\lambda_2^{\varphi_m}| \qquad (B.12)$$

because

$$\lambda_1^{\varphi_m} = \frac{1}{\lambda_2^{\varphi_m}} \quad \text{and}$$
$$\lambda_1^{\varphi_m} = 2\varphi^m - 1 + 2\sqrt{\varphi^m(\varphi^m - 1)} > 2\varphi^i - 1 \pm 2\sqrt{\varphi^i(\varphi^i - 1)}.$$

Thus, since $T^{-1} z_{1,2}^m$ has at least one positive coordinate, we deduce from (B.11) and (B.12) that

$$\alpha_1^{\varphi_m} \geq 0, \qquad \alpha_2^{\varphi_m} \geq 0. \qquad (B.13)$$

As in §6.3.2, for the case $\mathcal{B}$ is non-negative definite, let us calculate $\alpha_{1,2}^{\varphi_m}$. The vector $\tilde{z}_1^m$ conjugate to the vector $z_1^m$ is orthogonal to the vectors $z_{1,2}^2, z_{1,2}^3, \dots$. Let us show that $\tilde{z}_1^m$ is also orthogonal to $z_1^m$. Indeed, the vectors $z_1^m$ and $\tilde{z}_1^m$ can be expressed as follows:

$$z_1^m = \begin{pmatrix} x^m \\ -\dfrac{2}{\lambda_1^m + 1} D^t x^m \end{pmatrix}, \qquad \tilde{z}_1^m = \begin{pmatrix} x^m \\ \dfrac{2\lambda_1^m}{\lambda_1^m + 1} D^t x^m \end{pmatrix}, \qquad (B.14)$$

where for brevity we designate $\lambda_1^{\varphi_{max}}$ by $\lambda_1^m$, $z_1^{\varphi_{max}}$ by $z_1^m$, and $x^{\varphi_{max}}$ by $x^m$. Then,

$$\langle z_1^m, \tilde{z}_1^m \rangle =$$

$$\langle x^m, x^m \rangle - \frac{4\lambda_1^m}{(\lambda_1^m + 1)^2} \langle D^t x^m, D^t x^m \rangle = \tag{B.15}$$

$$\langle x^m, x^m \rangle - \frac{1}{\varphi_m} \langle x^m, DD^t x^m \rangle =$$

$$\langle x^m, x^m \rangle - \langle x^m, x^m \rangle = 0.$$

The conjugate vector $\tilde{z}_1^m$ is not orthogonal only to $z_2^m$ in decomposition (B.9), and similarly $\tilde{z}_2^m$ is not orthogonal only to $z_1^m$. From (B.9) we get

$$\langle Tz, \tilde{z}_1^m \rangle = \alpha_2^m \langle z_2^m, \tilde{z}_1^m \rangle, \qquad \langle Tz, \tilde{z}_2^m \rangle = \alpha_1^m \langle z_1^m, \tilde{z}_2^m \rangle. \tag{B.16}$$

Let us find $\langle z_2^m, \tilde{z}_1^m \rangle$ and $\langle z_1^m, \tilde{z}_2^m \rangle$. We have:

$$\langle z_2^m, \tilde{z}_1^m \rangle = \langle x^m, x^m \rangle - \frac{4}{(\lambda_2^m + 1)^2} \langle x^m, DD^t x^m \rangle =$$

$$\langle x^m, x^m \rangle (1 - \frac{4}{(\lambda_2^m + 1)^2} \frac{(\lambda_2^m + 1)^2}{4(\lambda_2^m)^2}) = \langle x^m, x^m \rangle (1 - (\lambda_1^m)^2). \tag{B.17}$$

Similarly,

$$\langle z_1^m, \tilde{z}_2^m \rangle = \langle x^m, x^m \rangle (1 - (\lambda_2^m)^2). \tag{B.18}$$

Thus, from (B.16), (B.17) and (B.18) we get

$$\langle Tz, \tilde{z}_1^m \rangle = \alpha_2^m \langle x^m, x^m \rangle (1 - (\lambda_1^m)^2),$$

$$\langle Tz, \tilde{z}_2^m \rangle = \alpha_1^m \langle x^m, x^m \rangle (1 - (\lambda_2^m)^2). \tag{B.19}$$

**Theorem B.3 ([SuSt75], [SuSt78]).** *If $z$ is a regular vector for the graph $\Gamma$ with indefinite Tits form $\mathcal{B}$ in a given orientation $\Omega$, then*

$$\langle Tz, \tilde{z}_1^m \rangle \ \leq \ 0, \qquad \langle Tz, \tilde{z}_2^m \rangle \ \geq \ 0. \tag{B.20}$$

*Proof.* Since $\lambda_1^m > 1$ and $\lambda_2^m < 1$, the theorem follows from (B.19) and (B.13). $\square$

We denote the linear form $\langle Tz, \tilde{z}_1^m \rangle$ (resp. $\langle Tz, \tilde{z}_2^m \rangle$) by $\rho_\Omega^1$ (resp. $\rho_\Omega^2$). Then, conditions (B.20) have the following form:

$$\rho_\Omega^1(z) \ \leq \ 0, \qquad \rho_\Omega^2(z) \ \geq \ 0. \tag{B.21}$$

Similar results were obtained by Y. Zhang in [Zh89, Prop.1.5], and by J. A. de la Peña, M. Takane in [PT90, Th.2.3].

For an application of (B.20) to the star graph, see §B.4.4.

# B.3 Transforming elements and sufficient regularity conditions

In this section we consider only extended Dynkin diagrams. We will show in Theorem B.10 that *the necessary regularity condition of the vector $z$ (6.53) coincides with the sufficient condition (Proposition B.4) only if the vector $z$ is a positive root in the corresponding root system*. For an arbitrary vector $z$, this is not true. Since dimension $\dim V$ of the indecomposable representation $V$ is a positive root (Kac's theorem for any diagrams, Th. 2.16), then *for the indecomposable representation $V$, the necessary regularity condition of the vector $\dim V$ coincides with the sufficient condition*. For an arbitrary decomposable representation, this is not true.

We start from the bicolored orientation $\Lambda$.

## B.3.1 The sufficient regularity conditions for the bicolored orientation

**Proposition B.4 ([St82]).** *Let $\Gamma$ be an extended Dynkin diagram, i.e., $B$ is non-negative definite. Let $z$ be a root in the root system associated with $\Gamma$. If the $\Lambda$-defect of the vector $z$ is zero:*

$$\rho_\Lambda(z) = 0, \tag{B.22}$$

*then $z$ is regular in the bicolored orientation $\Lambda$.*

*Proof.* Let

$$H = \{z \mid z > 0,\ z \text{ is a root},\ \rho_\Lambda(z) = 0\}. \tag{B.23}$$

It suffices to prove that

$$w_1 H \subset H \quad \text{and} \quad w_2 H \subset H. \tag{B.24}$$

Indeed, if (B.24) holds, then

$$\mathbf{C}_\Lambda^k H \subset H \quad \text{for all}\quad k \in \mathbb{N},$$

hence $\mathbf{C}_\Lambda^k z > 0$ if $z$ is a positive root satisfying the condition (B.22).

So, let us prove, for example, that $w_1 H \subset H$. Note that $z$ and $w_1 z$ are roots simultaneously. Thus, either $w_1 z > 0$ or $w_1 z < 0$. Suppose $w_1 z < 0$. Together with $z > 0$, by (3.4) we have $y = 0$. Hence

$$\rho_\Lambda(z) = \langle z, \tilde{z}^{1\vee} \rangle = \langle x, \tilde{x}^{1\vee} \rangle.$$

The coordinates of $\tilde{x}^{1\vee}$ are positive, the coordinates of $x$ are non-negative. If $\rho_\Lambda(z) = 0$, then $x = 0$; this contradicts to the condition $z > 0$. Therefore, $w_1 z > 0$. It remains to show that

$$\rho_\Lambda(w_1 z) = 0.$$

Again, by (3.4), (3.6) and (3.23) we have

$$
\begin{aligned}
\langle w_1 z, \tilde{z}^{1\vee} \rangle &= \left\langle \begin{pmatrix} -x - 2Dy \\ y \end{pmatrix}, \begin{pmatrix} x^{1\vee} \\ F^\vee x^{1\vee} \end{pmatrix} \right\rangle = \\
&- \langle x, x^{1\vee} \rangle - 2\langle y, D^t x^{1\vee} \rangle + \langle y, F^\vee x^{1\vee} \rangle = \\
&- \langle x, x^{1\vee} \rangle - \langle y, F^\vee x^{1\vee} \rangle = -\langle z, \tilde{z}^{1\vee} \rangle = 0.
\end{aligned}
\tag{B.25}
$$

Thus, $w_1 H \subset H$. Similarly, $w_2 H \subset H$.    □

To prove the sufficient regularity condition for arbitrary orientation, we need some properties of transforming elements $T$.

### B.3.2 A theorem on transforming elements

**Proposition B.5 ([St82]).** *Let $\Omega', \Omega''$ be two arbitrary orientations of the graph $\Gamma$ that differ by the direction of $k$ edges. Consider the chain of orientations, in which every two adjacent orientations differ by the direction of one edge:*

$$
\Omega' = \Lambda_0, \Lambda_1, \Lambda_2, \ldots, \Lambda_{k-1}, \Lambda_k = \Omega''.
\tag{B.26}
$$

*Then, in the Weyl group, there exist elements $P_i$ and $S_i$, where $i = 1, 2, ..., k$, such that*

$$
\begin{aligned}
\mathbf{C}_{\Lambda_0} &= P_1 S_1, \\
\mathbf{C}_{\Lambda_1} &= S_1 P_1 = P_2 S_2, \\
&\cdots \\
\mathbf{C}_{\Lambda_{k-1}} &= S_{k-1} P_{k-1} = P_k S_k, \\
\mathbf{C}_{\Lambda_k} &= S_k P_k.
\end{aligned}
\tag{B.27}
$$

*In addition, for each reflection and each $i = 1, 2, ..., k$, this reflection does not occur in the decomposition of either $P_i$ or $S_i^{-1}$.*

*Proof.* It suffices to consider the case $k = 1$. Let us consider the graph

$$
\Gamma_1 \cup \Gamma_2 = \Gamma \backslash l.
$$

The graph $\Gamma$ can be depicted as follows:

$$
\begin{array}{cccc}
\Omega' & \cdots & \longleftarrow & \cdots \\
& \Gamma_1 & & \Gamma_2 \\
\Omega'' & \cdots & \longrightarrow & \cdots \\
& \Gamma_1 & & \Gamma_2
\end{array}
\tag{B.28}
$$

The orientations $\Omega''$ and $\Omega'$ induce the same orientations on the graphs $\Gamma_1$ and $\Gamma_2$, and therefore they induce the same Coxeter transformations on subgraphs $\Gamma_1$ and $\Gamma_2$. Denote the corresponding Coxeter transformations by $\mathbf{C}_{\Gamma_1}$ and $\mathbf{C}_{\Gamma_2}$. Then

$$\mathbf{C}_{\Omega'} = \mathbf{C}_{\Gamma_2}\mathbf{C}_{\Gamma_1}, \qquad \mathbf{C}_{\Omega''} = \mathbf{C}_{\Gamma_1}\mathbf{C}_{\Gamma_2}.$$

Here,

$$P_1 = \mathbf{C}_{\Gamma_2}, \qquad S_1 = \mathbf{C}_{\Gamma_1}. \qquad \square$$

*Remark B.6.* Observe, in particular, that Proposition B.5 gives a simple proof of the fact that all the Coxeter transformations form one conjugacy class, cf. [Bo, Ch.5, §6].

**Theorem B.7 ([St82]).** *1) Under the condition of Proposition B.5,*

$$T^{-1}\mathbf{C}_{\Omega'}T = \mathbf{C}_{\Omega''}$$

*for the following $k+1$ transforming elements $T := T_i$:*

$$
\begin{aligned}
T_1 &= P_1 P_2 P_3 ... P_{k-2} P_{k-1} P_k, \\
T_2 &= P_1 P_2 P_3 ... P_{k-2} P_{k-1} S_k^{-1}, \\
T_3 &= P_1 P_2 P_3 ... P_{k-2} S_{k-1}^{-1} S_k^{-1}, \\
&\quad ... \\
T_{k-1} &= P_1 P_2 S_3^{-1} ... S_{k-2}^{-1} S_{k-1}^{-1} S_k^{-1}, \\
T_k &= P_1 S_2^{-1} S_3^{-1} ... S_{k-2}^{-1} S_{k-1}^{-1} S_k^{-1}, \\
T_{k+1} &= S_1^{-1} S_2^{-1} S_3^{-1} ... S_{k-2}^{-1} S_{k-1}^{-1} S_k^{-1}.
\end{aligned}
\tag{B.29}
$$

*In addition, for each reflection $\sigma_\alpha$, there exists a $T_i$ whose decomposition does not contain this reflection.*
*2) The following relation holds:*

$$T_p T_q^{-1} = \mathbf{C}_{\Omega'}^{q-p}. \tag{B.30}$$

*Proof.* 1) There are altogether $2^k$ transforming elements of the form

$$T = X_1 X_2 ... X_{k-1} X_k, \quad \text{where } X_i \in \{P_i, S_i^{-1}\}. \tag{B.31}$$

By Proposition B.5, for each reflection $\sigma_\alpha$ and for each $i$, we can select $X_i = P_i$ or $X_i = S_i^{-1}$ such that this reflection does not occur in $T$. Taking the product of all these elements $X_i$ we obtain the transforming element $T$ whose decomposition does not contain this reflection. It remains to show that every transforming element from the list (B.31) containing $2^k$ elements is of the form (B.29). By (B.27) we have

$$S_q^{-1} P_{q+1} = P_q S_{q+1}^{-1}.$$

Thus, all symbols $S_i^{-1}$ can be shifted to the right and all symbols $P_i$ can be shifted to the left.
2) It suffices to show (B.30) for $p < q$. By (B.29) we have

$$T_p T_q^{-1} =$$
$$(P_1 P_2 ... P_{k-p+1} S_{k-p+2}^{-1} ... S_k^{-1}) \times (S_k ... S_{k-q+2} P_1 P_{k-q+1}^{-1} ... P_1^{-1}) = \quad (\text{B.32})$$
$$P_1 P_2 ... P_{k-p+1} (S_{k-p+1} S_{k-p} ... S_{k-q+2}) P_{k-q+1}^{-1} ... P_1^{-1}.$$

Here, $k - p + 1 \geq k - q + 2$. In order to simplify eq. (B.32), we use eq. (B.27):

$$P_{k-p} P_{k-p+1} S_{k-p+1} S_{k-p} = P_{k-p} (S_{k-p} P_{k-p}) S_{k-p} = (P_{k-p} S_{k-p})^2,$$
$$\dots$$

$$P_{k-p-1} (P_{k-p} S_{k-p})^2 S_{k-p-1} =$$
$$P_{k-p-1} (S_{k-p-1} P_{k-p-1})^2 S_{k-p-1} = (P_{k-p-1} S_{k-p-1})^3, \quad (\text{B.33})$$
$$\dots$$

$$P_{k-q+2} P_{k-q+3} ... P_{k-p+1} S_{k-p+1} ... S_{k-q+2} S_{k-q+2} =$$
$$(P_{k-q+2} S_{k-q+2})^{q-p}.$$

Thus, from (B.32) and (B.33) we deduce:

$$T_p T_q^{-1} = P_1 P_2 ... P_{k-q+1} (P_{k-q+2} S_{k-q+2})^{q-p} P_{k-q+1}^{-1} ... P_1^{-1}. \quad (\text{B.34})$$

Again, by (B.27) we have

$$P_{k-q+1} (P_{k-q+2} S_{k-q+2})^{q-p} P_{k-q+1}^{-1} =$$
$$P_{k-q+1} (S_{k-q+1} P_{k-q+1})^{q-p} P_{k-q+1}^{-1} = (P_{k-q+1} S_{k-q+1})^{q-p},$$
$$\dots$$

$$P_{k-q} (P_{k-q+1} S_{k-q+1})^{q-p} P_{k-q}^{-1} = \quad (\text{B.35})$$
$$P_{k-q} (S_{k-q} P_{k-q})^{q-p} P_{k-q}^{-1} = (P_{k-q} S_{k-q})^{q-p},$$
$$\dots$$

$$P_1 (P_2 S_2)^{q-p} P_1^{-1} = P_1 (S_1 P_1)^{q-p} P_1^{-1} = (P_1 S_1)^{q-p}.$$

Therefore,

$$T_p T_q^{-1} = (P_1 S_1)^{q-p} = \mathbf{C}_{\Lambda_0}^{q-p} = \mathbf{C}_{\Omega'}^{q-p}. \quad \square \quad (\text{B.36})$$

*Remark B.8.* Theorem B.7 allows us to select a transforming element $T$ in such a way that its decomposition does not contain any given refection $\sigma_i$, and therefore $T$ does not change any given coordinate $i$. This fact was already used once in §B.2 for the proof of the necessary regularity conditions for diagrams with indefinite Tits form. Now, we will use Theorem B.7 to carry the sufficient regularity condition from a bicolored orientation $\Lambda$ in Proposition B.4 to an arbitrary orientation $\Omega'$, see Definition 6.30.

### B.3.3 The sufficient regularity conditions for an arbitrary orientation

**Proposition B.9 ([St82]).** *Let $\Gamma$ be an extended Dynkin diagram. Let $\Omega = \Lambda_0$, $\Omega' = \Lambda_k$. If $z$ is a positive root with zero $\Omega'$-defect, then $z$ is the $\Omega'$-regular vector.*

*Proof.* If $z > 0$, then there exists a positive coordinate $z_\alpha > 0$. Take a transforming element $T_i$ whose decomposition does not contain the reflection $\sigma_\alpha$. Then $(T_i z)_\alpha > 0$. Since $T_i z$ is the root, we have $T_i z > 0$. The equality $\rho_{\Omega'}(z) = 0$ means that $\rho_\Omega(T_i z) = 0$, see Definition 6.29. Since $T_i z > 0$ and $T_i z$ is the root, we see by Proposition B.4 that $T_i z$ is $\Omega$-regular and

$$\mathbf{C}_\Omega^m T_i z > 0 \text{ for all } m \in \mathbb{N}. \tag{B.37}$$

Suppose that
$$u = T_i^{-1} \mathbf{C}_\Omega^m T_i z < 0 \text{ for some } m = m_0.$$

Then by Theorem B.7 there exists a $T_j$ such that $T_j u < 0$, i.e.,

$$T_j T_i^{-1} \mathbf{C}_\Omega^m T_i z < 0 \text{ for } m = m_0. \tag{B.38}$$

Again by Theorem B.7 we have $T_j T_i^{-1} = \mathbf{C}_\Omega^{i-j}$ and by (B.38) the following relation holds:

$$\mathbf{C}_\Omega^{i-j} \mathbf{C}_\Omega^m T_i z < 0 \text{ for } m = m_0 \tag{B.39}$$

that contradicts (B.37). Therefore,

$$u = \mathbf{C}_{\Omega'}^m z = T_i^{-1} \mathbf{C}_\Omega^m T_i z > 0 \text{ for all } m \in \mathbb{N}.$$

Thus, $z$ is a $\Omega'$-regular vector.   $\square$

From Theorem 6.33 together with Proposition B.9 we get the following

**Theorem B.10.** *The indecomposable representation $V$ of the graph $\Gamma$ (which is an extended Dynkin diagram) with orientation $\Omega$ is regular in the orientation $\Omega$ if and only if*

$$\rho_\Omega(\dim V) = 0.$$

*Proof.* Indeed, the dimensions of the indecomposable representations of the extended Dynkin diagrams are roots, see [DR76], [Kac80]; however, there are indecomposable representations whose dimensions are not usual (real) roots but *imaginary roots*, see §2.2.1. They are vectors from the kernel of the Tits form, and are proportional to vectors $z^1$ which are fixed points of the Weyl group. In particular, $Tz^1 = z^1$, so $\langle z^1, \tilde{z}^{1\vee} \rangle = 0$ directly implies $\langle Tz^1, \tilde{z}^{1\vee} \rangle = 0$.
$\square$

### B.3.4 The invariance of the defect

We will show that the $\Omega$-defect $\rho_\Omega$ is invariant under the Coxeter transformation $\mathbf{C}_\Omega$ and $\rho_\Omega$ does not depend on the choice of the transforming element $T_i$ in the Weyl group, see Definition 6.29 and Theorem B.7. In other words, the following proposition holds.

**Proposition B.11.** *1) The Coxeter transformation* $\mathbf{C}_\Omega$ *preserves the linear form* $\rho_\Omega$:

$$\rho_\Omega(\mathbf{C}_\Omega z) = \rho_\Omega(z) \text{ for any vector } z. \tag{B.40}$$

*2) If* $T_i$ *and* $T_j$ *are transforming elements defined by (B.29), i.e.,*

$$T_i^{-1}\mathbf{C}_\Lambda T_i = \mathbf{C}_\Omega \text{ and } T_j^{-1}\mathbf{C}_\Lambda T_j = \mathbf{C}_\Omega, \tag{B.41}$$

*then we have*

$$\langle T_i z, \tilde{z}^{1\vee} \rangle = \langle T_j z, \tilde{z}^{1\vee} \rangle \text{ for any vector } z. \tag{B.42}$$

*Proof.* First, observe that (B.42) does not hold for an arbitrary matrix $T$, since the matrix $kT$, where $k \in \mathbb{R}$, also satisfies (B.41), but does not satisfy (B.42).

In (B.25) we showed that

$$\langle w_i z, \tilde{z}^{1\vee} \rangle = -\langle z, \tilde{z}^{1\vee} \rangle. \tag{B.43}$$

By (B.43) we have

$$\langle z, \tilde{z}^{1\vee} \rangle = \langle \mathbf{C}_\Lambda z, \tilde{z}^{1\vee} \rangle \text{ for any vector } z, \tag{B.44}$$

i.e., (B.40) holds for any bicolored orientation $\Lambda$. Since $z$ in (B.44) is an arbitrary vector, we have

$$\langle T z, \tilde{z}^{1\vee} \rangle = \langle \mathbf{C}_\Lambda T z, \tilde{z}^{1\vee} \rangle = \langle T \mathbf{C}_\Omega z, \tilde{z}^{1\vee} \rangle,$$

or

$$\rho_\Omega(z) = \rho_\Omega(\mathbf{C}_\Omega z),$$

i.e., (B.40) holds for an arbitrary orientation $\Omega$.

Further, from (B.44) we have

$$\rho_\Lambda(z) = \rho_\Lambda(\mathbf{C}_\Lambda z) = \rho_\Lambda(\mathbf{C}_\Lambda^{i-j} z).$$

By (B.30)

$$\rho_\Lambda(z) = \rho_\Lambda(T_j T_i^{-1} z).$$

Substituting $T_i z$ instead of $z$ we have

$$\rho_\Lambda(T_i z) = \rho_\Lambda(T_j z),$$

and (B.42) is proved. □

For examples of regularity conditions for extended Dynkin diagrams $\widetilde{D}_4$, $\widetilde{E}_6$, $\widetilde{G}_{12}$, $\widetilde{G}_{22}$ and the star $*_{n+1}$, see §B.4.

# B.4 Examples of regularity conditions

**Definition B.12.** Let $\Gamma$ be a graph, $x_0 \in \Gamma$ be a point of maximal branching degree and let all arrows of $\Gamma$ be directed to the point $x_0$. The corresponding orientation is said to be the *central orientation* and denoted by $\Lambda_0$.

We will consider the necessary regularity conditions for some diagrams in bicolored, central, and other orientations. If an orientation $\Lambda'$ is obtained from another orientation $\Lambda''$ by reversing of all arrows of a graph, then the corresponding regularity conditions coincide:

$$\rho_{\Lambda'} = \rho_{\Lambda''} \quad \text{if } \mathcal{B} \text{ is positive definite}$$

and

$$\rho^1_{\Lambda'} = \rho^1_{\Lambda''}, \quad \rho^2_{\Lambda'} = \rho^2_{\Lambda''} \quad \text{if } \mathcal{B} \text{ is indefinite.}$$

Let us introduce an equivalence relation $\mathcal{R}$ on the set of orientations of the graph $\Gamma$. Two orientations $\Lambda'$ and $\Lambda''$ will be called *equivalent*, if one orientation can be obtained from the other one by reversing all arrows or by an automorphism of the diagram; for equivalent orientations, we write:

$$\Lambda' \equiv \Lambda'' \mod \mathcal{R}.$$

Equivalent orientations have identical regularity conditions.

## B.4.1 The three equivalence classes of orientations of $\widetilde{D}_4$

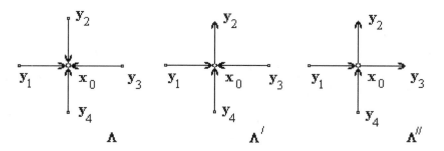

**Fig. B.1.**    For $\widetilde{D}_4$, the bicolored and central orientations coincide

In the case of $\widetilde{D}_4$, the bicolored orientation $\Lambda$ coincides with the central orientation $\Lambda_0$. The orientations $\Lambda$, $\Lambda'$, and $\Lambda''$ cover all possible equivalence classes.

a) The bicolored orientation $\Lambda$, see Fig. B.1.

Here, $T = I$ and by Theorem 6.33 and Proposition B.4 we have

$$z^1 = \begin{pmatrix} 2 \\ 1 \\ 1 \\ 1 \\ 1 \end{pmatrix} \begin{matrix} x_0 \\ y_1 \\ y_2 \\ y_3 \\ y_4 \end{matrix}, \quad \tilde{z}^1 = \begin{pmatrix} 2 \\ -1 \\ -1 \\ -1 \\ -1 \end{pmatrix}, \quad \rho_\Lambda(z) = y_1 + y_2 + y_3 + y_4 - 2x_0. \qquad (B.45)$$

Originally, the linear form $\rho_\Lambda(z)$ in (B.45) was obtained by I. M. Gelfand and V. A. Ponomarev in the work devoted to classifications of quadruples of linear subspaces of arbitrary dimension [GP72].

b) The orientation $\Lambda'$, see Fig. B.1. Here we have

$$\mathbf{C}_\Lambda = \sigma_{y_4}\sigma_{y_3}\sigma_{y_2}\sigma_{y_1}\sigma_{x_0}, \qquad \mathbf{C}_{\Lambda'} = \sigma_{y_4}\sigma_{y_3}\sigma_{y_1}\sigma_{x_0}\sigma_{y_2}, \qquad T = \sigma_{y_2}.$$

Then, by Theorem 6.33 and Proposition B.9 we get the following condition of the $\Lambda'$-regularity:

$$Tz = \begin{pmatrix} x_0 \\ y_1 \\ x_0 - y_2 \\ y_3 \\ y_4 \end{pmatrix}, \qquad \begin{array}{c} y_1 + (x_0 - y_2) + y_3 + y_4 - 2x_0 = 0, \text{ or} \\ y_1 + y_3 + y_4 = y_2 + x_0. \end{array}$$

c) The orientation $\Lambda''$, see Fig. B.1. Here, $T = \sigma_{y_2}\sigma_{y_3}$ (or $\sigma_{y_1}\sigma_{y_4}$), and we have the following condition of the $\Lambda''$-regularity:

$$\begin{array}{c} y_1 + (x_0 - y_2) + (x_0 - y_3) + y_4 - 2x_0 = 0, \text{ or} \\ y_1 + y_4 = y_2 + y_3. \end{array} \qquad (B.46)$$

## B.4.2 The bicolored and central orientations of $\widetilde{E}_6$

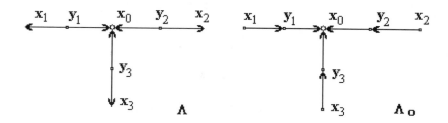

**Fig. B.2.**    For $\widetilde{E}_6$, the bicolored and central orientations

We consider only the bicolored orientation $\Lambda$ and the central orientation $\Lambda_0$ in Fig. B.2. The Coxeter transformations and transforming element $T$ are:

$$\mathbf{C}_\Lambda = \sigma_{y_3}\sigma_{y_2}\sigma_{y_1}\sigma_{x_3}\sigma_{x_2}\sigma_{x_1}\sigma_{x_0},$$
$$\mathbf{C}_{\Lambda_0} = \sigma_{x_3}\sigma_{x_2}\sigma_{x_1}\sigma_{y_3}\sigma_{y_2}\sigma_{y_1}\sigma_{x_0},$$
$$T = \sigma_{x_3}\sigma_{x_2}\sigma_{x_1}.$$

We have

$$
z^1 = \begin{pmatrix} 3 \\ 1 \\ 1 \\ 1 \\ 2 \\ 2 \\ 2 \end{pmatrix} \begin{matrix} x_0 \\ x_1 \\ x_2 \\ x_3 \\ y_1 \\ y_2 \\ y_3 \end{matrix}, \qquad
\tilde{z}^1 = \begin{pmatrix} 3 \\ 1 \\ 1 \\ 1 \\ -2 \\ -2 \\ -2 \end{pmatrix}, \qquad
Tz = \begin{pmatrix} x_0 \\ y_1 - x_1 \\ y_2 - x_2 \\ y_3 - x_3 \\ y_1 \\ y_2 \\ y_3 \end{pmatrix},
$$

and by Theorem 6.33 and Proposition B.4 a condition of $\Lambda$-regularity is

$$\rho_\Lambda(z) = 3x_0 + x_1 + x_2 + x_3 + 3x_0 - 2y_1 - 2y_2 - 2y_3.$$

The choice of element $T$ is ambiguous, but it was shown in Proposition B.11 that the regularity condition does not depend on this choice.

So, a condition of $\Lambda_0$-regularity is

$$\rho_{\Lambda_0}(z) =$$
$$3x_0 + (y_1 - x_1) + (y_2 - x_2) + (y_3 - x_3) - 2(y_1 + y_2 + y_3) = 0,$$

or

$$\rho_{\Lambda_0}(z) = x_1 + x_2 + x_3 + y_1 + y_2 + y_3 - 3x_0.$$

### B.4.3 The multiply-laced case. The two orientations of $\widetilde{G}_{21}$ and $\widetilde{G}_{22} = \widetilde{G}_{21}^\vee$

We have

$$
z^1 = \begin{pmatrix} 2 \\ 1 \\ 3 \end{pmatrix}, \tilde{z}^1 = \begin{pmatrix} 2 \\ -1 \\ -3 \end{pmatrix} \begin{matrix} x_1 \\ y_1 \\ y_2 \end{matrix}, \quad
z^{1\vee} = \begin{pmatrix} 2 \\ 1 \\ 1 \end{pmatrix}, \tilde{z}^{1\vee} = \begin{pmatrix} 2 \\ -1 \\ -1 \end{pmatrix} \begin{matrix} x_1 \\ y_1 \\ y_2 \end{matrix}. \quad \text{(B.47)}
$$

Take vectors $\tilde{z}^1$ and $\tilde{z}^{1\vee}$ from (B.47) and substitute them in (6.52). We get the following regularity condition $\rho_\Lambda$ (resp. $\rho_\Lambda^\vee$):

$$\rho_\Lambda(z) = y_1 + y_2 - 2x_1 \quad \text{for } \widetilde{G}_{21},$$
$$\rho_\Lambda^\vee(z) = y_1 + 3y_2 - 2x_1 \quad \text{for } \widetilde{G}_{22}.$$

Now consider the orientation $\Lambda'$, Fig. B.3. The Coxeter transformations can be expressed as follows:

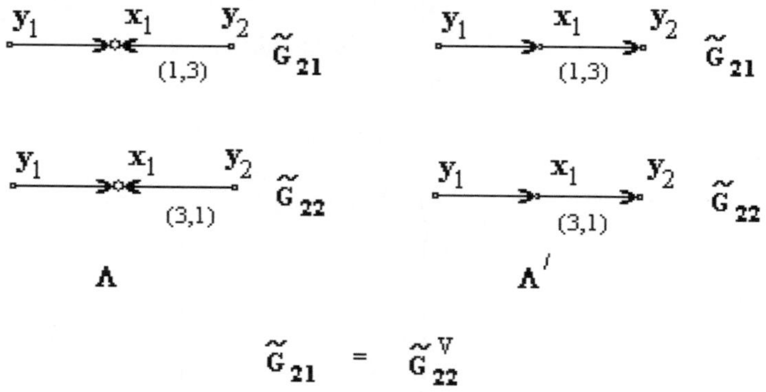

$$\widetilde{G}_{21} \;=\; \widetilde{G}_{22}^{\vee}$$

**Fig. B.3.**    The two orientations of $\widetilde{G}_{21}$ and $\widetilde{G}_{22} = \widetilde{G}_{21}^{\vee}$

$$\mathbf{C}_{\Lambda} = \sigma_{y_2}\sigma_{y_1}\sigma_{x_1}, \quad \mathbf{C}_{\Lambda}^{\vee} = \sigma_{y_2}^{\vee}\sigma_{y_1}^{\vee}\sigma_{x_1}^{\vee},$$
$$\mathbf{C}_{\Lambda'} = \sigma_{y_1}\sigma_{x_1}\sigma_{y_2}, \quad \mathbf{C}_{\Lambda'}^{\vee} = \sigma_{y_1}^{\vee}\sigma_{x_1}^{\vee}\sigma_{y_2}^{\vee},$$

Transforming elements $T$ and $T^{\vee}$ are:

$$T = \sigma_{y_2}, \quad T^{\vee} = \sigma_{y_2}^{\vee}.$$

and

$$Tz = \begin{pmatrix} x_1 \\ y_1 \\ 3x_1 - y_2 \end{pmatrix}, \quad T^{\vee}z = \begin{pmatrix} x_1 \\ y_1 \\ x_1 - y_2 \end{pmatrix}.$$

Finally, we have

$$\rho_{\Lambda'}(z) = y_1 - y_2 + x_1 \quad \text{for } \widetilde{G}_{21},$$
$$\rho_{\Lambda'}^{\vee}(z) = y_1 - 3y_2 + x_1 \quad \text{for } \widetilde{G}_{22}.$$

### B.4.4 The case of indefinite $\mathcal{B}$. The oriented star $*_{n+1}$

Consider the oriented star $*_{n+1}$ with a bicolored orientation.

According to Remark 3.7 the matrix $DD^t$ is a scalar. By (3.17) $DD^t = \dfrac{n}{4}$ and the maximal eigenvalue $\varphi^m = \dfrac{n}{4}$. By (3.16) we have

$$\lambda_{1,2}^m = \frac{n - 2 \pm \sqrt{n(n-4)}}{2}. \tag{B.48}$$

Let $x_m = 1$. Then

$$\frac{-2}{\lambda_{1,2}^m + 1} D^t x^m = \frac{-2}{\lambda_{1,2}^m + 1} [1, 1, ..., 1]^t.$$

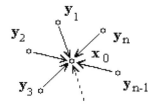

**Fig. B.4.**    A bicolored orientation of the star $*_{n+1}$

Thus, we have the following eigenvectors $z_1^m$, $z_2^m$, and their conjugate vectors $\tilde{z}_1^m$, $\tilde{z}_2^m$ (see Definition 6.31 and Proposition 6.32):

$$
z_1^m = \begin{pmatrix} \lambda_1^m + 1 \\ 1 \\ 1 \\ \cdots \\ 1 \end{pmatrix}, z_2^m = \begin{pmatrix} \lambda_2^m + 1 \\ 1 \\ 1 \\ \cdots \\ 1 \end{pmatrix}, \tilde{z}_1^m = \begin{pmatrix} \lambda_2^m + 1 \\ -1 \\ -1 \\ \cdots \\ -1 \end{pmatrix}, \tilde{z}_2^m = \begin{pmatrix} \lambda_1^m + 1 \\ -1 \\ -1 \\ \cdots \\ -1 \end{pmatrix}.
$$

According to (B.20) we obtain the following condition of $\Lambda$-regularity:

$$
(\lambda_2^m + 1)x_0 - \sum y_i \leq 0, \qquad (\lambda_1^m + 1)x_0 - \sum y_i \geq 0. \tag{B.49}
$$

From (B.48) and (B.49) we deduce

$$
\frac{n - \sqrt{n(n-4)}}{2} x_0 \leq \sum y_i \leq \frac{n + \sqrt{n(n-4)}}{2} x_0 \text{ or}
$$

$$
\frac{n - \sqrt{n(n-4)}}{2} \sum y_i \leq x_0 \leq \frac{n + \sqrt{n(n-4)}}{2} \sum y_i \text{ , or}
$$

$$
\left| x_0 - \frac{1}{2} \sum y_i \right| \leq \frac{\sqrt{n(n-4)}}{2n} \sum y_i \text{ or}
$$

$$
x_0^2 - x_0 \sum y_i + \frac{1}{4} \left( \sum y_i \right)^2 \leq \frac{1}{4} \left( \sum y_i \right)^2 - \frac{1}{n} \left( \sum y_i \right)^2 \text{ or}
$$

$$
x_0^2 - x_0 \sum y_i \leq -\frac{1}{n} \left( \sum y_i \right)^2 \text{ or}
$$

$$
x_0^2 - x_0 \sum y_i + \sum y_i^2 \leq \sum y_i^2 - \frac{1}{n} \left( \sum y_i \right)^2
$$

Since the left hand side of the latter inequality is the Tits form $\mathcal{B}$, we obtain the following condition of $\Lambda$-regularity:

$$
\mathcal{B}(z) \leq \frac{1}{n} \sum_{0 < i < j} (y_i - y_j)^2.
$$

# C

## Miscellanea

## C.1 The triangle groups and Hurwitz groups

The group generated by $X, Y, Z$ that satisfy the relations (A.6)

$$X^p = Y^q = Z^r = XYZ = 1$$

is said to be a *triangle group*. As it was mentioned in §A.2, the finite polyhedral groups from Table A.1 are triangle groups. Set

$$\mu(p, q, r) := \frac{1}{p} + \frac{1}{q} + \frac{1}{r}.$$

The triangle group is finite if and only if $\mu(p, q, r) > 1$. There are only three triangle groups

$$(2, 4, 4), \quad (2, 3, 6), \quad (3, 3, 3),$$

for which $\mu(p, q, r) = 1$. These groups are infinite and soluble. For references, see [Con90], [Mu01]. For all other triangle groups, we have

$$\mu(p, q, r) < 1.$$

These groups are infinite and insoluble. The value $1 - \mu(p, q, r)$ attains the minimum value $\frac{1}{42}$ at $(2, 3, 7)$. Thus, $(2, 3, 7)$ is, in a sense, the minimal infinite insoluble triangle group. The importance of the triangle group $(2, 3, 7)$ is revealed by the following theorem due to Hurwitz.

**Theorem C.1 (Hurwitz, [Hur1893]).** *If $X$ is a compact Riemann surface of genus $g > 1$, then $|AutX| \leq 84(g - 1)$, and moreover, the upper bound of this order is attained if an only if $|AutX|$ is a homomorphic image of the triangle group $(2, 3, 7)$.*

For further references, see [Con90], [Con03].

A *Hurwitz group* is any finite nontrivial quotient of the triangle group $(2, 3, 7)$. In other words, the finite group $G$ is the Hurwitz group if it has generators $X, Y, Z \in G$ such that

$$X^2 = Y^3 = Z^7 = XYZ = 1.$$

M. Conder writes that the significance of the Hurwitz groups "...is perhaps best explained by referring to some aspects of the theory of Fuchsin groups, hyperbolic geometry, Riemann surfaces...", see [Con90, p.359] and a bibliography cited there.

M. Conder [Con80] using the method of coset graphs developed by G. Higman has shown that the alternating group $A_n$ is a Hurwitz group for all $n \geq 168$.

Recently, A. Lucchini, M. C. Tamburini and J. S. Wilson showed that most finite simple classical groups of sufficiently large rank are Hurwitz groups, see [LuT99], [LuTW00]. For example, the groups $SL_n(q)$ are Hurwitz, for all $n > 286$ [LuTW00], and the groups $Sp_{2n}(q), SU_{2n}(q)$ are Hurwitz, for all $n > 371$ [LuT99]. (These mentioned groups act in the $n$-dimensional vector space over the field $F_q$ of the prime characteristic $q$.)

The sporadic groups have been treated in a series of papers by Woldar and others; for a survey and references, see [Wi01]. R. A. Wilson shows in [Wi01], that the Monster is also a Hurwitz group.

## C.2 The algebraic integers

If $\lambda$ is a root of the polynomial equation

$$a_n x^n + a_n x^{n-1} + \cdots + a_1 x + a_0 = 0, \tag{C.1}$$

where $a_i$ for $i = 0, 1, \ldots, n$ are integers and $\lambda$ satisfies no similar equation of degree $< n$, then $\lambda$ is said to be an *algebraic number* of degree $n$. If $\lambda$ is an algebraic number and $a_n = 1$, then $\lambda$ is called an *algebraic integer*.

A polynomial $p(x)$ in which the coefficient of the highest order term is equal to 1 is called the *monic polynomial*. The polynomial (C.1) with integer coefficients and $a_n = 1$ is the *monic integer polynomial*.

The algebraic integers of degree 1 are the ordinary integers (elements of $\mathbb{Z}$). If $\alpha$ is an algebraic number of degree $n$ satisfying the polynomial equation

$$(x - \alpha)(x - \beta)(x - \gamma) \cdots = 0,$$

then there are $n - 1$ other algebraic numbers $\beta$, $\gamma$, ... called the *conjugates* of $\alpha$. Furthermore, if $\alpha$ satisfies any other algebraic equation, then its conjugates also satisfy the same equation.

**Definition C.2.** An algebraic integer $\lambda > 1$ is said to be a *Pisot number* if all its conjugates (other then $\lambda$ itself) satisfy $|\lambda'| < 1$.

The smallest Pisot number,

$$\lambda_{Pisot} \approx 1.324717...,\tag{C.2}$$

is a root of $\lambda^3 - \lambda - 1 = 0$ (for details and references, see [McM02]). This number appears in Proposition 4.16 as a limit of the spectral radius $\rho(T_{2,3,n})$ as $n \to \infty$.

**Definition C.3.** Let $p(x)$ be a monic integer polynomial, and define its *Mahler measure* to be

$$\|p(x)\| = \prod_{\beta} |\beta|,\tag{C.3}$$

where $\beta$ runs over all (complex) roots of $p(x)$ outside the unit circle.

*Remark C.4.* 1) Thanks are due to C. J. Smyth who kindly informed me about Siegel's work [Si44]. It was Siegel who showed that two smallest Pisot number were the positive zero $\theta_1$ of $x^3 - x - 1$, and the positive zero $\theta_2$ of $x^4 - x^3 - 1$, where

$$\theta_1 = 1.324717..., \qquad \theta_2 = 1.380728....$$

Siegel also proved that any other Pisot number is larger than $\sqrt{2}$.

2) In his thesis [Sm71], C. J. Smyth proved that among nonreciprocal integer polynomials, the polynomial $x^3 - x - 1$ has the smallest Mahler measure.

3) The spectral radius $\rho(T_{2,3,n})$ as $n \to \infty$ (4.18) was obtained by Y. Zhang [Zh89] and used in the study of regular components of an Auslander-Reiten quiver. The spectral radius (4.18) coincides with the smallest *Pisot number* (C.2).

It is well known that $\|p(x)\| = 1$ if and only if all roots of $p(x)$ are roots of unity. In 1933, Lehmer [Leh33] asks whether, for each $\varepsilon \geq 1$, there exists an algebraic integer such that

$$1 < \|\alpha\| < 1 + \varepsilon\tag{C.4}$$

Lehmer found polynomials with smallest Mahler measure for small degrees and stated in [Leh33, p.18] that the polynomial with minimal root $\alpha$ (in the sense of C.4) he could find is the polynomial of degree 10:

$$1 + x - x^3 - x^4 - x^5 - x^6 - x^7 + x^9 + x^{10},\tag{C.5}$$

see [Hir02], [McM02]; cf. Remark 4.15. Outside the unit circle, the polynomial (C.5) has only one root

$$\lambda_{Lehmer} \approx 1.176281...\tag{C.6}$$

The number (C.6) is called *Lehmer's number*; see Proposition 4.16, Remark 4.15 and Table 4.4.

**Definition C.5.** A *Salem number* is a real algebraic integer $\lambda > 1$, whose other conjugates all have modulus at most 1, with at least one having modulus exactly 1.

It is known that every Pisot number is a limit of Salem numbers; for details and references, see [MRS99], [MS05]. Conjecturally, Lehmer's number (C.6) is the smallest Salem number, [Leh33], [GH01]. McKee and Smyth [MS05] introduced notions of a *Salem graph* and *Pisot graph* whose spectral radii are respectively the Salem number and Pisot number, see Remark 4.21, heading 4).

The positive root of the quadratic equation $\lambda^2 - \lambda - 1 = 0$ is a well-known constant

$$\lambda_{Golden} \approx 1.618034..., \tag{C.7}$$

called the *Golden mean* or *Divine proportion*. This number appears in Proposition 4.17 as a limit of the spectral radius $\rho(T_{3,3,n})$ as $n \to \infty$.

The smallest Mahler measure among reciprocal polynomials of degree at most 6 is

$$M_6 = ||x^6 - x^4 + x^3 - x^2 + 1|| \approx 1.401268..., \tag{C.8}$$

see [Mos98, p.1700]. This number appears in Proposition 4.17 as a root of

$$\mathcal{X}(T_{3,3,4}) = x^8 + x^7 - 2x^5 - 3x^4 - 2x^3 + x + 1.$$

This is the polynomial of minimal degree among polynomials $\mathcal{X}(T_{3,3,n})$, where $n = 4, 5, 6, \ldots$ with indefinite Tits form, see Proposition 4.17 and Table 4.5.

## C.3 The Perron-Frobenius Theorem

We say that a matrix is *positive* (resp. *non-negative*) if all its entries are positive (resp. non-negative). We use the notation $A > 0$ (resp. $A \geq 0$) for positive (resp. non-negative) matrix. A square $n \times n$ matrix $A$ is called *reducible* if the indices 1, 2, ..., $n$ can be divided into the disjoint union of two nonempty sets $\{i_1, i_2, \ldots, i_p\}$ and $\{j_1, j_2, \ldots, j_q\}$ (with $p + q = n$) such that

$$a_{i_\alpha j_\beta} = 0, \text{ for } \alpha = 1, \ldots, p \text{ and } \beta = 1, \ldots, q.$$

In other words, $A$ is *reducible* if there exists a permutation matrix $P$, such that

$$PAP^t = \begin{pmatrix} B & 0 \\ C & D \end{pmatrix},$$

where $B$ and $D$ are square matrices. A square matrix which is not reducible is said to be *irreducible*.

**Theorem C.6 (Perron-Frobenius).** *Let $A$ be an $n \times n$ non-negative irreducible matrix. Then the following holds:*

*1) There exists a positive eigenvalue $\lambda$ such that*

$$|\lambda_i| \leq \lambda, \text{ where } i = 1, 2, \ldots, n.$$

*2) There is a positive eigenvector $z$ corresponding to the eigenvalue $\lambda$:*

$$Az = \lambda z, \text{ where } z = (z_1, \ldots, z_n)^t \text{ and } z_i > 0 \text{ for } i = 1, 2, \ldots, n.$$

*Such an eigenvalue $\lambda$ is called the* dominant eigenvalue *of $A$.*

*3) The eigenvalue $\lambda$ is a simple root of the characteristic equation of $A$.*

The following important corollary from the Perron-Frobenius theorem holds for the eigenvalue $\lambda$:

$$\lambda = \max_{z \geq 0} \min_i \frac{(Az)_i}{z_i} \qquad (z_i \neq 0),$$

$$\lambda = \min_{z \geq 0} \max_i \frac{(Az)_i}{z_i} \qquad (z_i \neq 0).$$

For details, see [MM64], [Ga90].

## C.4 The Schwartz inequality

Let $\mathcal{B}$ be the quadratic Tits form associated with a tree graph (simply or multiply laced), and let $\mathbf{B}$ be the matrix of $\mathcal{B}$. Let $\mathcal{B}$ be positive definite or non-negative definite, and

$$\ker \mathbf{B} = \{x \mid \mathbf{B}x = 0\}. \tag{C.9}$$

Since $(x, y) = \langle x, \mathbf{B}y \rangle = \langle \mathbf{B}x, y \rangle$, it follows that

$$\ker \mathbf{B} = \{x \mid (x, y) = 0 \text{ for all } y \in \mathcal{E}_\Gamma\} \subseteq \{x \mid \mathcal{B}(x) = 0\}. \tag{C.10}$$

Let $x, y \in \mathbb{R}^n$, where $n$ is the number of vertices in $\Gamma_0$. Then the following *Schwartz inequality* is true:

$$(x, y)^2 \leq \mathcal{B}(x)\mathcal{B}(y). \tag{C.11}$$

To prove (C.11), it suffices to consider the inequality $(x + \alpha y, x + \alpha y) \geq 0$ which is true for all $\alpha \in \mathbb{R}$. Then the discriminant of the polynomial

$$(x, x) + 2\alpha(x, y) + \alpha^2(y, y)$$

should be non-positive, whence (C.11).

Over $\mathbb{C}$, there exists $x \notin \ker \mathbf{B}$ such that $\mathcal{B}(x) = 0$. For example, the eigenvectors of the Coxeter transformation with eigenvalues $\lambda \neq \pm 1$ satisfy this condition because

$$\mathcal{B}(x) = \mathcal{B}(Cx) = \lambda^2 \mathcal{B}(x).$$

If $\mathcal{B}(x) = 0$ and $x \in \mathbb{R}^n$, then from (C.11) we get $(x, y) \leq 0$ and $(x, -y) \leq 0$, i.e., $(x, y) = 0$ for all $y \in \mathbb{R}^n$. In other words, $x \in \ker \mathbf{B}$. Taking (C.10) into account we see that if $\mathcal{B}$ is non-negative definite then

$$\ker \mathbf{B} = \{x \mid \mathcal{B}(x) = 0\}. \tag{C.12}$$

## C.5 The complex projective line and stereographic projection

The $n$-dimensional *complex projective space* $\mathbb{C}P^n$ is the set of all complex lines in $\mathbb{C}^{n+1}$ passing through the origin. Two points

$$z_1 = (z_1^0, z_1^1, z_1^2, \ldots, z_1^n), \quad z_2 = (z_2^0, z_2^1, z_2^2, \ldots, z_2^n) \in \mathbb{C}^{n+1}$$

lie on the same line, if

$$
\begin{aligned}
&z_2 = w z_1 \text{ for some complex factor } w \in \mathbb{C} \setminus \{0\}, \text{ i.e.} \\
&z_2^i = w z_1^i \text{ for } i = 0, 1, \ldots, n.
\end{aligned}
\tag{C.13}
$$

The points (C.13) constitute an equivalence class denoted by $[z^0 : z^1 : \cdots : z^n]$.

Clearly, the *complex projective line* $\mathbb{C}P^1$ is the set of all lines in $\mathbb{C}^2$. By (C.13), the points of $\mathbb{C}P^1$ are classes of complex pairs $z = [z^0, z^1]$ up to a factor $w \in \mathbb{C}$.

The correspondence

$$[z^0 : z^1] \mapsto z^0/z^1 \tag{C.14}$$

sets the following bijection maps:

$$
\begin{aligned}
&\mathbb{C}P^1 \setminus \{[1 : 0]\} \Longleftrightarrow \mathbb{C}, \text{ and} \\
&\mathbb{C}P^1 \Longleftrightarrow \mathbb{C} \cup \infty.
\end{aligned}
\tag{C.15}
$$

Let $z_1, z_2$ be two vectors from $\mathbb{C}^2$. Define a map

$$F : \mathbb{C}^2 \longrightarrow \mathbb{R}^3 \tag{C.16}$$

by setting

$$F(z_1, z_2) := \left( \frac{z_1 \overline{z}_2 + \overline{z}_1 z_2}{z_1 \overline{z}_1 + \overline{z}_2 z_2}, \frac{z_1 \overline{z}_2 - \overline{z}_1 z_2}{i(z_1 \overline{z}_1 + \overline{z}_2 z_2)}, \frac{z_1 \overline{z}_1 - \overline{z}_2 z_2}{z_1 \overline{z}_1 + \overline{z}_2 z_2} \right), \tag{C.17}$$

(see, for example, [Alv02]).

Let $z_2 = wz_1$, then

$$F(z_1, z_2) = \left( \frac{\overline{w} + w}{1 + |w|^2}, \frac{\overline{w} - w}{i(1 + |w|^2)}, \frac{1 - |w|^2}{1 + |w|^2} \right). \tag{C.18}$$

If $w = u + iv$, then

$$F(z_1, z_2) = \left( \frac{2u}{1 + u^2 + v^2}, \frac{2v}{1 + u^2 + v^2}, \frac{1 - u^2 - v^2}{1 + u^2 + v^2} \right). \tag{C.19}$$

It is easily to see that

$$(2u)^2 + (2v)^2 + (1 - u^2 - v^2)^2 = (1 + u^2 + v^2)^2,$$

so $F(z_1, z_2)$ maps every vector $[z_1, z_2]$ to a point (C.19) on the unit sphere $S^2$ in $\mathbb{R}^3$. Another vector $[z_3, z_4]$ defines the same line in the $\mathbb{C}P^1$ if and only if

$$\frac{z_4}{z_3} = \frac{z_2}{z_1} = w,$$

i.e., the map F (C.18) defines a bijection from the complex projective line $\mathbb{C}P^1$ and the unit sphere in $\mathbb{R}^3$:

$$F : \mathbb{C}P^1 \Longleftrightarrow S^2. \tag{C.20}$$

Let $(x, y, z)$ be a point on $S^2$ distinct from the *north pole* $N = \{0, 0, 1\}$. The *stereographic projection* is the map

$$S : S^2 \backslash N \longrightarrow \mathbb{C} \tag{C.21}$$

defined by

$$S(x, y, z) := \frac{x}{1 - z} + i \frac{y}{1 - z}. \tag{C.22}$$

see Fig. C.1. For details, see, for example, [Jen94, §2.9].

Consider the composition $S(F(z_1, z_2))$. By (C.17), we have

$$1 - z = \frac{2\overline{z}_2 z_2}{z_1 \overline{z}_1 + \overline{z}_2 z_2}, \tag{C.23}$$

and

$$\frac{x}{1 - z} = \frac{z_1 \overline{z}_2 + \overline{z}_1 z_2}{2 \overline{z}_2 z_2} = \frac{1}{2} \left( \frac{z_1}{z_2} + \frac{\overline{z}_1}{\overline{z}_2} \right),$$

$$\tag{C.24}$$

$$\frac{y}{1 - z} = \frac{z_1 \overline{z}_2 - \overline{z}_1 z_2}{2i \overline{z}_2 z_2} = \frac{1}{2i} \left( \frac{z_1}{z_2} - \frac{\overline{z}_1}{\overline{z}_2} \right).$$

Let

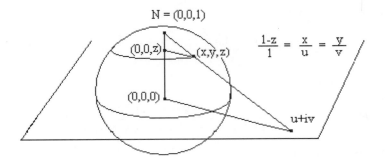

**Fig. C.1.**    The stereographic projection

$$\frac{z_1}{z_2} = w = u + iv.$$

Then

$$\frac{x}{1-z} = \frac{1}{2}(w + \overline{w}) = u, \qquad \frac{y}{1-z} = \frac{1}{2i}(w - \overline{w}) = v, \tag{C.25}$$

and

$$S(F(z_1, z_2)) = u + iv = w = \frac{z_1}{z_2}. \tag{C.26}$$

## C.6 The prime spectrum, the coordinate ring, the orbit space

### C.6.1 Hilbert's Nullstellensatz (Theorem of zeros)

From now on, we assume that $R$ is a commutative ring with unit.

A proper ideal $\mathfrak{m}$ of $R$ is said to be *maximal* if $\mathfrak{m}$ is not a proper subset of any other proper ideal of $R$. An ideal $\mathfrak{m} \subset R$ is maximal if and only if the quotient ring $R/\mathfrak{m}$ is a field. For example, every ideal $p\mathbb{Z}$ is maximal in the ring of integers $\mathbb{Z}$ if $p$ is a prime number, and, in this case, the quotient ring $\mathbb{Z}/p\mathbb{Z}$ is a field.

A proper ideal $\mathfrak{p}$ of a commutative ring $R$ is called a *prime ideal* if the following condition holds:

for any $a, b \in R$, if $a \cdot b \in \mathfrak{p}$, then either $a \in \mathfrak{p}$ or $b \in \mathfrak{p}$. \tag{C.27}

An ideal $\mathfrak{p} \subset R$ is prime if and only if the quotient ring $R/\mathfrak{p}$ is an *integral domain* (i.e., the commutative ring which has no divisors of 0). Examples of prime ideals (recall that $(x_1, \ldots, x_n)$ is the ideal generated by $x_1, \ldots, x_n$):

the ideal $(5, \sqrt{6})$ in the ring $\mathbb{Z}[\sqrt{6}]$,

the ideal $(x)$ in the ring $\mathbb{Z}[x]$, \tag{C.28}

the ideal $(y + x + 1)$ in the ring $\mathbb{C}[x, y]$.

The maximal ideals are prime since the fields are integral domains, but not conversely:

$$\boxed{\text{maximal ideals}} \subset \boxed{\text{prime ideals}} \, . \tag{C.29}$$

The ideal $(x) \in \mathbb{Z}[x]$ is prime, but not maximal, since, for example:

$$(x) \subset (2, x) \subset \mathbb{Z}[x]. \tag{C.30}$$

Let $k$ be an algebraically closed field (e.g., the complex field $\mathbb{C}$), and let $I$ be an ideal in $k[x_1, \ldots, x_n]$. Define $\mathbb{V}_k(I)$, the *zero set* of $I$ , by

$$\mathbb{V}_k(I) = \{(a_1, \ldots, a_n) \in k \mid f(a_1, \ldots . a_n) = 0 \text{ for all } f \in I\}. \tag{C.31}$$

Denote by $\mathbb{A}_k^n$ (or just $\mathbb{A}^n$) the $n$-dimensional *affine space* over the field $k$. The *Zariski topology* on $\mathbb{A}_k^n$ is defined to be the topology whose closed sets are the zero sets $\mathbb{V}_k(I)$.

For any ideal $I$ of the commutative ring $R$, the radical $\sqrt{I}$ of $I$ is the set

$$\{a \in R \mid a^n \in I \text{ for some integer } n > 0\}. \tag{C.32}$$

The radical of an ideal $I$ is always an ideal of $R$. If $I = \sqrt{I}$, then $I$ is called a *radical ideal*. The prime ideals are radical:

$$\boxed{\text{prime ideals}} \subset \boxed{\text{radical ideals}} \, . \tag{C.33}$$

**Theorem C.7.** (Hilbert's Nullstellensatz, [Re88, Ch.2], or [Mum88, pp.9-11])
*Let $k$ be an algebraically closed field.*
*1) The maximal ideals in the ring $A = k[x_1, \ldots, x_n]$ are the ideals*

$$\mathfrak{m}_P = (x_1 - a_1, \ldots, x_n - a_n) \tag{C.34}$$

*for some point $P = (a_1, \ldots, a_n)$. The ideal $\mathfrak{m}_P$ coincides with the ideal $I(P)$ of all functions which vanish at $P$.*
*2) If $\mathbb{V}_k(I) = \emptyset$, then $I = k[x_1, \ldots, x_n]$.*
*3) For any ideal $J \subset A$, we have*

$$I(\mathbb{V}_k(J)) = \sqrt{J}, \tag{C.35}$$

The set $X \subset \mathbb{A}_k^n$ is called an *affine variety* if $X = \mathbb{V}_k(I)$ for some ideal $I \subset A$, see (C.31). An affine variety $X \subset \mathbb{A}_k^n$ is said to be *irreducible* if there does not exist a decomposition into the disjoint union

$$X = X_1 \coprod X_2,$$

where $X_1$, $X_2$ are two proper subsets of $X$. For example, the affine variety

$$X = \{(x, y) \subset \mathbb{A}_\mathbb{C}^2 \mid xy = 0\}$$

is decomposed into the sum of

$$X_1 = \{(x, y) \subset \mathbb{A}_\mathbb{C}^2 \mid x = 0\} \text{ and } X_2 = \{(x, y) \subset \mathbb{A}_\mathbb{C}^2 \mid y = 0\}.$$

**Proposition C.8.** *Let $X \subset \mathbb{A}_k^n$ be an affine variety and $I = I(X)$ the corresponding ideal, i.e., $X = \mathbb{V}_k(I)$, see (C.31). Then*

$$X \text{ is irreducible } \Longleftrightarrow I(X) \text{ is the prime ideal.}$$

For a proof of this proposition, see, e.g., [Re88, §3.7].

One of the important corollaries of Hilbert's Nullstellensatz is the following one-to-one correspondence [Re88, §3.10] between subvarieties $X \subset \mathbb{A}_k^n$ and ideals $I \subset A$:

$$
\begin{array}{ccc}
\{ \text{ radical ideals } \} & \Longleftrightarrow & \{ \text{ affine varieties } \} \\
\cup & & \cup \\
\{ \text{ prime ideals } \} & \Longleftrightarrow & \{ \text{ irreducible affine varieties } \} \qquad \text{(C.36)} \\
\cup & & \cup \\
\{ \text{ maximal ideals } \} & \Longleftrightarrow & \{ \text{ points } \}
\end{array}
$$

### C.6.2 The prime spectrum

The *prime spectrum* $\mathrm{Spec}(R)$ of a given commutative ring $R$ is defined to be the set of proper prime ideals of $R$:

$$\{\mathfrak{p} \subset R \mid \mathfrak{p} \text{ is a prime ideal of } R\}. \qquad \text{(C.37)}$$

The ring $R$ itself is not counted as a prime ideal, but $(0)$, if prime, is counted.

A topology is imposed on $\mathrm{Spec}(R)$ by defining the sets of *closed sets*. For any subset $J$ of $R$, the *closed set* $V(J)$ is defined to be the set of the prime ideals containing $J$:

$$V(J) := \{\mathfrak{p} \mid \mathfrak{p} \supseteq J\} \subset \mathrm{Spec}(R). \qquad \text{(C.38)}$$

The *closure* $\overline{\mathcal{P}}$ of the subset $\mathcal{P}$ is the intersection of all closed sets containing $\mathcal{P}$:

$$\overline{\mathcal{P}} = \bigcap_{V(J) \supset \mathcal{P}} V(J). \qquad \text{(C.39)}$$

Consider the closure of points of the topological space $\mathrm{Spec}(R)$. For a point of $\mathrm{Spec}(R)$ which is the prime ideal $\mathfrak{p}$, we see that

$$\overline{\mathfrak{p}} = \bigcap_{V(J) \supset \mathfrak{p}} V(J) = V(\mathfrak{p}) \qquad \text{(C.40)}$$

which consists of all prime ideals $\mathfrak{p}' \supset \mathfrak{p}$. In particular, the closure $\overline{\mathfrak{p}}$ consists of one point if and only if the ideal $\mathfrak{p}$ is maximal. Any point which coincides with its closure is called a *closed point*. Thus, closed points in $\mathrm{Spec}(R)$ correspond one-to-one to maximal ideals.

In the conditions of Hilbert's Nullstellensatz every maximal ideal $J$ is $\mathfrak{m}_P$ (C.34) defined by some point $P = (a_1, \ldots, a_n) \in k^n$. Thus, in the finitely generated ring $A = k[x_1, \ldots, x_n]$ over an algebraically closed field $k$, the closed points in $\mathrm{Spec}(A)$ correspond to the maximal ideals, hence to the points $P = (a_1, \ldots, a_n) \in k^n$.

In the topological space $\mathrm{Spec}(R)$ there exist non-closed points . Let $R$ have no divisors of 0. Then the ideal $(0)$ is prime and is contained in all other prime ideals. Thus, the closure $\overline{(0)}$ consists of all prime ideals, i.e., coincides with the space $\mathrm{Spec}(R)$. The point $(0)$ is an everywhere dense point in $\mathrm{Spec}(R)$. Any everywhere dense point is called a *generic point*.

**Definition C.9.** Let $\mathcal{P}$ be an irreducible closed subset of $\mathrm{Spec}(R)$. A point $a \in \mathcal{P}$ is said to be a *generic point of* $\mathcal{P}$ if the closure $\bar{a}$ coincides with $\mathcal{P}$.

**Proposition C.10.** ([Mum88, p.126]) *If* $x \in \mathrm{Spec}(R)$, *then the closure* $\overline{\{x\}}$ *of* $\{x\}$ *is irreducible and* $x$ *is a generic point of* $\{x\}$. *Conversely, every irreducible closed subset* $\mathcal{P} \subset \mathrm{Spec}(R)$ *is equal to* $V(J)$ *for some prime ideal* $J \subset R$ *and* $J$ *is its unique generic point.*

A *distinguished open set* of $\mathrm{Spec}(R)$ is defined to be an open set of the form

$$\mathrm{Spec}(R)_f := \{\mathfrak{p} \in \mathrm{Spec}(R) \mid f \notin \mathfrak{p}\}. \tag{C.41}$$

for any element $f \in R$.

*Example C.11.* 1) If $R$ is a field, then $\mathrm{Spec}(R)$ has just one point $(0)$.

2) Let $R = k[X]$ be a polynomial ring in one variable $x$. Then $\mathrm{Spec}(R)$ is the affine line $\mathbb{A}_k^1$ over $k$. There exist two types of prime ideals: $(0)$ and $(f(X))$, where $f$ is an irreducible polynomial. For any algebraically closed $k$, the closed points are all of the form $(X - a)$. The point $(0)$ is generic.

3) $\mathrm{Spec}(\mathbb{Z})$ consists of closed points for every prime ideal $(\mathrm{p})$, plus the ideal $(0)$.

### C.6.3 The coordinate ring

Let $V \subset \mathbb{A}_k^n$ be an affine variety and $I(V)$ an ideal of $V$. The *coordinate ring* of the affine variety $V$ is defined to be:

$$k[V] := k[x_1, \ldots, x_n]/I(V). \tag{C.42}$$

A *regular function* on the affine variety $V$ is the restriction to $V$ of a polynomial in $x_1, \ldots, x_n$ modulo $I(V)$ (i.e., modulo functions vanishing on $V$). Thus, the regular functions on the affine variety $V$ are elements of the coordinate ring $k[V]$.

*Example C.12.* 1) Let $f(x, y)$ be a complex polynomial function on $\mathbb{C}^2$. The coordinate ring of a plane curve defined by the polynomial equation $f(x, y) = 0$ in $\mathbb{A}_\mathbb{C}^2$ is

$$\mathbb{C}[x, y]/(f(x, y)). \tag{C.43}$$

2) Consider two algebraic curves: $y = x^r, r \in \mathbb{N}$, and $y = 0$. The coordinate rings of these curves are isomorphic:

$$y = x^r \quad \text{corresponds to} \quad \mathbb{C}[x, y]/(y - x^r) \simeq \mathbb{C}[x, x^r] \simeq \mathbb{C}[x],$$
$$y = 0 \quad \text{corresponds to} \quad \mathbb{C}[x, y]/(y) \simeq \mathbb{C}[x, x] \simeq \mathbb{C}[x].$$

Thus, in the sense of algebraic geometry, the curves $y = x^r$ and $y = 0$ are equivalent.

3) On the other hand,

$$\mathbb{C}[x, y]/(y^2 - x^3) \not\simeq \mathbb{C}[x],$$

and the curves $y^2 = x^3$ and $y = 0$ are not equivalent (in the sense of algebraic geometry). Indeed, there exists the isomorphism

$$x \longmapsto T^2, \qquad y \longmapsto T^3,$$

and

$$\mathbb{C}[x, y]/(y^2 - x^3) \simeq \mathbb{C}[T^2, T^3] \subset \mathbb{C}[T].$$

The affine variety $X = \{(x, y) \mid y^2 = x^3\}$ is called *Neile's parabola*, or the *semicubical parabola*.

*Remark C.13.* Consider the coordinate ring $R = k[V]$ of the affine variety $V$ over an algebraically closed $k$. Then,

*the prime spectrum $\mathrm{Spec}(R)$ contains exactly*
*the same information as the variety $V$.*

Indeed, according to Hilbert's Nullstellensatz, the maximal ideals of $k[V]$ correspond one-to-one to points of $V$:

$$v \in V \Longleftrightarrow \mathfrak{m}_v \subset k[V].$$

Besides, according to Proposition C.8 every other prime ideal $\mathfrak{p} \subset k[V]$ is the intersection of maximal ideals corresponding to the points of some irreducible subvariety $Y \subset V$:

$$\mathfrak{p}_Y = \bigcap_{v \in Y} \mathfrak{m}_v.$$

### C.6.4 The orbit space

Let $G$ be a finite group acting on affine variety $V$ and $k[V]$ the coordinate ring of $V$. The problem here is that the quotient space $V/G$ might not exist, even for very trivial group actions, see Example C.14. To find how this problem is resolved in geometric invariant theory (GIT), see [Dol03] or [Kr85].

*Example C.14.* Consider the multiplicative group $G = k^\times$ acting on the affine line $\mathbb{A}_k^1$. The orbit space for this action consists of two orbits: $\{0\}$ and $\mathbb{A}_k^1 \backslash \{0\}$. The second orbit $\mathbb{A}_k^1 \backslash \{0\}$ is not a closed subset in the Zariski topology, and the first orbit $\{0\}$ is contained in the closure of the orbit $\mathbb{A}_k^1 \backslash \{0\}$. Thus, the point $\{0\}$ is the generic point (§C.6.2).

The following two definitions of the *categorical quotient* and the *orbit space* can be found, e.g., in P. E. Newstead's textbook [Newst78, p.39].

**Definition C.15.** Let $G$ be an algebraic group acting on a variety $V$. A *categorical quotient* of $V$ by $G$ is a pair $(Y, \varphi)$, where $Y$ is a variety and $\varphi$ is a morphism $\varphi : V \longrightarrow Y$ such that
(i) $\varphi$ is constant on the orbits of the action;
(ii) for any variety $Y'$ and morphism $\varphi' : V \longrightarrow Y'$ which is constant on orbits, there is a unique morphism $\psi : Y \longrightarrow Y'$ such that $\psi \circ \varphi = \varphi'$.

**Definition C.16.** A categorical quotient of $V$ by $G$ is called an *orbit space* if $\varphi^{-1}(y)$ consists of a single orbit for all $y \in Y$. The *orbit space* is denoted by $V/G$.

The *orbit space* $X = V/G$ is an affine variety whose points correspond one-to-one to orbits of the group action.

**Proposition C.17.** *Let $G$ be a finite group of order $n$ acting on an affine variety $V$. Assume that the characteristic of $k$ does not divide $n$. Set*

$$X := \operatorname{Spec} k[V]^G. \tag{C.44}$$

*Then $X$ is an orbit space for the action of $G$ on $V$.*

For a proof of this theorem, see, e.g., [Shaf88, Ch.1 §2, Ex.11] or [Rom00, Th.1.6].

# C.7 Fixed and anti-fixed points of the Coxeter transformation

## C.7.1 The Chebyshev polynomials and the McKay-Slodowy matrix

We continue to use block arithmetic for $(m + k) \times (m + k)$ matrices as in §3.1.3, namely:

$$K = \begin{pmatrix} 2I & 2D \\ 2F & 2I \end{pmatrix}, \quad w_1 = \begin{pmatrix} -I & -2D \\ 0 & I \end{pmatrix}, \quad w_2 = \begin{pmatrix} I & 0 \\ -2F & -I \end{pmatrix}. \tag{C.45}$$

Let

$$\Theta = \frac{1}{2}K - I = \begin{pmatrix} 0 & D \\ F & 0 \end{pmatrix}. \tag{C.46}$$

The matrix $\Theta$ coincides (up to the factor $-\frac{1}{2}$) with the McKay matrix from §A.4 or with the Slodowy matrix in the multiply-laced case, (A.56), (A.63). Introduce two recurrent series $f_p(\Theta)$ and $g_p(\Theta)$ of polynomials in $\Theta$:

$$
\begin{aligned}
f_0(\Theta) &= 0, \\
f_1(\Theta) &= K = 2\Theta + 2I, \\
f_{p+2}(\Theta) &= 2\Theta f_{p+1}(\Theta) - f_p(\Theta) \quad \text{for } p \geq 0
\end{aligned}
\tag{C.47}
$$

and

$$
\begin{aligned}
g_0(\Theta) &= 2I, \\
g_1(\Theta) &= K - 2I = 2\Theta, \\
g_{p+2}(\Theta) &= 2\Theta g_{p+1}(\Theta) - g_p(\Theta) \quad \text{for } p \geq 0.
\end{aligned}
\tag{C.48}
$$

For example,

$$
\begin{aligned}
f_2(\Theta) &= 2\Theta(2\Theta + 2I) = (K - 2I)K = 4\begin{pmatrix} DF & D \\ F & FD \end{pmatrix}, \\
f_3(\Theta) &= (K - 2I)(K - 2I)K - K = 2\begin{pmatrix} 4DF - I & 4DFD - D \\ 4FDF - F & 4FD - I \end{pmatrix},
\end{aligned}
\tag{C.49}
$$

$$
\begin{aligned}
g_2(\Theta) &= 2(2\Theta^2 - I) = 2\begin{pmatrix} 2DF - I & 0 \\ 0 & 2FD - I \end{pmatrix}, \\
g_3(\Theta) &= 4\Theta(2\Theta^2 - I) - 2\Theta = 2\begin{pmatrix} 0 & 8DFD - 6D \\ 8FDF - 6F & 0 \end{pmatrix}.
\end{aligned}
\tag{C.50}
$$

*Remark C.18.* Let us formally substitute $\Theta = \cos t$. Then
   1) The polynomials $g_p(\Theta)$ are Chebyshev polynomials of the first kind (up to the factor $\frac{1}{2}$):

$$
g_p(\cos t) = 2 \cos pt.
$$

Indeed,

$$
g_1(\Theta) = 2\Theta = 2\cos t, \quad g_2(\Theta) = 2(2\Theta^2 - 1) = 2\cos 2t,
$$

and

$$
g_{p+2}(\Theta) = 2(2\cos t \cos(p+1)t - \cos pt) = 2\cos(p+2)t.
$$

   2) The polynomials $f_p(\Theta)$ are Chebyshev polynomials of the second kind (up to the factor $2\Theta + 2$):

$$
f_p(\cos t) = 2(\cos t + 1)\frac{\sin pt}{\sin t}.
$$

Indeed,

$$f_1(\Theta) = 2\Theta + 2I = 2\cos t + 2, \quad f_2(\Theta) = 2\Theta(2\Theta + 2I) = 4\cos t(\cos t + 1),$$

and

$$f_{p+2}(\Theta) = 2\frac{\cos t + 1}{\sin t}(2\cos t \sin(p+1)t - \sin pt) = 2(\cos t + 1)\frac{\sin(p+2)t}{\sin t}.$$

## C.7.2 A theorem on fixed and anti-fixed points

The purpose of this section is to prove the following

**Theorem C.19 ([SuSt82]).** *1) The fixed points of the powers of the Coxeter transformation satisfy the following relations*

$$\mathbf{C}^p z = z \quad \Longleftrightarrow \quad f_p(\Theta)z = 0.$$

*2) The anti-fixed points of the powers of the Coxeter transformation satisfy the following relations*

$$\mathbf{C}^p z = -z \quad \Longleftrightarrow \quad g_p(\Theta)z = 0.$$

In the proof of Theorem C.19 we will need some properties of block matrices. Define the involution $\sigma$ by setting

$$\sigma : A = \begin{pmatrix} P & Q \\ S & T \end{pmatrix} \quad \mapsto \quad \overline{A} = \begin{pmatrix} P & Q \\ -S & -T \end{pmatrix}. \tag{C.51}$$

The map $\sigma$ is a linear operator because $\overline{\alpha_1 A_1 + \alpha_2 A_2} = \alpha_1 \overline{A_1} + \alpha_2 \overline{A_2}$.
The involution $\sigma$ does not preserve products:

$$\overline{AR} \neq \overline{A}\,\overline{R},$$

but

$$\overline{AR} = \overline{A}\,R \quad \text{for any } R = \begin{pmatrix} X & Y \\ U & V \end{pmatrix}, \tag{C.52}$$

and

$$R\,\overline{A} = -\,\overline{RA} \quad \text{for any } R \text{ of the form } \begin{pmatrix} 0 & Y \\ U & 0 \end{pmatrix}. \tag{C.53}$$

The following relations are easy to check:

$$\mathbf{C} = w_1 w_2 = \begin{pmatrix} 4DF - I & 2D \\ -2F & -I \end{pmatrix},$$

$$\mathbf{C}^{-1} = w_2 w_1 = \begin{pmatrix} -I & -2D \\ 2F & 4FD - I \end{pmatrix}, \tag{C.54}$$

$$w_2 \mathbf{C} = \begin{pmatrix} 4DF - I & 2D \\ -8FDF + 4F & -4FD + I \end{pmatrix},$$

$$w_1 \mathbf{C}^{-1} = \begin{pmatrix} -4DF + I & 4D - 8DFD \\ 2F & 4FD - I \end{pmatrix},$$

(C.55)

$$w_1 - w_2 = 4 \begin{pmatrix} -2I & -2D \\ 2F & 2I \end{pmatrix} = -\overline{K}, \quad w_1 + w_2 = -(K - 2I), \quad \text{(C.56)}$$

$$\mathbf{C} - \mathbf{C}^{-1} = 4 \begin{pmatrix} DF & D \\ -F & -FD \end{pmatrix} = 4 \overline{\begin{pmatrix} DF & D \\ F & FD \end{pmatrix}},$$

(C.57)

$$\mathbf{C} + \mathbf{C}^{-1} = 2 \begin{pmatrix} 2DF - I & 0 \\ 0 & 2FD - I \end{pmatrix},$$

$$w_2 \mathbf{C} - w_1 \mathbf{C}^{-1} = 2 \begin{pmatrix} 4DF - I & 4DFD - D \\ -(4FDF - F) & -(4FD - I) \end{pmatrix} =$$

(C.58)

$$2 \overline{\begin{pmatrix} 4DF - I & 4DFD - D \\ 4FDF - F & 4FD - I \end{pmatrix}},$$

$$w_2 \mathbf{C} + w_1 \mathbf{C}^{-1} = \begin{pmatrix} 0 & -8DFD + 6D \\ -8FDF + 6F & 0 \end{pmatrix}, \quad \text{(C.59)}$$

$$2\Theta w_2 + I = -\mathbf{C}, \quad 2\Theta w_1 + I = -\mathbf{C}^{-1}. \quad \text{(C.60)}$$

**Proposition C.20.** *The following relations hold*

$$w_2 - w_1 = \overline{f_1(\Theta)}, \quad \text{(C.61)}$$

$$\mathbf{C} - \mathbf{C}^{-1} = \overline{f_2(\Theta)}, \quad \text{(C.62)}$$

$$w_2\mathbf{C} - w_1\mathbf{C}^{-1} = \overline{f_3(\Theta)}, \tag{C.63}$$

$$w_2 + w_1 = -g_1(\Theta), \tag{C.64}$$

$$\mathbf{C} + \mathbf{C}^{-1} = g_2(\Theta), \tag{C.65}$$

$$w_2\mathbf{C} + w_1\mathbf{C}^{-1} = -g_3(\Theta). \tag{C.66}$$

*Proof.* The proposition follows from comparing relations (C.47)–(C.48) and relations (C.57)–(C.59).
(C.61) follows from (C.47) and (C.56).
(C.62) follows from (C.49) and (C.57).
(C.63) follows from (C.49) and (C.58).
(C.64) follows from (C.48) and (C.56).
(C.65) follows from (C.50) and (C.57).
(C.66) follows from (C.50) and (C.59). □

**Proposition C.21.** *The following relations hold:*

$$\overline{f_{2m}(\Theta)} = \mathbf{C}^m - \mathbf{C}^{-m}, \qquad \overline{f_{2m+1}(\Theta)} = w_2\mathbf{C}^m - w_1\mathbf{C}^{-m}, \tag{C.67}$$

$$g_{2m}(\Theta) = \mathbf{C}^m + \mathbf{C}^{-m}, \qquad g_{2m+1}(\Theta) = -(w_2\mathbf{C}^m + w_1\mathbf{C}^{-m}). \tag{C.68}$$

*Proof.* 1) Relations (C.61)–(C.63) realize the basis of induction. By (C.53), by the induction hypothesis, and by (C.60) we have

$$\begin{aligned}
\overline{f_{2m+2}(\Theta)} &= \overline{2\Theta f_{2m+1}(\Theta)} - \overline{f_{2m}(\Theta)} = -2\Theta\,\overline{f_{2m+1}(\Theta)} - \overline{f_{2m}(\Theta)} = \\
&- 2\Theta(w_2\mathbf{C}^m - w_1\mathbf{C}^{-m}) - (\mathbf{C}^m - \mathbf{C}^{-m}) = \\
&- (2\Theta w_2 + I)\mathbf{C}^m + (2\Theta w_1 + I)\mathbf{C}^{-m} = \mathbf{C}^{m+1} - \mathbf{C}^{-(m+1)}.
\end{aligned}$$

In the same way we have

$$\begin{aligned}
\overline{f_{2m+3}(\Theta)} &= -2\Theta\,\overline{f_{2m+2}(\Theta)} - \overline{f_{2m+1}(\Theta)} = \\
&- 2\Theta(\mathbf{C}^{m+1} - \mathbf{C}^{-(m+1)}) - w_2\mathbf{C}^m - w_1\mathbf{C}^{-m} = \\
&- (2\Theta w_1 + I)w_2\mathbf{C}^m + (2\Theta w_2 + I)w_1\mathbf{C}^{-m} = \\
&\mathbf{C}^{-1}w_2\mathbf{C}^m - \mathbf{C}w_1\mathbf{C}^{-m} = w_2\mathbf{C}^{m+1} - w_1\mathbf{C}^{-(m+1)}.
\end{aligned}$$

2) Relations (C.64)–(C.66) realize the basis of induction. By the induction hypothesis and by (C.60) we have

$$g_{2m+2}(\Theta) = 2\Theta g_{2m+1}(\Theta) - g_{2m}(\Theta) =$$
$$- 2\Theta(w_2\mathbf{C}^m + w_1\mathbf{C}^{-m}) - (\mathbf{C}^m + \mathbf{C}^{-m}) =$$
$$- (2\Theta w_2 + I)\mathbf{C}^m - (2\Theta w_1 + I)\mathbf{C}^{-m} = \mathbf{C}^{m+1} + \mathbf{C}^{-(m+1)}.$$

By analogy, we have

$$g_{2m+3}(\Theta) = 2\Theta g_{2m+2}(\Theta) - g_{2m+1}(\Theta) =$$
$$- 2\Theta(\mathbf{C}^{m+1} + \mathbf{C}^{-(m+1)}) - (w_2\mathbf{C}^m + w_1\mathbf{C}^{-m}) =$$
$$- (2\Theta w_1 + I)w_2\mathbf{C}^m - (2\Theta w_2 + I)w_1\mathbf{C}^{-m} = w_2\mathbf{C}^{m+1} + w_1\mathbf{C}^{-(m+1)}.\square$$

Theorem C.19 now follows from Proposition C.21.

# References

[ArGV86]  V. I. Arnold, S. M. Gussein-Zade, A. N. Varchenko. *Singularities of differentiable maps* I, II. Birkhäuser, Basel, 1986, 1988.

[ASS06]  I. Assem, D. Simson, A. Skowroński, *Elements of the Representation Theory of Associative Algebras Vol.1 Techniques of Representation Theory* London Math. Soc. Student Texts (No. 65), Cambridge, 2006. x+458 pp.

[AtMa69]  M. F. Atiyah, I. G. Macdonald, *Introduction to Commutative Algebra.* Reading, Addison-Wesley, 1969.

[AuRS95]  M. Auslander, I. Reiten, S. O. Smalø, *Representation Theory of Artin Algebras.* Cambridge Studies in Advanced Mathematics 36, Cambridge, 1995.

[AuPR79]  M. Auslander, M. I. Platzeck, I. Reiten, *Coxeter Functors without Diagrams.* Trans. Amer. Math. Soc. 250 (1979), 1–46.

[A'C75]  N. A'Campo, *Le groupe de monodromie du deploiement des singularites isolees de courbes planes.* I. (French) Math. Ann. 213 (1975), 1–32.

[A'C76]  N. A'Campo, *Sur les valeurs propres de la transformation de Coxeter.* Invent. Math. 33 (1976), no. 1, 61–67.

[Alv02]  J. C. Álvarez Paiva *Interactive Course on Projective Geometry. Geometry of the Complex Pojective Line,* 2000,
http://www.math.poly.edu/courses/projective_geometry .

[Bak04]  A. Baker, *Representations of Finite Groups,* 2004,
http://www.maths.gla.ac.uk/~ajb/dvi-ps/groupreps.pdf .

[Blu06]  M. Blume, *McKay correspondence over non algebraically closed fields,* 2006, arXiv: math.RT/0601550v1

[Bol96]  A. Boldt, *Two Aspects of Finite-Dimensional Algebras: Uniserial Modules and Coxeter Polynomials.* University of California, Ph.D. thesis, 1996.

[BT97]  A. Boldt, M. Takane, *The spectral classes of unicyclic graphs.* J. Pure Appl. Algebra 133 (1998), no. 1-2, 39–49.

[Bo]  N. Bourbaki, *Groupes et agebres de Lie, Chaptires 4,5,6.* Paris, Hermann, 1968.

[Ben93]  D. J. Benson, *Polynomial invariants of finite groups.* London Math. Soc. Lecture Note Series, vol. 304, Cambridge University Press, Cambridge, 1993. x+118 pp.

[BLM89]  S. Berman, Y. S. Lee, R. V. Moody, *The spectrum of a Coxeter transformation, affine Coxeter transformations and the defect map.* J. Algebra 121 (1989), no. 2, 339–357.

[Brn01]  D. Berenstein, V. Jejjala, R .G. Leigh, *D-branes on singularities: new quivers from old.* Phys. Rev. D (3) 64 (2001), no. 4, 046011, 13 pp.

[BGP73]  I. N. Bernstein, I. M. Gelfand, V. A. Ponomarev, *Coxeter functors, and Gabriel's theorem.* Uspehi Mat. Nauk 28 (1973), no. 2(170), 19–33. English translation: Russian Math. Surveys 28 (1973), no. 2, 17–32.

[BS06]  J. Bernstein, O. Schwarzman, *Complex crystallographic Coxeter groups and affine root systems.* J. Nonlinear Math. Phys. 13 (2006), no. 2, 163–182.
J. Bernstein, O. Schwarzman, *Chevalley's theorem for the complex crystallographic groups.* J. Nonlinear Math. Phys. 13 (2006), no. 3, 323–351

[BS06a]  J. Bernstein, O. Schwarzman, *Complex crystallographic Coxeter groups and affine root systems,* Preprint 82/2006, Max Planck Institute for Mathematics in the Sciences, http://www.mis.mpg.de/preprints/2006/prepr2006_82.html .

[Br35]  R. Brauer, *Sur les invariants intégraux des variétés représentatives des groupes de Lie simples clos,* C. R. Acad. Sci. Paris, 201,(1935), 419–421.

[Br03]  J. Brundan, *Topics in Representation Theory,* Lecture Notes, University of Oregon, http://darkwing.uoregon.edu/~brundan/math607winter03/ch2.pdf, 2003.

[Car70]  R. Carter, *Conjugacy classes in the Weyl group.* 1970 Seminar on Algebraic Groups and Related Finite Groups (The Institute for Advanced Study, Princeton, N.J., 1968/69) 297–318, Springer, Berlin.

[Cip00]  M. Çiperjani, *The McKay Correspondence,* University of Massachusetts, http://www.math.princeton.edu/~mciperja/main.ps, 2000.

[Col58]  A. J. Coleman, *The Betti numbers of the simple Lie groups.* Canad. J. Math. 10 (1958), 349–356.

[Col89]  A. J. Coleman, *Killing and the Coxeter transformation of Kac-Moody algebras.* Invent. Math. 95 (1989), no. 3, 447–477.

[Col89a]  A. J. Coleman, *The greatest mathematical paper of all time.* Math. Intelligencer 11 (1989), no. 3, 29–38.

[Con80]  M. Conder, *Generators for alternating and symmetric groups.* J. London Math. Soc. (2) 22 (1980), no. 1, 75–86.

[Con90]  M. Conder, *Hurwitz groups: a brief survey.* Bull. Amer. Math. Soc. (N.S.) 23 (1990), no. 2, 359–370.

[Con03]  M. Conder, *Group actions on graphs, maps and surfaces with maximum symmetry.* invited paper in: Groups St Andrews 2001 in Oxford, London Math. Soc. Lecture Note Series, vol. 304, Cambridge University Press, 63–91, 2003.

[CCS72]  J. H. Conway, H. S. M. Coxeter, G. C. Shephard, *The centre of a finitely generated group.* Commemoration volumes for Prof. Dr. Akitsugu Kawaguchi's seventieth birthday, Vol. II. Tensor (N.S.) 25 (1972), 405–418; erratum, ibid. (N.S.) 26 (1972), 477.

[Cox34]  H. S. M. Coxeter, *Discrete groups generated by reflections.* Ann. of Math. (2) 35 (1934), no. 3, 588–621.

[Cox40]  H. S. M. Coxeter, *The binary polyhedral groups, and other generalizations of the quaternion group.* Duke Math. J. 7, (1940), 367–379.

[Cox49]  H. S. M. Coxeter, *Regular polytopes.* New York (1949).

[Cox51]  H. S. M. Coxeter, *The product of the generators of a finite group generated by reflections.* Duke Math. J. 18, (1951), 765–782.

[CoxM84]  H. S. M. Coxeter and W. O. J. Moser, *Generators and relations for discrete groups.* 4th ed., Springer-Verlag, NY, 1984.

[Ch50]  C. Chevalley, *The Betti numbers of the exceptional simple Lie groups.* Proceedings of the International Congress of Mathematicians, Cambridge, Mass., 1950, vol. 2, Amer. Math. Soc., Providence, R. I., 1952, 21–24.

[Ch55]  C. Chevalley, *Invariants of finite groups generated by reflections.* Amer. J. Math. 77 (1955), 778–782.

[Cr01]  A. Craw, *The McKay correspondence and representations of the McKay quiver.* Warwick Ph.D. thesis, 2001.

[CrW93]  W. Crawley-Boevey, *Geometry of representations of algebras.* A graduate course given at Oxford University, 1993.

[CrW99]  W. Crawley-Boevey, *Representations of quivers, preprojective algebras and deformations of quotient singularities.* Lectures from a DMV Seminar "Quantizations of Kleinian singularities", Oberwolfach, 1999, http://www.amsta.leeds.ac.uk/~pmtwc/dmvlecs.pdf .

[CrW01]  W. Crawley-Boevey, *Geometry of the moment map for representations of quivers.* Compositio Math. 126 (2001), no. 3, 257–293.

[CR62]  C. W. Curtis and I. Reiner, *Representation theory of finite groups and associative algebras.* Wiley (1962).

[CDS95]  D. M. Cvetković, M. Doob, H. Sachs, *Spectra of graphs. Theory and applications.* Third edition. Johann Ambrosius Barth, Heidelberg, (1995), ii+447 pp.

[CRS95]  D. M. Cvetković, P. Rowlinson, S. Simić, *Spectral generalizations of line graphs. On graphs with least eigenvalue −2.* London Mathematical Society Lecture Note Series, 314. Cambridge University Press, Cambridge, (2004). xii+298 pp.

[Dol03]  I. Dolgachev, *Lectures on invariant theory.* London Mathematical Society Lecture Note Series, 296. Cambridge University Press, Cambridge, (2003), xvi+220 pp.

[DL03]  V. Dlab, P. Lakatos, *On spectral radii of Coxeter transformations.* Special issue on linear algebra methods in representation theory. Linear Algebra Appl. 365 (2003), 143–153.

[DR74]  V. Dlab, C. M. Ringel, *Représentations des graphes valués.* (French) C. R. Acad. Sci. Paris Sér. A 278 (1974), 537–540.

[DR74a]  V. Dlab, C. M. Ringel, *Representations of graphs and algebras.* Carleton Mathematical Lecture Notes, No. 8. Department of Mathematics, Carleton University, Ottawa, Ont., 1974. iii+86 pp.

[DR76]  V. Dlab, C. M. Ringel, *Indecomposable representations of graphs and algebras.* Mem. Amer. Math. Soc. 6 (1976), no. 173.

[DR81]  V. Dlab and C. M. Ringel, *Eigenvalues of Coxeter transformations and the Gelfand-Kirillov dimension of preprojective algebras.* Proc. Amer. Math. Soc. 83 (2) (1981), 228–232.

[DF73]  P. Donovan, M. R. Freislich, *The representation theory of finite graphs and associated algebras.* Carleton Mathematical Lecture Notes, No. 5. Carleton University, Ottawa, Ont., 1973. iii+83 pp.

[Drz74]  Yu. A. Drozd, *Coxeter transformations and representations of partially ordered sets.* Funkcional. Anal. i Priložen. 8 (1974), no. 3, 34–42. English translation: Functional Anal. Appl. 8 (1974), 219–225 (1975).

[Drz80]  Yu. A. Drozd, *Tame and wild matrix problems.* Representation theory, II (Proc. Second Internat. Conf., Carleton Univ., Ottawa, Ont., 1979), pp. 242–258, Lecture Notes in Math., 832, Springer, Berlin-New York, 1980.

[DrK04]  Yu. A. Drozd, E. Kubichka, *Dimensions of finite type for representations of partially ordered sets*. (English. English summary) Algebra Discrete Math. 2004, no. 3, 21–37.

[DuVal34]  P. Du Val, *On isolated singularities which do not affect the condition of adjunction*. Proc. Cambridge Phil. Soc. 30 (1934), 453–465.

[Ebl02]  W. Ebeling, *Poincaré series and monodromy of a two-dimensional quasi-homogeneous hypersurface singularity*. Manuscripta Math. 107 (2002), no. 3, 271–282.

[EbGu99]  W. Ebeling, S. M. Gussein-Zade, *On the index of a vector field at an isolated singularity*. The Arnoldfest (Toronto, ON, 1997), 141–152, Fields Inst. Commun., 24, Amer. Math. Soc., Providence, RI, 1999.

[Fr51]  J. S. Frame, *Characteristic vectors for a product of n reflections*. Duke Math. J. 18, (1951), 783–785.

[FSS96]  J. Fuchs, B. Schellekens, C. Schweigert, *From Dynkin diagram symmetries to fixed point structures*. Comm. Math. Phys. 180 (1996), no. 1, 39–97.

[Gab72]  P. Gabriel, *Unzerlegbare Darstellungen I*. Manuscripta Math. 6 (1972), 71–103.

[Gb02]  A. Gabrielov, *Coxeter-Dynkin diagrams and singularities*. In: Selected Papers of E.B. Dynkin, AMS, 2000, p.367–369.

[GH01]  E. Ghatem, E. Hironaka, *The arithmetic and geometry of Salem numbers*. Bull. Amer. Math. Soc. 38 (2001), 293–314.

[Ga90]  F. Gantmakher, *The theory of matrices*. Chelsea, NY, 1990.

[GorOnVi94]  V. V. Gorbatsevich, A. L. Onishchik, È. B. Vinberg. *Structure of Lie groups and Lie algebras*. (Russian) Current problems in mathematics. Fundamental directions, Vol. 41, 5–259, Itogi Nauki i Tekhniki, Akad. Nauk SSSR, Moscow, 1990. Translation by V. Minachin. Translation edited by A. L. Onishchik and È. B. Vinberg. Encyclopaedia of Mathematical Sciences, 41. Springer-Verlag, Berlin, 1994. iv+248 pp.

[GP69]  I. M. Gelfand, V. A. Ponomarev, *Remarks on the classification of a pair of commuting linear transformations in a finite-dimensional space*. Funkcional. Anal. i Priložen. 3 1969 no. 4, 81–82. English translation: Functional Anal. Appl. 3, 325–326, 1969.

[GP72]  I. M. Gelfand, V. A. Ponomarev, *Problems of linear algebra and classification of quadruples of subspaces in a finite-dimensional vector space*. Hilbert space operators and operator algebras (Proc. Internat. Conf., Tihany, 1970), pp. 163–237. Colloq. Math. Soc. Janos Bolyai, 5, North-Holland, Amsterdam, 1972.

[GP74]  I. M. Gelfand, V. A. Ponomarev, *Free modular lattices and their representations*. Collection of articles dedicated to the memory of Ivan Georgievic Petrovskii (1901-1973), IV. Uspehi Mat. Nauk 29 (1974), no. 6(180), 3–58. English translation: Russian Math. Surveys 29 (1974), no.6, 1–56.

[GP76]  I. M. Gelfand, V. A. Ponomarev, *Lattices, representations, and their related algebras. I*. Uspehi Mat. Nauk 31 (1976), no. 5(191), 71–88. English translation: Russian Math. Surveys 31 (1976), no.5, 67–85.

[Gu76]  S. M. Gussein-Zade, *The characteristic polynomial of the classical monodromy for series of singularities*. Funkcional. Anal. i Priložen. 10 (1976), no. 3, 78–79. English translation: Functional Anal. Appl. 3, 229–231, 1976.

[GP79]  I. M. Gelfand, V. A. Ponomarev, *Model algebras and representations of graphs*. Funkts.Anal. i Prilozhen. 13 (1979) no.3, 1–12, English translation: Funct.Anal. Appl.13 (1980),no.3,157–166.

[GV83]  G. Gonzalez-Sprinberg, J. L. Verdier, *Construction geometrique de la corre-spondance de McKay*. Ann. Sci. Ecole Norm. Sup. (4) 16, (1983), no.3 409–449.

[Ha89]  M. Hamermesh, *Group Theory and Its Application to Physical Problems*. New York: Dover, 1989.

[Har95]  J. Harris, *Algebraic Geometry. A First Course*. Springer-Verlag, 1995.

[HPR80]  D. Happel, U. Preiser, C. M. Ringel, *Binary polyhedral groups and Eu-clidean diagrams*. Manuscripta Math. 31 (1980), no. 1-3, 317–329.

[Hil1890]  D. Hilbert, *Über die Theorie der Algebraischen Formen*. Math. Annalen 36. (1890), 473–534.

[Hir02]  E. Hironaka, *Lehmer's problem, McKay's correspondence, and* 2, 3, 7. Topics in algebraic and noncommutative geometry (Luminy/Annapolis, MD, 2001), 123–138, Contemp. Math., 324, Amer. Math. Soc., Providence, RI, 2003.

[Hob02]  J. van Hoboken, *Platonic solids, binary polyhedral groups, Kleinian singu-larities and Lie algebras of type $A, D, E$*. Master's Thesis, University of Ams-terdam, http://home.student.uva.nl/joris.vanhoboken/scriptiejoris.ps, 2002.

[How82]  R. Howlett, *Coxeter groups and M-matrices*. Bull. London Math. Soc. 14 (1982), no. 2, 137–141.

[Ho41]  H. Hopf, *Über die Topologie der Gruppenmannigfaltigkeiten und ihre Verall-gemeinerungen*. Annals of Math. 42 (1941), 22–52.

[Hu75]  J. E. Humphreys, *Representations of* SL(2, $p$). Amer. Math. Monthly, 82 (1975), 21–39.

[Hu78]  J. E. Humphreys, *Introduction to Lie algebras and representation theory*. Second printing, revised. Graduate Texts in Mathematics, 9. Springer-Verlag, New York-Berlin, 1978. xii+171 pp.

[Hu90]  J. E. Humphreys, *Reflection groups and Coxeter groups*. Cambridge Studies in Advanced Mathematics, 29. Cambridge University Press, Cambridge, 1990. xii+204 pp.

[Hu97]  J. E. Humphreys, *Comparing modular representations of semisimple groups and their Lie algebras*. Modular interfaces (Riverside, CA, 1995), 69–80, AMS/IP Stud. Adv. Math., 4, Amer. Math. Soc., Providence, RI, 1997.

[Hur1893]  A. Hurwitz, *Über algebraische Gebilde mit eindeutigen Transformationen in sich*. Math. Ann., 41 (1893), 403–442.

[Hur1897]  A. Hurwitz, *Über die Erzeugung der Invarianten durch Integration*. Göttingen. Nachrichten, (1897), pp. 71–90.

[Il87]  G. G. Ilyuta, *On the Coxeter transform of an isolated singularity*. Uspekhi Mat. Nauk 42 (1987), no. 2(254), 227–228; English translation: Russian Math. Surveys 42 (1987), no. 2, 279–280.

[Il95]  G. G. Ilyuta, *Characterization of simple Coxeter-Dynkin diagrams*. Funkt. Anal. i Prilozhen. 29 (1995), no. 3, 72–75; English translation: Funct. Anal. Appl. 29 (1995), no. 3, 205–207 (1996)

[IN99]  Y. Ito, I. Nakamura, *Hilbert schemes and simple singularities*. New trends in algebraic geometry (Warwick, 1996), 151–233, London Math. Soc. Lecture Note Ser., 264, Cambridge Univ. Press, Cambridge, 1999.

[JL2001]  G. James, M. Leibeck, *Representations and characters of groups*, 2d edi-tion, Cambridge University Press, 2001.

[Jen94]  G. Jennings, *Modern Geometry with Applications*, *Springer-Verlag*. New York, 1994.

[Kac68]  V. G. Kac, *Simple irreducible graded Lie algebras of finite growth*. Izv. Akad. Nauk SSSR Ser. Mat. 32 (1968) 1323–1367, English translation: Math. USSR-Izv. 2 (1968), 1271–1311.

[Kac69] V. Kac, *Automorphisms of finite order of semisimple Lie algebras.* Funk. Anal. i Priložen. 3 1969, 3, 94–96. English translation: Functional Anal. Appl. 3 (1969), 252–254.

[Kac80] V. Kac, *Infinite root systems, representations of graphs and invariant theory.* Invent. Math. 56 (1980), no. 1, 57–92

[Kac82] V. Kac, *Infinite root systems, representations of graphs and invariant theory. II.* J. Algebra 78 (1982), no. 1, 141–162.

[Kac83] V. Kac, *Root systems, representations of quivers and invariant theory.* Invariant theory (Montecatini, 1982), 74–108, Lecture Notes in Math., 996, Springer, Berlin, 1983.

[Kac93] V. Kac, *Infinite-Dimensional Lie Algebras.* 3d edition, Cambridge University Press, 1993.

[Kar92] G. Karpilovsky, *Group Representations: Introduction to Group Representations and Characters*, Vol 1 Part B, North-Holland Mathematics Studies 175, Amsterdam, 1992.

[Kir04] A. Kirillov, Jr. *Introduction to Lie Groups and Lie Algebras*, Department of Mathematics, SUNY at Stony Brook, Stony Brook, NY, 2004, http://www.math.sunysb.edu/~kirillov/mat552/liegroups.pdf

[Kir05] A. Kirillov, Jr. *Infinite Dimensional Lie Algebras*, Department of Mathematics, SUNY at Stony Brook, Stony Brook, NY, 2005, http://www.math.sunysb.edu/~blafard/notes

[Kl1884] F. Klein, *Lectures on the Ikosahedron and the Solution of Equations of the Fifth Degree.* Birkhäuser publishing house Basel, 1993 (reproduction of the book of 1888).

[Kn85] H. Knörrer, *Group representations and the resolution of rational double points.* in: Finite groups - coming of age. Cont. Math. 45, (1985) 175–222.

[KMSS03] V. A. Kolmykov, V. V. Menshikh, V. F. Subbotin, M. V. Sumin. *Four coefficients of the characteristic polynomial of the Coxeter transformation.* Mat. Zametki 73 (2003), no. 5, 788–792; English translation: Math. Notes 73 (2003), no. 5-6, 742–746.

[Kos59] B. Kostant, *The principal three-dimensional subgroup and the Betti numbers of a complex simple Lie group.* Amer. J. Math. 81 (1959) 973–1032.

[Kos84] B. Kostant, *The McKay correspondence, the Coxeter element and representation theory.* The mathematical heritage of Elie Cartan (Lyon, 1984), Asterisque 1985, Numero Hors Serie, 209–255.

[Kos04] B. Kostant, *The Coxeter element and the branching law for the finite subgroups of* SU(2), preprint, 2004, **arXiv: math.RT/0411142**.

[Kos06] B. Kostant, *The Coxeter element and the branching law for the finite subgroups of* SU(2). The Coxeter legacy, 63–70, Amer. Math. Soc., Providence, RI, 2006.

[Kosz50] J. L. Koszul, *Homologie et cohomologie des algèbres de Lie*, Bull. Soc. Math. France 78, (1950). 65–127.

[Kr85] H. Kraft, *Geometrische Methoden in der Invariantentheorie.* (German) [Geometrical methods in invariant theory], Friedr. Vieweg & Sohn, Braunschweig, 1984. x+308 pp.

[KR86] H. Kraft, Ch. Riedtmann, *Geometry of representations of quivers.* Representations of algebras (Durham, 1985), 109–145, London Math. Soc. Lecture Note Ser., 116, Cambridge Univ. Press, Cambridge, 1986.

[Lak99a] P. Lakatos, *On the spectral radius of Coxeter transformations of trees.* Publ. Math. Debrecen 54 (1999), no. 1-2, 181–187.

[Lak99b]  P. Lakatos, *On the Coxeter polynomials of wild stars.* Linear Algebra Appl. 293 (1999), no. 1-3, 159–170.

[Lev66]  J. Levine, *Polynomial invariants of knots of codimension two.* Ann. of Math. (2) 84, 1966, 537–554.

[Leh33]  D. H. Lehmer, *Factorization of certain cyclotomic functions.* Ann. of Math. 34 (1933), 461–469.

[LuTW00]  A. Lucchini, M. C. Tamburini and J. S. Wilson, *Hurwitz groups of large rank.* J. London Math. Soc. (2) 61 (2000), no. 1, 81–92.

[LuT99]  A. Lucchini, M. C. Tamburini, *Classical groups of large rank as Hurwitz groups.* J. Algebra 219 (1999), no. 2, 531–546.

[Lus83]  G. Lusztig, *Some examples of square integrable representations of semisimple p-adic groups.* Trans. Amer. Math. Soc. 277 (1983), no. 2, 623–653.

[Lus99]  G. Lusztig, *Subregular nilpotent elements and bases in K-theory.* Dedicated to H. S. M. Coxeter on the occasion of his 90th birthday. Canad. J. Math. 51 (1999), no. 6, 1194–1225.

[Mac72]  I. G. Macdonald, *Affine root systems and Dedekind's η-function.* Invent. Math. 15 (1972), 91–143

[Max98]  G. Maxwell, *The normal subgroups of finite and affine Coxeter groups.* Proc. London Math. Soc. (3) 76 (1998), no. 2, 359–382

[McM02]  C. McMullen, *Coxeter groups, Salem numbers and the Hilbert metric.* Publ. Math. Inst. Hautes Études Sci. no. 95 (2002), 151–183.

[McK80]  J. McKay, *Graphs, singularities, and finite groups.* The Santa Cruz Conference on Finite Groups (Univ. California, Santa Cruz, Calif., 1979), pp. 183-186, Proc. Sympos. Pure Math., 37, Amer. Math. Soc., Providence, R.I., 1980.

[McK81]  J. McKay, *Cartan matrices, finite groups of quaternions, and Kleinian singularities.* Proc. Amer. Math. Soc. 81 (1981), no. 1, 153–154.

[McK99]  J. McKay, *Semi-affine Coxeter-Dynkin graphs and $G \subseteq SU_2(C)$.* Dedicated to H. S. M. Coxeter on the occasion of his 90th birthday. Canad. J. Math. 51 (1999), no. 6, 1226–1229.

[McK01]  J. McKay, *A Rapid Introduction to ADE Theory*, January 1, 2001, http://math.ucr.edu/home/baez/ADE.html .

[Men85]  V. V. Menshikh, *Conjugacy in the Weyl group and the regularity of the graph representation.* Applications of topology in modern calculus, Voronezh Univ. Press, Voronezh, 1985, 144–150 (in Russian).

[MM64]  M. Markus, H. Minc, *A Survey o Matrix Theory and Matrix Inequalities.* Allyn and Bacon, Boston, 1964.

[Mohr04]  S. Mohrdieck *A Steinberg Cross-Section for Non-Connected Affine Kac-Moody Groups*, 2004, arXiv: math.RT/0401203.

[Mo68]  R. V. Moody, *A New Class of Lie Algebras.* J. Algebra 10 1968 211–230.

[Mon07]  S. Montarani, *On some finite dimensional representations of symplectic reflection algebras associated to wreath products*, 2007, arXiv: math.RT/0411286.

[Mos98]  M. Mossinghoff, *Polynomials with small Mahler measure.* Mathematics of Computation, 67 (1998), no. 224, 1697–1705.

[MRS99]  J. F. McKee, P. Rowlinson, C. J. Smyth, *Salem numbers and Pisot numbers from stars.* Number theory in progress, Vol. 1 (Zakopane-Kościelisko, 1997), 309–319, de Gruyter, Berlin, 1999.

[MS05]  J. F. McKee, C. J. Smyth, *Salem numbers, Pisot numbers, Mahler measure and graphs*, preprint, 2005, arXiv: math.NT/0503480

[Mu01] Q. Mushtaq, *Parametrization of triangle groups as subgroups of* PSL$(2, q)$. Southeast Asian Bull. Math. 25 (2001), no. 2, 309–312.

[Mum88] D. Mumford, *Introduction to algebraic geometry (preliminary version of first 3 chapters)* Reissued as The red book of varieties and schemes, Springer-Verlag, (1988), Lecture Notes in Mathematics, 1358.

[MWZ99] P. Magyar, J. Weyman, A. Zelevinsky, *Multiple flag varieties of finite type.* Adv. Math. 141 (1999), no. 1, 97–118.

[Nag59] M. Nagata. *On the 14th problem of Hilbert.* Amer. J. Math. 81 (1959), 766–772.

[Naz73] L. A. Nazarova, *Representations of quivers of infinite type.* Izv. Akad. Nauk SSSR Ser. Mat. 37 (1973), 752–791, English translation: Math. USSR-Izv. 7 (1973), 749–792.

[NR72] L. A. Nazarova, A. V. Roiter, *Representations of partially ordered sets.* (Russian) Investigations on the theory of representations. Zap. Naučn. Sem. Leningrad. Otdel. Mat. Inst. Steklov. (LOMI) 28 (1972), 5–31.

[NR73] L. A. Nazarova, A. V. Roiter, *Polyquivers and Dynkin schemes.* (Russian) Funkcional. Anal. i Priložen. 7 (1973), no. 3, 94–95.

[NaOR77] L. A. Nazarova, S. A. Ovsienko, A. V. Roiter, *Polyquivers of a finite type. (Russian) Quadratic forms and polyquivers.* Akad. Nauk Ukrain. SSR Inst. Mat. Preprint No. 23 (1977), 17–23.

[NaOR78] L. A. Nazarova, S. A. Ovsienko, A. V. Roiter, *Polyquivers of finite type.* (Russian) Algebra, number theory and their applications. Trudy Mat. Inst. Steklov. 148 (1978), 190–194, 277 English translation: Functional Anal. Appl. 7 (1973), 252–253 (1974).

[Newst78] P. E. Newstead. *Introduction to moduli problems and orbit spaces.* Tata Institute Lecture Note, 51, Springer, (1978).

[OvTa77] V. Ovsienko, S. Tabachnikov, *Projective Differential Geometry Old and New From the Schwarzian Derivative to the Cohomology of Diffeomorphism Groups.* Series: Cambridge Tracts in Mathematics, 2005, No. 165.

[OnVi90] A. L. Onishchik, È. B. Vinberg, *Lie groups and algebraic groups.* Translated from the Russian and with a preface by D. A. Leites. Springer Series in Soviet Mathematics. Springer-Verlag, Berlin, 1990. 328 pp.

[Pie82] Pierce R. S., *Associative algebras.* Graduate Texts in Mathematics, 88. Studies in the History of Modern Science, 9. Springer-Verlag, New York-Berlin, 1982. xii+436

[Pr94] V. V. Prasolov, *Problems and theorems in linear algebra.* Translated from the Russian manuscript by D. A. Leites, American Mathematical Society, Providence, RI, 1994. xviii+225 pp.

[PT90] J. A. de la Peña, M. Takane, *Spectral properties of Coxeter transformations and applications.* Arch. Math. 55 (1990), no. 2, 120–134.

[PV94] V. L. Popov, È. B. Vinberg, *Invariant Theory*, Encycl. of Math. Sci., Algebraic Geometry. IV, Springer Verlag, Vol. 55, 1994, 123–284.

[Re88] M. Reid, *Undergraduate Algebraic Geometry.* London Mathematical Society Student Texts 12, Cambridge Univ. Press, Cambridge, 1988

[Re01] M. Reid, *Surface cyclic quotient singularities and Hirzebruch–Jung resolutions.* 2001,
http://www.maths.warwick.ac.uk/~miles/surf/more/cyclic.pdf .

[Rie02] O. Riemenschneider, *Special representations and the two-dimensional McKay correspondence*, Hokkaido Math. J. 32 (2003), no. 2, 317-333.

[Rin76] C. M. Ringel, *Representations of K-species and bimodules.* J. Algebra 41 (1976), no. 2, 269–302.

[Rin80] C. M. Ringel, *The rational invariants of the tame quivers.* Invent. Math. 58 (1980), no. 3, 217–239.

[Rin94] C. M. Ringel, *The spectral radius of the Coxeter transformations for a generalized Cartan matrix.* Math. Ann. 300 (1994), no. 2, 331–339.

[Rob06] S. Roberts, *King of Infinite Space: Donald Coxeter, the Man Who Saved Geometry,* Walker & Company, 2006.

[Ro77] A. V. Roiter, *Roots of integer quadratic forms. (Russian) Quadratic forms and polyquivers.* Akad. Nauk Ukrain. SSR Inst. Mat. Preprint No. 23 (1977), 3–16.

[Rom00] T. Romstad, *Non-Commutative Algebraic Geometry Applied to Invariant Theory of Finite Group Actions.* Cand. Scient. Thesis. Institute of Mathematics, University of Oslo (2000), http://folk.uio.no/romstad/thesis.pdf .

[Ros04] W. Rossmann, *McKay's correspondence and characters of finite subgroups of SU(2).* Noncommutative harmonic analysis, 441–458, Progr. Math., 220, Birkhäuser Boston, Boston, MA, 2004.

[Sam41] H. Samelson, *Beiträge zur Topologie der Gruppen-Mannigfaltigkeiten.* (German) Ann. of Math. (2) 42, (1941). 1091–1137.

[Sat60] I. Satake, *On representations and compactifications of symmetric Riemannian spaces.* Ann. of Math. (2) 71 (1960), 77–110.

[Sc24] I. Schur, *Neue Anwendung der Integralrechnung auf Probleme der Invariantentheorie.* Sitzungsber. Preuss. Akad. (1924), 189, 297, 346.

[Ser05] V. Serganova, *Reflection functors.* Representation Theory, Lecture Notes, 2005, University of California, http://math.berkeley.edu/~serganov/math252/notes11.pdf.

[Shaf88] I. Shafarevich, *Osnovy algebraicheskoi geometrii.* (Russian) [Fundamentals of algebraic geometry.] Second edition. Nauka, Moscow, (1988) 352 pp.

[ShT54] G. C. Shephard, J. A. Todd, *Finite unitary reflection groups.* Canad. J. Math. 6 (1954), 274–304.

[Shi00] Jian-yi Shi, *Conjugacy relation on Coxeter elements.* Adv. Math. 161 (2001), no. 1, 1–19.

[Si44] C. L. Siegel, *Algebraic integers whose conjugates lie in the unit circle.* Duke Math. J. 11, (1944), 597–602.

[SW00] A. Skowroński, J. Weyman, *The algebras of semi-invariants of quivers.* Transform. Groups 5 (2000), no. 4, 361–402.

[Sln77] N. J. A. Sloane, Error-correcting codes and invariant theory: new applications of a nineteenth-century technique. Amer. Math. Monthly 84 (1977), no. 2, 82–107.

[Sl80] P. Slodowy, *Simple singularities and simple algebraic groups.* Lecture Notes in Mathematics, 815. Springer, Berlin, 1980.

[Sl83] P. Slodowy, *Platonic solids, Kleinian singularities, and Lie groups.* Algebraic geometry (Ann Arbor, Mich., 1981), 102-138, Lecture Notes in Math., 1008, Springer, Berlin, 1983.

[Sm71] C. J. Smyth, On the product of the conjugates outside the unit circle of an algebraic integer. Bull. London Math. Soc. 3 (1971) 169–175.

[Sp77] T. A. Springer, *Invariant Theory,* Lect. Notes in Math. 585, Springer-Verlag, NY, 1977.

230    References

[Sp87]  T. A. Springer, *Poincaré series of binary polyhedral groups and McKay's correspondence*. Math. Ann. 278 (1987), no. 1-4, 99–116.

[Sp94]  T. A. Springer, *Linear Algebraic Groups*, Encycl. of Math. Sci., Algebraic Geometry. IV, Springer Verlag, Vol. 55, 1994, 1–123.

[Stm04]  J. R. Stembridge, *Tight quotients and double quotients in the Bruhat order*. Electron. J. Combin. 11 (2004/06), no. 2, Research Paper 14, 41 pp.

[Stn79]  R. P. Stanley, *Invariants of finite groups and their applications to combinatorics*, Bull. Amer. Math. Soc., 1 (1979), 475–511.

[Stb85]  R. Steinberg, *Finite subgroups of $SU_2$, Dynkin diagrams and affine Coxeter elements*. Pacific J. Math. 118 (1985), no. 2, 587–598.

[Stb59]  R. Steinberg, *Finite reflection groups*. Trans. Amer. Math. Soc. 91 (1959), 493–504.

[SuSt75]  V. F. Subbotin, R. B. Stekolshchik, *The spectrum of the Coxeter transformation and the regularity of representations of graphs*. In: Transactions of department of mathematics of Voronezh University. 16, Voronezh Univ. Press, Voronezh (1975), 62–65, (in Russian).

[SuSt78]  V. F. Subbotin, R. B. Stekolshchik, *The Jordan form of the Coxeter transformation, and applications to representations of finite graphs*. Funkcional. Anal. i Priložen. 12 (1978), no. 1, 84–85. English translation: Functional Anal. Appl. 12 (1978), no. 1, 67–68.

[SuSt79]  V. F. Subbotin, R. B. Stekolshchik, *Sufficient conditions of regularity of representations of graphs*. Applied analysis, Voronezh, (1979), 105–113, (in Russian).

[SuSt82]  V. F. Subbotin, R. B. Stekolshchik, *The relation between the Coxeter transformation, the Cartan matrix and Chebyshev polynomials*. In: A. I. Perov et al (eds.) Operator methods in nonlinear analysis, 118–121, Voronezh. Gos. Univ., Voronezh, 1982, (in Russian).

[St75]  R. B. Stekolshcik, *Conditions for the regularity of representations of graphs*. In: Transactions of department of mathematics of Voronezh University. 16, Voronezh Univ. Press, Voronezh, 1975, 58–61, (in Russian).

[St81]  R. B. Stekolshchik, *Coxeter numbers of the extended Dynkin diagrams with multiple edges*. Abstracts of the XVI National (the USSR) algebraic conference. Part 2, Leningrad, p.187, 1981, (in Russian).

[St82]  R. B. Stekolshchik, *Transforming elements connecting the Coxeter transforms, and conditions for the regularity of roots*. Funkt. Anal. i Prilozhen. 16 (1982), no. 3, 84–85. English translation: Functional Anal. Appl. 16 (1982), no. 3, 229–230 (1983).

[St84]  R. B. Stekolshchik, *Invariant elements in a modular lattice*. Funktsional. Anal. i Prilozhen. 18 (1984), no. 1, 82–83. English translation: Functional Anal. Appl. 18 (1984), no. 1, 73–75.

[St85]  R. B. Stekolshchik, *Study of Coxeter transformations and regularity problems associated with representation theory of graphs*. Ph.D.Thesis, Kiev, 95p, 1985, (in Russian).

[St05]  R. B. Stekolshchik, *Notes on Coxeter Transformations and the McKay correspondence*, arXiv: math.RT/0510216, 1–154.

[St06]  R. B. Stekolshchik, *Kostant's generating functions, Ebeling's theorem and McKay's observation relating the Poincare series*, arXiv: math.RT/0608500, 1-22.

[St07]  R. B. Stekolshchik, *Gelfand–Ponomarev and Herrmann constructions for quadruples and sextuples*, J. Pure Appl. Algebra, 211 (2007) no. 1, 95–202.

[Ti66]  J. Tits, *Classification of algebraic semisimple groups*. 1966 Algebraic Groups and Discontinuous Subgroups (Proc. Sympos. Pure Math., Boulder, Colo., 1965) pp. 33–62 Amer. Math. Soc., Providence, R.I., 1966.

[Vin85]  È. B. Vinberg, *Hyperbolic reflection groups*. Uspekhi Mat. Nauk 40 (1985), no. 1(241), 29–66, 255 English translation: Russian Math. Surveys 40 (1985), no.1, 31–75.

[Wa98]  A. Wassermann, *Lecture Notes on Kac-Moody and Virasoro Algebras*. Michaelmas 1998, Cambridge, Part III, http://iml.univ-mrs.fr/~wasserm/

[Weib]  C. A. Weibel, *History of Homological Algebra*. http://www.math.rutgers.edu/~weibel/history.dvi .

[Wi01]  R. A. Wilson, *The Monster is a Hurwitz group*. J. Group Theory 4 (2001), no. 4, 367–374.

[Zh89]  Y. Zhang, *Eigenvalues of Coxeter transformations and the structure of regular components of an Auslander-Reiten quiver*. Comm. Algebra 17 (1989), no. 10, 2347–2362.

[Zhe73]  D. Zhelobenko, *Compact Lie groups and their representations*. AMS. Translations of Math. Monographs, 40 (1973).

## Papers published in difficult to access VINITI depositions

[KMSS83]  V. A. Kolmykov, V. V. Menshikh, V. F. Subbotin, M. V. Sumin, *Certain properties of acyclic graphs and polynomials related to them*. Voronezh, VINITI deposition, 12.12.1983, n. 6707-B83, 21 p, (in Russian).

[KMSS83a]  V. A. Kolmykov, V. V. Menshikh, V. F. Subbotin, M. V. Sumin, *Study of spectral properties of trees*. Voronezh, VINITI deposition, 12.12.1983, n. 6709-B83, 30 p, (in Russian).

[MeSu82]  V. V. Menshikh, V. F. Subbotin, *Orientations of graphs and Coxeter transformations they generate*. Voronezh, VINITI deposition, 30.12.1982, n. 6479-B82, 26p, (in Russian).

[St82a]  R. B. Stekolshchik, *The Coxeter transformation associated with the extended Dynkin diagrams with multiple edges*. Kishinev, VINITI deposition, 02.09.1982, n. 5387-B82, 17p, (in Russian).

[SuSum82]  V. F. Subbotin, M. V. Sumin, *Study of polynomials related with graph representations*. Voronezh, VINITI deposition, 30.12.1982, n. 6480-B82, 37p, (in Russian).

[SuU85]  V. F. Subbotin, N. N. Udodenko, *On certain properties of characteristic polynomials, of the Coxeter transformation*. Voronezh, VINITI deposition, 07.08.1985, n. 5922-B85, 34p, (in Russian).

[SuU88]  V. F. Subbotin, N. N. Udodenko, *On certain results and problems of the representation theory of graphs and the spectral theory of the Coxeter transformation*. Voronezh, VINITI deposition, 12.07.1988, n. 6295-B88, 60p, (in Russian).

# Index

Printing: Krips bv, Meppel, The Netherlands
Binding: Stürtz, Würzburg, Germany